WEED CONTROL
IN TURF AND ORNAMENTALS

WEED CONTROL
IN TURF AND ORNAMENTALS

A. J. Turgeon
Penn State University

L. B. McCarty
Clemson University

Nick Christians
Iowa State University

Prentice Hall
Upper Saddle River, New Jersey
Columbus, Ohio

Library of Congress Cataloging-in-Publication Data

Turgeon, A. J. (Alfred J.), 1943–
Weed control in turf and ornamentals / A. J. Turgeon, L. B. McCarty, Nick Christians. — Ist ed.
p. cm.
Includes bibliographical references.
ISBN-13: 978-0-13-159122-6
ISBN-10: 0-13-159122-3
1. Turfgrasses—Weed control. 2. Ornamental grasses—Weed control. 3. Turf management.
I. McCarty, L. B. (Lambert Blanchard), 1958- II. Christians, Nick Edward, 1949- III. Title.
SB608.T87T89 2009
635. 9′64295—dc22
2008021580

Vice President and Executive Publisher: Vernon R. Anthony
Acquisitions Editor: William Lawrensen
Editorial Assistant: Nancy Kesterson
Production Coordination: Nitin Agarwal, Aptara®, Inc.
Project Manager: Maren L. Miller
AV Project Manager: Janet Portisch
Operations Specialist: Laura Weaver
Art Director: Candace Rowley
Interior and Cover Design: Anne DeMarinis
Cover Image: Sandra Cunningham/Shutterstock
Director of Marketing: David Gesell
Senior Marketing Coordinator: Alicia Dysert

Photo credits: All photos courtesy of authors, unless otherwise noted.

This book was set in AGaramond by Aptara®, Inc., and was printed and bound by Quebecor, Dubuque. The cover was printed by Phoenix Color Corp.

Pearson Education Ltd.
Pearson Education Singapore Pte. Ltd.
Pearson Education Canada, Ltd.
Pearson Education—Japan
Pearson Education Australia Pty. Limited
Pearson Education North Asia Ltd.
Pearson Educación de Mexico, S.A. de C.V.
Pearson Education Malaysia Pte. Ltd.

Prentice Hall
is an imprint of

www.pearsonhighered.com

10 9 8 7 6 5 4 3 2 1
Paper Bound ISBN-13: 978-0-13-159122-6
ISBN-10: 0-13-159122-3
Loose Leaf ISBN-13: 978-0-13-505482-6
ISBN-10: 0-13-505482-6

contents

7 Turfgrass Plant Growth Regulators

8 Weed Control in Landscapes and Nonturf Areas

Pearson AG Is Going Green

Issues of sustainability and preserving our natural resources consistently rank among the most important concerns to our customers. To help do our part, Pearson AG is implementing the following eco-friendly initiatives to our publishing program:

- Printing this book, as well as all future Pearson AG titles, with paper fiber from managed forests certified by the Sustainable Forestry Initiative (SFI).
- Integrating the use of vegetable-based ink products that contain a minimum of 45% of renewable resource content and more than 5% by weight of petroleum distillates.
- Offering alternative versions to traditional printed textbooks, such as our "Student Value Edition" and the e-book version of the text in the CourseSmart platform.
- Providing electronic versions of supplemental material, such as Power-Point Presentations, Test Banks, and Instructor Manuals, which can be found by registering with our Instructor Resource Center on the web at www.pearsoned.com.

For more information regarding the Sustainable Forestry Initiative, visit **www.sfiprogram.org.**

Preface

This book was designed to support an undergraduate or graduate course in weed control for turfgrass or ornamental/environmental horticulture majors. It can also be useful as a reference text for field professionals working on golf courses or sports fields, or in the landscape maintenance or lawn care industries. It differs from other weed control books in that it deals specifically with controlling weeds in turfgrass and landscape-plant communities. And it differs from other turfgrass or landscape-plant books that include information on weed control by providing a more comprehensive coverage of the subject.

The book is divided into three units. Unit I, "Weeds of Turf and Ornamentals," covers a broad array of weed species that occur in turfgrass or landscape-plant communities throughout the United States. Unit II, "Herbicides," includes detailed information on herbicide action and metabolism in plants, their fate in the environment, and herbicide formulations and application methods. Unit III, "Weed Control," covers various methods for controlling weeds and includes comprehensive lists of herbicides and plant growth regulators and their use in controlling weeds in turfgrass and landscape-plant communities.

A. J. Turgeon
L. B. McCarty
N. Christians

About the Authors

A. J. Turgeon, Ph.D., is a professor of turfgrass management at Penn State University. He came to Penn State in 1986 as head of the Agronomy Department and served in that position until 1994, when he moved into an academic position in the department to pursue his turfgrass interests full time. He is involved primarily in case-based teaching, the development and evaluation of Web-accessible educational resources, and research in turfgrass morphogenesis and management systems. Dr. Turgeon completed his B.S. degree at Rutgers University in 1965. After a three-year tour in the U.S. Army, he pursued graduate study at Michigan State University, where he completed his Ph.D. degree in 1971. He then served as assistant—followed by associate—professor of turfgrass management at the University of Illinois from 1971 to 1979. He moved to Texas A&M University, where he served as resident director of research at the Dallas Research and Extension Center from 1980 to 1983. His next assignment was vice president for research and technical services with the Tru Green Corporation in East Lansing, Michigan, from 1983 to 1986. Dr. Turgeon has received numerous national and local awards for leadership and teaching, including the National ADEC Award for Excellence in Distance Education in 2000. He is the author of *Turfgrass Management,* currently in its eighth edition, the coauthor of Poa Annua: *Physiology, Culture, and Control of Annual Bluegrass,* and *The Turf Problem Solver: Case Studies and Solutions for Environmental, Cultural, and Pest Problems,* and is the editor of *Turf Weeds and Their Control.*

Bert McCarty, Ph.D., is a professor of horticulture specializing in turfgrass science and management at Clemson University in Clemson, South Carolina. A native of Batesburg, South Carolina, he received a B.S. degree in agronomy and soils from Clemson University, an M.S. in crop science from North Carolina State University, and a Ph.D. in plant physiology and plant pathology from Clemson University. Dr. McCarty spent nine years as a turfgrass specialist at the University of Florida in Gainesville. While there, he oversaw the design and construction of the state-of-the-art research and education turfgrass facility "The Envirotron." He also was author or coauthor of the books *Best Management Practices for Florida Golf Courses, Weeds of Southern Turfgrasses,* and *Florida Lawn Handbook.* In 1996, he moved to Clemson University, where he is currently involved in research, extension, and teaching activities. He has published over 500 articles dealing with all phases of turfgrass management and has given over 1,000 presentations. He is coauthor of the books *Color Atlas of Turfgrass Weeds, Best Golf Course Management Practices, Southern Lawns, Managing Bermudagrass Turf, Sod Production in the Southern United States, Fundamentals of Turfgrass and Agricultural Chemistry,* and *Designing and Maintaining Bermudagrass Sports Fields in the United States;* is a coauthor of the GCSAA seminars *Weed Control, Advanced Weed Management, Managing Bentgrass in Heat-Stress Environments,* and *Managing Ultradwarf Bermudagrass Golf Greens;* and is active in a number of professional societies. He also teaches a course in advanced turfgrass management. His research focuses on weed management, PGR use, and environmental stress physiology issues affecting turfgrass managers.

Nick E. Christians, Ph.D., is a university professor of horticulture at Iowa State University, where he has been involved in teaching and research of turfgrass management since 1979. He received a B.S. from Colorado State University and M.S. and Ph.D. degrees from the Ohio State University. He is the author of *Fundamentals of Turfgrass Management* and coauthor of *The Mathematics of Turfgrass Maintenance* and *Mathematics for the Green Industry.* He is also coauthor of *Scotts Lawns: Your Guide to a Beautiful Yard* and *Scotts Southern Lawns: Your Guide to a Beautiful Yard.* He has authored or coauthored more than 900 refereed papers, popular publications, abstracts, and published research reports. He has received a number of national and local awards for teaching and advising, including the American Society for Horticultural Science Outstanding Undergraduate Educator Award in 1991.

unit one
Weeds of Turf and Ornamentals

CHAPTER 1 **Grasses and Grass-Like Weeds** CHAPTER 2 **Broadleaf Weeds**

The classical definitions of a weed are "a plant out of place" and "a plant growing where it is not wanted." In turfgrass communities, the definition can be expanded to "a plant that is undesirable because of its disruptive effect on the esthetic appearance, stabilizing capacity, or overall utility of a turf." For example, creeping bentgrass may be considered a weed when grown in a stand of Kentucky bluegrass, but it is desirable when grown as a monoculture in a golf course green or fairway. Weeds can also detract from the esthetic appearance of woody and herbaceous ornamentals in ornamental planting beds, as well as in populations of ground cover plants.

Reasons for classifying a plant as a weed are numerous. In addition to being unsightly, weeds compete with desirable plants for light, oxygen, soil nutrients, soil moisture, carbon dioxide, and space. Weeds also act as hosts for pests such as plant pathogens, nematodes, and insects. Certain weeds cause allergic reactions in humans, due to their pollen or their volatile chemicals. When seed is sold, the definition of a weed can become a legal matter.

Probably the most undesirable characteristic of weeds in turf and ornamentals is their disruption of plant stand uniformity. Different leaf width and/or shape, different growth habit, and/or different color contribute to unsightliness. For example, many broadleaf weeds, such as dandelion, plantains, and pennywort, have leaf widths different from turf or ornamentals. Goosegrass, pathrush, smutgrass, and dallisgrass tend to form clumps or patches that also disrupt plant stand uniformity. In addition, large clumps are difficult to mow effectively, and they increase maintenance problems. Plant stand uniformity also is disrupted by weed seed heads. Annual bluegrass, for example, is largely unnoticed on putting surfaces until prolific seed heads appear in late winter and spring. Seed heads also disrupt the smoothness and trueness of the turf's playing surface.

Note: Color photos of the weeds discussed in Chapters 1 and 2 are found in Color Plates 1 through 16.

Plant color is another factor in determining the potential of a weed problem in turf. The lighter green color typically associated with annual sedge often distracts the golfer who is focused on the playing surface. Bahiagrass and *Poa trivialis* often have different colors when grown in combination with other turf species.

Because weeds include species from many different families of plants, they have been divided in this text into two chapters, one covering all grasses and grass-like weeds and the other covering all broadleaved species.

*1
Grasses and Grass-Like Weeds

ANNUAL GRASSES

The annual grasses are common weeds in turf and landscape areas. Annuals complete their life cycle in a 12-month period, not necessarily in a calendar year. There are both **summer annuals** and **winter annuals**. Summer annuals germinate in the spring and die in the fall, generally after the first frost. They must come back from seed each spring. Winter annuals germinate in the late summer and the fall, live through the winter as mature species, produce a seed crop in the spring, and then die during the stress periods of summer. Some grasses are annuals in cooler areas, but may persist as perennials in milder climates. There are both preemergence and postemergence herbicides for the control of annual grasses. Generally, properly timed preemergence herbicides provide the best approach to effective control.

Annual Bluegrass (*Poa annua* L.)

Synonyms: This grass is also widely known in the turfgrass industry by its genus and species name (*Poa annua*), or simply, *Poa.*

Annual bluegrass has a boat-shaped leaf tip and a folded vernation. It is also generally classed as a bunch grass, although some biotypes do produce short stolons. It is usually considered to be a winter annual, although it can also persist as a perennial.

Annual bluegrass is one of the most widely distributed turf and ornamental weeds. It can be found throughout most of the world in both warm and cool climatic regions. Annual bluegrass is very tolerant of low mowing heights and can be found on golf course greens at mowing heights as low as 0.08 inch (2 mm). It can become a serious weed problem in golf course greens, tees, and fairways. It can also become a problem in intensely managed sports fields and in lawns in cool and wet regions. Annual bluegrass is often found in perennial flower beds and other landscape areas where soil is exposed. It is a profuse seed former, and much of its competitive advantage is due to its ability to form seed throughout the season, even at the lowest mowing heights on golf greens.

There are currently no effective chemical controls for annual bluegrass in most close-mown turf areas. It is not very competitive with turf maintained at mowing heights of 2 inches (5 cm) and above, and simply increasing the mowing height of lawn grasses can prevent it from becoming a weed problem in most climatic zones. Properly timed preemergence herbicide applications before seed germination in the late summer can provide reasonably good control in ornamental plantings.

Barnyardgrass (*Echinochloa crus-galli* [L.] Beauv.)

Barnyardgrass has a rolled vernation. Stems are compressed and may have a purple coloration on the sheaths near the stem base. Its broad, coarse-textured leaves and stems may grow flat on the ground's surface in lawns. Its seedhead is a distinct panicle that protrudes upright above the canopy. Mature seedheads often have a purple coloration, particularly on lower seed stalks.

Barnyardgrass is a warm-season summer annual that germinates in the spring in temperate climatic regions. It usually germinates at about the same time as crabgrass. Barnyardgrass is not very competitive with well-maintained turf. Proper fertilization, irrigation, and mowing will typically eliminate this species. Standard preemergence herbicides applied before seed germination in the spring will generally provide satisfactory control in both turf and landscape areas.

CRABGRASS (*Digitaria* spp.)

The term "crabgrass" is familiar to most people in the United States and is often associated with any weed that disrupts the uniformity of the turf in midsummer. Specifically, crabgrass refers to several plant species in the genus *Digitaria.* Members of this genera are widely distributed around the world and do constitute the primary weed problem in some locations.

Large Hairy Crabgrass (*Digitaria sanguinalis* [L.] Scop.)

Large hairy crabgrass is covered with fine hairs on both sides of the leaves and on the sheaths. Its seedhead is quite distinctive and looks like protruding fingers—thus its Latin name *Digitaria* (digits, or fingers). The seedhead branches from multiple locations on the seed stalk. This weed has a rolled vernation. It grows to 10 inches (25 cm) or more in height, but can also tolerate low mowing heights in turf.

Large hairy crabgrass is widely distributed throughout the United States and can be found in nearly every state. It is a particular problem in the upper Midwest and in the northeastern states, where it is among the most hated weeds in the landscape. Hairy crabgrass is a summer annual that germinates in the spring when soil temperatures reach approximately 55°F (13°C). The timing of its germination varies from as early as February in warmer climates to late May in cooler regions. It dies in the fall after producing an abundant seed crop. Being a warm-season species allows it to outcompete the cool-season lawn grasses, like Kentucky bluegrass, in the high-temperature stress periods of midsummer. There are both preemergence and postemergence herbicides that can selectively control large hairy crabgrass in turf. Preemergence control is generally preferred, and postemergence control is recommended only where needed.

Smooth Crabgrass (*Digitaria ischaemum* [Schreb.] Schreb. ex Muhl.)

Synonym: Small Crabgrass

Smooth crabgrass is very similar to large hairy crabgrass except that it lacks the hairs and generally grows to only a few inches in height. A few long hairs may appear at the collar, and the stem and sheath may take on a reddish to purple color. It also has a rolled vernation. It has the same finger-like seedheads, and its seed germinates at about the same time, as large hairy crabgrass. Large hairy and smooth crabgrasses are often found together in the same lawn and are controlled in the same way. The range of smooth crabgrass is similar to that of large hairy crabgrass, although higher populations of the large hairy crabgrass are generally observed in northern regions.

Southern Crabgrass (*Digitaria ciliaris* [Retz.] Koel.)

Southern crabgrass is very similar to large hairy crabgrass in its general appearance, including dense hairs on leaves and sheaths. It is better adapted to warmer regions, however, and extends into Florida, Mexico, and Central America. Its region of adaptation overlaps with that of large hairy crabgrass through the central United States into Kansas and Nebraska. Preemergence control methods are similar to those used for large hairy and smooth crabgrass.

Tropical Crabgrass (*Digitaria bicornis* [Lam.] Roemer & J.A. Schultes ex Loud.)

Synonym: Asia Crabgrass

Tropical crabgrass is a warm-season summer annual found in lawns along the Gulf Coast of the United States and throughout the state of Florida. It does not persist in colder areas. It has hairs on the leaf and sheath, similar to large hairy crabgrass, but can be distinguished from that species by the fact that its seedheads branch from a single central point.

Blanket Crabgrass (*Digitaria serotina* [Walt.] Michx.)

Synonyms: Dwarf Crabgrass, Rabbit Crabgrass

Blanket crabgrass is found in the central United States and into the tropics. In temperate climates, it is a summer annual, but it may be a perennial in warmer climates. It has hairy leaves and sheaths and can form a dense mat of stolons.

Crowfootgrass (*Dactyloctenium aegyptium* [L.] Willd.)

Synonym: Egyptian Grass

Crowfootgrass has a rolled vernation and a membranous ligule with a fringe of hairs along its edge. The leaf blades have hairs along their margins near the base. It looks a little like crabgrass, but is in a different genus. The seedhead is very distinct and easily identified. Crowfootgrass is composed of multiple spikelets that generate from a central growing point and radiate out like fingers.

This summer annual species is most common in the southern states and extends down into Central and South America. It does not compete well with highly maintained turf and is usually found on poorly maintained areas where the grass is thin. It is also found in flower beds, gardens, and other landscape areas.

Fall Panicum (*Panicum dichotomiflorum* Michx.)

Fall panicum produces a very large (several inches across), pointed panicle seedhead with multiple branches. The seedhead often has a purple coloration at maturity. Its coarse textured leaves are glossy on the underside and have a distinct midrib. Individual plants can grow quite large, spreading 1 foot (0.3 m) or more in diameter.

Fall panicum is generally considered to be a field weed in the Midwest, but can spread into flower beds and into poorly maintained areas in lawns. It is often found along curbsides in compacted areas and may be observed growing over the edge of the curb and down into the street. It can be controlled with preemergence herbicides, but proper lawn management will generally prevent it from becoming established. It does not tolerate continuous mowing. Fall panicum is similar in appearance to witchgrass (*Panicum capillare*), although witchgrass is covered with hairs. Also, the panicle of witchgrass remains partially covered by the sheath, whereas the panicle of fall panicum fully emerges from the sheath.

FOXTAIL (*Setaria* spp.)

The foxtails, like crabgrass, are very well known to most people who have lawns and landscape areas to maintain. They are known by their distinct seedhead that looks like a fox's tail. They are warm-season summer annuals that germinate in spring and die in fall after the first frost. Foxtail often takes over vacant lots and may be a problem in new lawns the first year or two following establishment.

Giant Foxtail (*Setaria faberi* Herrm.)

Giant foxtail is the largest of the foxtails and may reach several feet in height. It is best identified by its nodding seedhead, unlike yellow and green foxtail, which have upright seedheads. It is generally considered to be a field weed in most areas of the United States, and it will not tolerate mowing in turf areas. It does, however, readily establish in disturbed areas in the landscape.

Yellow Foxtail (*Setaria pumila* [Poir.] Roem. & J.A. Schult.)

Yellow foxtail has a rolled vernation and has a ligule that is a fringe of hairs. Leaves are hairy on the upper surface. It has a lighter green color than the other foxtails and takes on a yellow-green color at maturity.

Yellow foxtail is a smaller plant than giant foxtail and generally reaches a height of 1 to 2 feet (0.3 m). It is widely distributed throughout the United States and can be found as far south as Florida. It will tolerate mowing better than giant foxtail and will persist for a period in turf. It is less tolerant of mowing than crabgrass and will generally disappear if the lawn is properly maintained. Yellow foxtail germinates later than crabgrass and generally requires soil temperatures in the range of 65°F (18°C).

Green Foxtail (*Setaria viridis* [L.] Beauv.)

Green foxtail is the smallest of the three foxtails and has a darker green color than the yellow foxtail. Its leaves are smooth, unlike those of yellow foxtail, which has hairs on the upper side of the leaves. Its seedhead is generally smaller than the seedhead of yellow foxtail.

Its range is generally the cooler areas of the United States, and it is rarely found in the Deep South. Its seed germinates at approximately the same time as yellow foxtail's.

Goosegrass (*Eleusine indica* [L.] Gaertn.)

Synonym: Silver Crabgrass, Indian Goosegrass

Goosegrass has a silvery stem base and a seedhead that at first appearance looks like crabgrass. In some parts of the United States it is known as "silver crabgrass." There are, however, clear differences between goosegrass and crabgrass. The individual seed stalks on goosegrass appear somewhat like a zipper, with the individual seeds protruding in two directions from the central stalk. The individual seeds are also larger than those on crabgrass. Crabgrass has a rolled vernation, whereas goosegrass has a folded vernation. Goosegrass can be found throughout northern Iowa and Illinois, but is much more common through the central part of the United States, from Kansas City to Washington, D.C.

Goosegrass is a warm-season summer annual that germinates in the spring and dies in the fall, like crabgrass. Goosegrass generally germinates two to three weeks later than crabgrass, however. This makes goosegrass a difficult problem to control, because the preemergence herbicides used to control crabgrass may lose their efficacy before goosegrass germinates. Goosegrass is hard to control even with proper application timing of preemergence herbicides. The relatively expensive preemergence herbicide "oxadiazon" may be needed to achieve control of this difficult grass.

Lovegrass (*Eragrostis pilosa* [L.] Beauv.)

Synonym: India Lovegrass

Lovegrass has a rolled vernation, with narrow pointed leaves that have a rough upper surface and a smooth underside. Lovegrass forms a tufted bunch with protruding panicle seedheads.

Lovegrass is generally a sign of compacted soil. It is often observed in gravel driveways and along the edges of poorly maintained lawns. It does not compete well with healthy turf. It can be controlled with preemergence herbicides, but core aerification followed by proper fertilization and irrigation will generally eliminate it. There are many species in the genus *Eragrostis.* The species *pilosa* is often found in landscapes and in compacted areas around lawns in the Midwest and eastern United States. It may be found as far south as Texas.

SANDBUR (*Cenchrus*)

Sandburs are quite familiar to anyone who has stepped on one or who has spent 30 minutes pulling them from a dog's coat and foot pads. This species produces seed with very sharp burs that can easily penetrate the skin. There are several species of the genus *Cenchrus* found throughout the world.

Field Sandbur (*Cenchrus spinifex* Cav.)

Field sandbur is a summer annual species. It has a folded vernation and a ligule that is a fringe of hairs. Its most outstanding feature is a spike of up to 20 burs with very sharp points that can penetrate the skin. These spurs are easily detached from the mature inflorescence and tend to be a real problem for barefooted people and for dogs that pick them up on their fur.

Field sandbur is found throughout most of the United States, but increases in the central and southern parts of the country. It forms dense mats in lawns, particularly on sandy hillsides where the grass is thin. It is also commonly found on bare soil and may infest flower beds. It is not easily controlled by standard preemergence herbicides. The best control is a combination of proper lawn management practices and preemergence herbicides. Benefin (Balan) is one of the more effective of the preemergence herbicides on this species.

Southern Sandbur (*Cenchrus echinatus* L.)

Southern sandbur is similar to field sandbur, but it is found farther south in the United States and extends down into Central America. It is a summer annual and also produces its seed with sharp burs. Its leaf is rough to the touch, and the sheath is smooth. It is often found on sandy, infertile soil areas.

Witchgrass (*Panicum capillare* L.)

Witchgrass is very similar to fall panicum, with the exception that its leaves and sheaths are covered with hairs. It has a large, rounded panicle of seeds that generally stays partially hidden in the sheath until it matures. It is a summer annual.

Witchgrass is a major field weed in crop-production areas of the Midwest and eastern United States, but it is also quite common in landscapes and in compacted areas at the edges of lawns. It is commonly observed along street curbs and may grow over the edge of the curb and down into the street. It does not compete well with properly maintained turf and is intolerant of regular mowing.

PERENNIAL GRASSES

Perennial grass weeds also commonly infest lawns and other landscape areas. These weeds are the most difficult species to control selectively in turf because they are so similar to the lawn grasses that they infest. The management practices that favor the lawn grasses also favor these species, and the herbicides generally affect the perennial grass weeds and the lawn grasses in the same way. The solution to most of these will be nonselective herbicides that kill all species in the lawn, followed by reestablishment with the desirable lawn grass. In many cases, perennial grasses that are themselves useful turfgrasses in one situation become weeds when they infest other turf species.

Bahiagrass (*Paspalum notatum* Flueggé)

Bahiagrass is a warm-season grass that spreads by very coarse-textured rhizomes which grow along the surface of the ground. The sheath base of the individual plant often has a reddish to purple coloration. It produces a distinct, branched seedhead that may extend a foot (0.3 m) or more above the turf. Bahiagrass can spread by vegetative growth of rhizomes or by seed.

Bahiagrass is used as a turf along the Gulf Coast of the United States, particularly in Florida. It is quite intolerant of cold temperatures and is not found in the central or northern states. It is generally used in low-maintenance turf, but, with proper management, can produce a reasonably good-quality turf on lawns. It becomes a weed when it encroaches into flower beds and other landscape areas. It may also disrupt the uniformity of St. Augustinegrass and bermudagrass lawns.

Bermudagrass (*Cynodon dactylon* [L.] Pers.)

Bermudagrass is a warm-season grass that has rhizomes and stolons on the same plant. Its leaves have a folded vernation and a ligule that is a fringe of hairs. It produces a branched seedhead that looks like a smaller version of the crabgrass' seedhead. It is an aggressive sod former and readily spreads over the soil surface.

Bermudagrass is the primary turfgrass of much of the warmer climatic areas of the world. In the United States, it extends from the central states throughout the southern states, with Kansas as its northern limit. It readily spreads into flower beds and can spread under and over weed barriers. The rhizomes can penetrate bulb species, such as tulips, and will kill them in some situations. On golf courses, bermudagrass can spread into creeping bentgrass greens and disrupt the playing surface. It is a useful grass species for golf course tees and fairways, but may become a weed on golf courses where other species constitute the primary turf.

Broomsedge (*Andropogon virginicus* L.)

Synonyms: Broomgrass, Sagegrass

Broomsedge is a warm-season perennial grass with a folded vernation and a membranous ligule. Leaf blades have hairs near the base, and sheaths are compressed.

Broomsedge is generally found in the eastern United States, but has also been reported in Florida and several of the western states. It produces upright seedheads that have cotton-like hairs. The seedheads persist in winter as brown tufts. Broomsedge is generally found in poorly maintained areas, and it does not compete well with properly maintained turf.

Creeping Bentgrass (*Agrostis stolonifera* L.)

Creeping bentgrass is a cool-season grass that has a rolled vernation, a pointed leaf tip, prominent veins on the upper side of its leaves, and a long membranous ligule. It spreads by extensive stolons.

Creeping bentgrass is a turfgrass used on golf course greens, tees, and fairways and on other areas where close mowing is necessary. It is well adapted to climates as far north as Alaska and can be found as far south as Florida. It can become a weed problem in Kentucky bluegrass lawns and golf course roughs, where it forms a dense, low-growing mat of stolons. It may occasionally be found in flower beds and other landscape areas. The recently released herbicide mesotrione (Tenacity) provides good control of creeping bentgrass in Kentucky bluegrass turf, with multiple applications.

Dallisgrass (*Paspalum dilatatum* Poir.)

Dallisgrass is a warm-season weed that has a rolled vernation and a tall membranous ligule. It has coarse-textured blades, with a prominent midvein and hairs around the collar area. It is a coarse-textured bunch grass with short rhizomes. Its finger-like seedhead is quite distinct, with three to six slender, flattened spikes.

Dallisgrass infests bermudagrass and other warm-season lawns. It is also found in landscape areas such as flower beds. It extends through the southern states and into the tropics and is commonly found in Hawaii. It can grow from seed or from rhizomes left in the soil after the above-ground plant parts have been removed. It is difficult to control and generally requires a nonselective systemic herbicide to kill it.

Johnsongrass (*Sorghum halepense* [L.] Pers.)

Johnsongrass is a warm-season perennial which spreads by large rhizomes that can be up to 1/2 inch (13 mm) in diameter. It has a rolled vernation and a distinct membranous ligule with jagged edges and hairs. It has coarse-textured leaves with hairs near their base. Its seedhead is a panicle.

Johnsongrass is a problem weed in many field crops and may reach 6 feet (1.8 m) in height. In lawns and landscape areas, it can adapt to defoliation and may persist under lawn-mowing heights, although proper turf management techniques can help to eliminate it. It is generally found in the central and southeastern states in the United States.

Kikuyugrass (*Pennisetum clandestinum* Hochst. ex Chiov.)

Kikuyugrass is a warm-season perennial that spreads by rhizomes and stolons. It has a ligule that is a fringe of hairs. Blades are relatively smooth with a few hairs, but sheaths are covered by fine hairs.

Kikuyugrass is one of the most aggressive of the turf and ornamental weeds. It can take over golf-course fairways and is often observed grown up over large shrubs, mailboxes, and taller objects such as electrical-wire poles. Fortunately, it is intolerant of cold and is limited to coastal areas of the United States, particularly the California coast and Hawaii. It is difficult to control, although repeated applications of triclopyr have met with some success.

Nimblewill (*Muhlenbergia schreberi* J.F. Gmel.)

Synonym: Wiregrass

Nimblewill is a warm-season perennial grass with a fine texture and a gray-green color. It spreads by stolons and forms a dense mat in cool-season lawns. It has a rolled vernation and a membranous ligule with jagged edges. It forms a narrow spike-like panicle of seeds, with each seed having a fine awn (hair) at its tip. It can be mistaken for creeping bentgrass in lawns; however, bentgrass has a longer membranous ligule, and nimblewill generally has hairs at the collar. Creeping bentgrass rarely produces seed in most of the central and southern United States. Where bentgrass does produce seed, the seeds will not have the awn present, as nimblewill does.

Nimblewill is one of the hardest species to control once it has become established in the lawn. It can be killed by nonselective herbicides, but it is also a good seed producer and often returns from seed. Because it is a warm-season species, it loses its chlorophyll in the fall and takes on a bleached, straw-like appearance that makes it stand out in cool-season lawns. Its thick stolon growth has given it the name "wire grass" in parts of the Midwest. Lawns that are to be renovated to have nimblewill removed should generally be sodded to prevent the reemergence of this difficult weed. It is found throughout the southeastern states and is also prevalent in Indiana, Ohio, and Pennsylvania. However, in recent years, it has become more common in the central and western states, including Iowa. The new herbicide mesotrione (Tenacity) has been found to provide postemergence control of nimblewill in Kentucky bluegrass turf.

Orchardgrass (*Dactylis glomerata* L.)

Orchardgrass is a cool-season perennial grass with a broad, gray-green leaf and a flat sheath. It has a very long membranous ligule, a folded vernation, and a bunch-type growth habit. The seedhead is a panicle that produces a fan-shaped, gray-green spike of seeds.

Orchardgrass often shows up as a weed in Kentucky bluegrass. It has a very fast growth rate and may grow an inch or more above a Kentucky bluegrass canopy, within a few days of mowing. There is no selective control in cool-season lawns, and it must be killed with nonselective herbicides. Small patches can be removed by digging followed by replacement of the soil in the area from which it was removed.

Perennial Ryegrass (*Lolium perenne* L.)

Perennial ryegrass is a cool-season perennial grass with a bunch-type growth habit. It has a folded vernation and may have a short membranous ligule, although the ligule is often absent. The leaves have a shiny backside. It produces a narrow spike of seeds in late spring that radiate laterally from the shiny clump of tillers in mown turf areas.

Perennial ryegrass is a useful turf and is often a part of seed mixes for lawns and sports fields. There are turf types and common, or "forage," types. The coarser-textured common types may become a weed in fine-textured grasses. This is particularly a problem in warm-season turf, such as bermudagrass and zoysiagrass. Ryegrass is a cool-season grass, and it stands out when the warm-season grasses are dormant in spring and fall. It can generally be controlled by spot treating with nonselective herbicides.

Quackgrass (*Elymus repens* [L.] Gould)

Quackgrass is a perennial, cool-season grass that is best known for its very extensive underground rhizome system that may extend for several feet in the soft soil of gardens and flower beds. It has long, clasping auricles that wrap around the stem at the collar area and a gray-green color. The texture is highly variable; leaves may range from fine to very coarse.

Quackgrass is a hated weed in field crops and landscape areas. It is a primary noxious weed in several midwestern states. The rhizome system makes this species very hard to kill. If the rhizome is not completely killed, it will come back and reinfest the area. Tilling the area only chops up the rhizome and spreads it through the soil. Pulling or cutting off the top of the weed only causes the buds on rhizomes to begin to grow and form new plants aboveground. Glyphosate (Roundup) can be used to nonselectively kill it, but the inability of glyphosate to translocate to the end of the extensive rhizome system can be a problem. Repeat applications of glyphosate following recovery from rhizomes, followed by sodding, will generally provide good control. Trying to reseed turf, even after several glyphosate applications, is usually not successful. In sod fields, fallowing the area for a season, with repeat glyphosate applications and repeat tilling, will usually eliminate quackgrass. Quackgrass is usually limited to the northern states, but has been found in California, North Carolina, and Oklahoma.

Rough Bluegrass (*Poa trivialis* L.)

Synonyms: Roughstalk Bluegrass. Golf course superintendents often shorten the genus species name and refer to this species as Poa triv.

Rough bluegrass has a folded vernation and a boat-shaped leaf tip. It may have a long membranous ligule, although types lacking the ligule are also found in the Midwest. It is a stolon former and can produce a spreading mat of stolons at low mowing heights.

Rough bluegrass is well adapted to shade, but can be found in full sun in close-mown turf, such as creeping bentgrass fairways on golf courses. It can be distinguished from creeping bentgrass by the rolled vernation and pointed leaf tip of the bentgrass. Rough bluegrass is used as a lawn grass in the Midwest for wet, shaded areas. It will also spread into full-sun areas in lawns, however. During dry weather it will turn brown while the other lawn grasses are still green. In recent years, it has become a serious weed in creeping bentgrass fairways on golf courses. In the southern United States, it is used for winter overseeding of bermudagrass and may become a weed problem when it persists into the late spring. Control is generally limited to nonselective spot treating in cool-season turf.

Smooth Bromegrass (*Bromus inermis* Leyss.)

Smooth bromegrass is a cool-season, perennial grass with a rolled vernation and a coarse-textured leaf. One of the best features used to identify it is its closed sheath that has the appearance of a V-necked sweater. It may also have a W-shaped constriction on the leaf blades called a water mark, although that feature is not always present. It has short rhizomes and is commonly found in Kentucky bluegrass turf.

Smooth bromegrass is a common part of the seed mixes used for roadside establishment in the central and northern parts of the Midwest. It is a good seed producer, and birds or wind may spread it to neighboring lawns. It is known for its rapid growth in early spring and late fall, at times when Kentucky bluegrass and perennial ryegrass are growing very slowly. Through most of the mid-season, it blends in well with the other lawn grasses. There is no selective control for smooth bromegrass, and nonselective herbicides must be used to kill it. It does have short rhizomes, but it is much easier to kill than quackgrass, and a single application of a nonselective herbicide is usually sufficient.

Tall Fescue (*Festuca arundinacea* Schreb.)

Tall fescue is a cool-season grass with a rolled vernation and a pointed leaf tip. It is a bunch grass, although some types have short rhizomes. The leaves are coarse textured and have prominent vernation on their upper sides. Short auricles may be present on common types.

Tall fescue can be maintained as a quality turf and is widely used on lawns and sports fields through the central part of the United States. Its poor cold tolerance limits its use in far northern states. Because of its rapid growth rate in the spring and its coarse texture, it may become a weed in fine-textured turf such as Kentucky bluegrass. These coarse-textured clumps are often present in lawns, even in states such as Minnesota, which is too far north to maintain tall fescue as a turf. It can be controlled nonselectively with glyphosate. There is also a herbicide called Corsair (chlorsulfuron) that can selectively remove tall fescue from Kentucky bluegrass turf. Corsair will kill perennial ryegrass, and the herbicide should not be used in Kentucky bluegrass/perennial ryegrass turf.

Thin Paspalum (*Paspalum setaceum* Michx.)

Synonym: Bull Paspalum

Thin paspalum is a warm-season grass with a rolled vernation and a membranous ligule that spreads by short rhizomes. It often has hairs on the blades and sheaths. The seedhead is a distinct spike with up to six separate spikelets.

Thin paspalum is generally found in poorly maintained turf, disturbed areas, and open soil sites in the garden and landscape. It is limited to the east central states and south into Central America.

Timothy (*Phleum pratense* L.)

Timothy is a cool-season bunch grass with a membranous ligule and leaves with a gray-green color. It often has a broad, bulb-like structure at its base. Its seedhead is a dense spike that resembles foxtail. The seedhead is more compact than that of foxtail, and the plant stands distinctly upright, whereas foxtail has longer awns and a seedhead that is generally nodding.

Timothy is a forage grass and is occasionally included in roadside seed mixtures in the Midwest. It stands out in lawns because of its coarse texture. There is no selective control for Timothy, but it is easily removed by the application of nonselective herbicides followed by reseeding.

Smutgrass (*Sporobolus indicus* [L.] R. Br.)

Smutgrass is a warm-season grass with a rolled vernation and a ligule that is a fringe of hairs. Its seedhead is a spike-like panicle that often develops a black smutty fungus.

Smutgrass forms coarse-textured clumps in lawns and other areas in the landscape. It is common in the southeastern and south central states, with its range extending into the tropics.

Torpedograss (*Panicum repens* L.)

Torpedograss is a coarse-textured, warm-season grass with a folded vernation and a membranous ligule that has a fringe of hairs along its upper edge. The upper sides of the leaves are also hairy. The seedhead is a spike with a whitish color. It spreads by sharply pointed rhizomes and also forms spreading stolons.

Torpedograss is generally a problem in the southern states, although it does extend along the western coast to Oregon and Washington. It is also found in Hawaii and Central America. It forms coarse-textured clumps in fine-textured turf and is common in gardens and flower beds. Spot treating with glyphosate is generally recommended for control.

Tropical Signalgrass (*Urochloa subquadripara* [Trin.] R.D. Webster)

Synonyms: Smallflowered Alexandergrass

Tropical signalgrass is a warm-season, coarse-textured plant with hairy blades and sheaths. The ligule is a fringe of hairs. It roots from the nodes. The seedhead is a raceme with branches that emerge at an angle at intervals from the main stalk. The branches give tropical signalgrass the appearance of a signal flag—thus its name "signalgrass."

Tropical signalgrass is very intolerant of cold and generally turns brown from frost, leaving coarse-textured brown clumps in the turf. It is a perennial in southern Florida and acts as an annual in northern areas with regular frost. It is generally found in coastal states, particularly Florida and Texas, but can be found along the Atlantic Coast as far north as South Carolina. It is a particular problem in St. Augustinegrass sod fields, but can also be found in bermudagrass on golf courses and lawns.

Windmillgrass (*Chloris verticillata* Nutt.)

Windmillgrass is a warm-season grass with a folded vernation and a prominent membranous ligule with hairs along the upper edge. The sheaths are flat, and blades are narrow, with hairs along the mid-vein. The seedhead is composed of multiple spikes that radiate out in a circular pattern from the central stalk in a windmill-like pattern. Individual seeds have long awns. At maturity, the seedhead will detach from the plant and roll over the ground's surface in the wind, like a tumbleweed, dropping seed as it rolls.

Windmillgrass is generally a problem in compacted areas and is often observed along street curbs. It may also infest any turf area that is weak or thin and can become a problem in flower beds and other landscape areas. It is generally associated with the warmer regions of the central and southern states, but in recent years has been spreading north and is now quite common as far north as central Iowa.

Zoysiagrass (*Zoysia japonica* [Hack.] Forbes)

Synonyms: Japanese lawngrass; Chinese lawngrass; Korean lawngrass

Zoysiagrass is a warm-season perennial with a rolled vernation and a membranous ligule. It spreads by rhizomes and stolons and forms a dense sod. Leaves can range from coarse textured to relatively fine in texture and are quite stiff and bristly. Blades are generally covered by fine, whitish hairs on both sides. The nodes generally have a brown husk that persists through the season.

Zoysiagrass can form a high-quality turf and is widely used on golf courses and lawns in the central United States. It can also be a weed, particularly when it infests cool-season lawns like Kentucky bluegrass. It goes dormant and turns brown in the fall and generally remains brown well into the spring, thereby making its presence quite apparent. It is very difficult to kill, and repeated glyphosate treatments, followed by reseeding of Kentucky bluegrass, is usually not effective. To completely remove it, repeated applications of glyphosate followed by sodding over the area is generally required. Zoysiagrass treated in this way usually will not emerge through the bluegrass sod.

SEDGES

Sedges are a group of monocotyledonous plants that look somewhat like grasses. They have triangular (three-ranked) stems, however, that differ from grasses that are two-ranked (meaning that their leaves emerge 180 degrees from one another). Sedges are generally associated with low, wet areas, but may occur almost anywhere in the landscape. There are selective controls for sedges in both warm-season and cool-season turf, although they are often limited to professional applicators. In gardens and flower beds, sedges can be pulled by hand or treated with nonselective herbicides. They often will regrow from underground structures, and persistence may be needed to control them by hand pulling.

Annual Sedge (*Cyperus compressus* L.)

Synonym: Water Sedge

Annual sedge spreads by seeds that germinate in the spring and dies with the first frost. The seedhead is a cluster of flat spikes that generally have a whorl of leaves just below the seed spikes.

Annual sedge is typically found in wet areas and is common on sandy soils. It tolerates low mowing heights and may be found in golf course greens. It grows throughout much of the northeastern United States and down into the southern states as far as Florida.

Cylindric Sedge (*Cyperus retrorsus* Chapm.)

The seedhead of cylindric sedge is quite distinct, with seed arranged in tight, rounded clusters. The clusters are generally subtended by up to seven bracts growing laterally from the central stalk. It may have short rhizomes. The stems form a dense clump that may grow 18 to 20 inches (46 to 51 cm) in height. Leaves are bright green and smooth.

Cylindric sedge is distributed throughout most of the United States. It is common on sandy soils and may be found in both wet and dry areas.

False Nutsedge (*Cyperus strigosus* L.)

Synonyms: September Sedge

False nutsedge spreads by rhizomes. It has yellow-green seedheads that are subtended by long bracts.

False nutsedge ranges from the northeastern Canadian provinces to the southern states in the United States. It is generally found in low wet areas in the landscape and may be found in turf. It is often found in wet ditches along roads.

Green Kyllinga (*Kyllinga brevifolia* Rottb.)

Green kyllinga has dark green stems and underground rhizomes that take on a purple color. The seed stalk generally has three very long leaves just below the round, 1/4-inch (6.4-mm)-diameter seedhead.

Green kyllinga ranges from Georgia on the north, down to Florida and Central America. It is generally found in wet, saturated areas in the landscape.

Hurricanegrass (*Fimbristylis cymosa* R. Br.)

Hurricanegrass spreads by rhizomes and seed. The seedhead is a brownish cluster of seeds borne at the end of a triangular stem that extends a few inches above the soil.

Hurricanegrass is a tufted perennial with a triangular stem that can form a dense mat in turf and landscape areas in both wet and dry soil conditions. It is found in southern Florida and down into tropical regions.

Purple Nutsedge (*Cyperus rotundus* L.)

Purple nutsedge has dark green, pointed blades. Seedheads are located at the end of triangular stems and are reddish brown to purple. Propagation may take place from oblong underground tubers that are covered with hairs. Tubers are often found in chains along underground rhizomes.

Purple nutsedge is common to the southern United States, and its range extends down into Central America. It is typically not found in the northern states, where yellow nutsedge generally predominates.

Tufted Kyllinga (*Cyperus tenuifolius* [Stevd.] Dandy)

Tufted kyllinga is similar to green kyllinga, but lacks rhizomes. It generally appears as tufted clumps in turf and flower beds. It has an oval-shaped inflorescence, with long bracts at its base. It is typically a summer annual, but may be a perennial in warmer areas.

Tufted kyllinga is found throughout the southeastern United States and extends westward into the south central states and down into Mexico.

Yellow Nutsedge (*Cyperus esculentus* L.)

Yellow nutsedge has yellow-green blades that taper to a thin tip. Blades can be fine in close-mown turf, but can be up to 1/2 inch (13 mm) wide when allowed to grow to full maturity in unmown landscape areas. Seedheads are yellow to brown and emerge from erect, triangular stems. It spreads by round, underground, hairless tubers that may dislodge from the plant at maturity and live for years before emerging.

Yellow nutsedge is a common weed in lawns throughout the United States and into Canada. It is generally associated with wet areas in the landscape, but may also be found in drier soil conditions. It has a distinctive branched seedhead, which is yellow to straw colored. It can be selectively controlled with halosufuron-methyl (Manage) or bentazon (Basagran). Repeated applications over several weeks may be required because of regrowth from underground tubers (nutlets). Persistent hand weeding may also be effective, but it will most likely require several weeks because of reemergence from tubers.

SPIDERWORT FAMILY (*Commelinaceae*)

The spiderwort family is characterized by flowers with three petal and six stamens. The flowers are generally blue or purple. The family has 50 genera and 700 species. They have grass-like leaves, and jointed stems with swollen nodes. Doveweed and spreading dayflower are examples of spiderworts that are significant weeds in the landscape.

Doveweed (*Murdannia nudiflora* [L.] Brenan)

Synonym: Naked Stem Dewflower

Doveweed is a summer annual with creeping stems. Its leaves are grass-like, but its purple or blue flowers are much different from grass flowers. Leaves may grow up to 4 inches (10 cm), but are typically shorter. Leaf sheaths generally have fine hairs.

Doveweed usually grows in turf. It is found as far north as Virginia, but is more commonly found in the Gulf Coast states.

Spreading Dayflower (*Commelina diffusa* Burm. f.)

Spreading dayflower is a summer annual, or weak perennial in tropical regions. It has wide lance-shaped leaves with parallel veins. The sheaths are closed and have fine hairs on the upper margin. It has blue flowers with three petals.

Spreading dayflower is found from the southern parts of the midwestern states, throughout the South and into Central America. It is generally found in moist, protected areas in the landscape.

LILY FAMILY (*Liliaceae*)

The lily family is a large family with as many as 4000 species. The plants in this family are highly variable, but they generally have parallel veins and an alternate leaf arrangement. Many members of this family are ornamentals and houseplants. A few, however, may be weed problems in the landscape. The most common of these weeds in the United States are Star of Bethlehem, wild garlic, and wild onion.

Star of Bethlehem (*Ornithogalum umbellatum* L.)

Star of Bethlehem is a cool-season perennial that spreads by seed and underground bulbs. Its leaves are grass-like or may look like wild onion or garlic; however, it lacks the distinctive odor of these plants. Leaves, which have a white stripe down the midrib, are generally a few inches long, but may range up to 12 inches (30 cm). It produces white flowers with six petals that give the plant a star-like appearance.

Star of Bethlehem was originally introduced into the United States as an ornamental. It has become a common weed throughout most of the country as well as Canada.

Wild Garlic (*Allium vineale* L.)
Wild Onion (*Allium canadense* L.)

Wild garlic and wild onion look a little like sedges in the lawn, but they are members of the lily family and are not sedges. They are cool-season plants. The garlic has hollow leaves that are round in cross section. Both species are common in Canada and through much of the United States. They are most common in the eastern United States and down into the southern states. The two are similar in appearance. The onion, however, has a fibrous coat on its bulb, whereas the garlic has a membranous coat. The leaf of the onion is also more flat and is not hollow.

Both species are widely distributed through the eastern United States and through the central southern states. They often appear in thin turf and open, disturbed sites.

*2 Broadleaf Weeds

ANNUAL BROADLEAVES

Broadleaf (dicotyledonous) weeds may be annuals in some climatic regions and perennials in other regions. The following weeds are generally annual through much of the northern and central regions of the United States. As was the case with the annual grasses, there are both summer annual and winter annual types. For annual broadleaves, both preemergence and postemergence controls are available. While annual grasses are generally best controlled with preemergence herbicides, postemergence herbicides are often the best choice for annual broadleaves. The reason is that annual broadleaves often germinate before or after annual grasses, making additional applications necessary if a preemergence approach is chosen.

Black Medic (*Medicago lupulina* L.)

Black medic has alternate trifoliate leaves with a long stalk on the central leaflet. Each leaf is jagged (toothed) at the upper edge. It has a spreading growth habit from a central taproot. It forms an open mat that can be a foot (0.3 m) or more in diameter. Its flower is yellow, and in late summer it produces a black, coiled seed pod. It is generally a summer annual, but may be a winter annual or biennial in some areas. It is found throughout the United States.

Black medic is often found in turf, but may also be found in open areas in the landscape. Its presence in turf is usually an indication that the grass is not as competitive as it could be with proper management. Black medic is often found on hillsides where water does not readily penetrate the soil or in compacted areas. Preemergence or postemergence herbicides will provide satisfactory control of this species, but proper lawn management will generally prevent it from becoming established.

Carpetweed (*Mollugo verticillata* L.)

Carpetweed is a summer annual with multiple creeping stems that have whorls of five to six shiny spoon-shaped leaves at regular intervals along their length. White flowers arise on stalks from the leaf axils. Tiny reddish seeds are produced from midsummer until the first frost.

Carpetweed is found throughout the United States in turf and open areas in the landscape. It is often associated with shade, but may also be found in full-sun areas. It is a common problem in spring seeded lawns. It forms a mat (carpet) of prostrate, branched stems that cover the surface of the ground. Preemergence and postemergence herbicides can be used to control it. Hand pulling may also be effective if the problem is not widespread.

Catchweed Bedstraw (*Galium aparine* L.)

Synonym: Stickywilly

Catchweed bedstraw forms a whorl of six to eight leaves along multiple creeping stems, similar to that of carpetweed. Unlike carpetweed, which is smooth, catchweed bedstraw is covered with stiff bristles that cause it to attach to clothes and dog hair. It is a winter annual that produces white flowers in the spring. The flowers are located on stems arising from the leaf axil.

Catchweed bedstraw is found throughout the United States and in Canada. It is commonly found in shaded areas and along wood lots, but may also be found in full sun in the landscape. Postemergence broadleaf-weed controls will kill it, but hand weeding can also be effective.

Common Chickweed (*Stellaria media* [L.] Vill.)

Common chickweed has spreading stems with smooth, shiny, oblong leaves. Its white flowers form in clusters at the end of the stems. It is a winter annual that germinates in late fall and generally becomes apparent in spring. It usually dies during stress periods of midsummer. It is commonly found in shaded areas and at the bases of trees, but may also be found in more open, sunny areas in the landscape.

Common chickweed forms runners that produce an intertwined mat that covers open areas in the turf and bare-soil areas in the landscape. It also appears in graveled areas along driveways and walks. It is not very competitive with well-maintained turf, and proper management will usually prevent its emergence. Preemergence and postemergence herbicides can control it, but as is the case with most annual broadleaves, a dense turf will prevent it from becoming a problem.

Common Groundsel (*Senecio vulgaris* L.)

Synonym: Ragwort

Common groundsel is a winter annual with upright stems that grow up to 2 feet (0.6 m) high. Its leaves range from deeply lobed to rounded (generally in young leaves. Upper leaves are usually attached directly to the stem. Common groundsel produces multiple yellow flowers that open into white, dandelion-like puffballs which disperse the seed.

Groundsel is found in eastern Canada and throughout most of the northern states of the United States and can be found in California and as far south as Arizona. It does not tolerate mowing and is generally not a serious problem in turf. Open, wet areas in the landscape may readily become infested. It can be killed by most postemergence broadleaf controls.

Common Lespedeza (*Kummerowia striata* [Thunb.] Schindler)

Synonym: Japanese clover

Common lespedeza is a summer annual legume that produces pink to purple flowers late in the season. It has smooth trifoliate leaves with a short spur at the end of each leaflet. It forms a spreading mat that can be a foot (0.3 m) or more in diameter. It may be mistaken for other spreading plants with trifoliate leaves in the landscape, such as black medic or white clover. The spur at the tip of the leaf distinguishes it from the others. Black medic also has a yellow flower, as opposed to the purplish flower of common lespedeza.

Common lespedeza grows in Ohio and Pennsylvania and other central eastern states and extends to Oklahoma and Texas. It is usually found in disturbed sites and bare-soil areas in the landscape. Like clover and black medic, common lespedeza can also form spreading mats in the turf. It can be controlled with selective broadleaf weed controls.

Common Mallow (*Malva neglecta* Wallr.)

Synonym: Round leaf mallow

Common mallow is generally considered to be a summer annual, but may be a biennial in some regions. Its rounded leaves are sometimes mistaken for ground ivy (a perennial). Ground ivy, however, has a triangular stem and a strong mint odor, while mallow has a rounded stem and lacks the mint odor. Common mallow has upright-growing stems, whereas ground ivy is an aggressive spreading species. Mallow produces large flowers that are purple to white and have five petals.

Common mallow is usually found in shaded areas in the landscape, but may also be found in open areas on bare soil or where the turf is thin. It is found throughout the United States, although it is more common in the Northeast and Midwest.

Common Ragweed (*Ambrosia artemisiifolia* L.)

Common ragweed is a summer annual which forms a deep taproot. The plant grows up to 6 feet (1.8 m) or more in height. Its leaves are deeply dissected and fine in texture and take on a ragged appearance. Leaves are opposite on the lower part of the stem and alternate above. Male and female flowers are produced on the same plant.

Common ragweed is well known to those with allergies. Its pollen is widely distributed in late spring through summer. It does not tolerate defoliation and does not persist in lawns, although it will readily germinate in turf and form small plants early in the season. It is commonly found in disturbed areas and along wooded areas in the landscape. It can be controlled by preemergence or postemergence herbicides.

Corn Speedwell (*Veronica arvensis* L.)

The upper leaves of corn speedwell are small and oblong, whereas lower leaves are rounded with small lobes. Blue flowers form in the spring, followed by the formation of a heart-shaped leaf. Its stems, leaves, and seeds have hairs.

Corn speedwell is a winter annual that often appears in the spring, following fall seeding of cool-season grasses. It is a common problem throughout the landscape and is often found on bare-soil areas or where the turf is thin. It is found throughout most of the United States and into Central America.

Dog Mustard (*Erucastrum gallicum* [Willd.] O. E. Schulz.)

Dog mustard is generally found in the northern parts of the midwestern United States and into Canada. It is commonly found in low maintenance areas such as along railroad tracks and roads, but may also be found in landscape areas such as perennial flower beds and gardens. It generally lives as an annual or winter annual. It has multibranched stems with hairs along the margin of the stem that point in a downward direction. Dog mustard's leaves are deeply lobed and opposite, and the plant has yellow flowers generally clustered at the top of the stem.

Field Dodder (*Cuscuta pentagona* Engelm.)

Field dodder is a parasitic plant that lacks chlorophyll. It spreads by seeds. Its stem is yellow and thread-like. It also forms scale-like leaves. Rather than growing in the soil, it attaches to the host plant and draws its energy from the infected plant. It produces light-pink to cream-colored, bell-shaped flowers and tiny seeds.

Field dodder is found throughout most of the United States. It is often associated with field crops like alfalfa and clover, but may also be found on some landscape plants, such as chrysanthemum, English ivy, and petunia. Dodder is generally not controlled with herbicides. It can be removed by hand. Vinegar (acetic acid) has also been reported to kill it (http://www.ipm.ucdavis.edu/PMG/PESTNOTES/pn7496.html).

Florida Pusley (*Richardia scabra* L.)

Synonym: Rough Mexican-clover

Florida pusley has oval leaves with an opposite arrangement on branched hairy stems. It is a summer annual that produces clusters of three to four white flowers at the end of its stems in midsummer.

Florida pusley is found from the northeastern states, through the Midwest, and down into the southern states. It is generally found in bare soil and open disturbed areas.

Hairy Galinsoga (*Galinsoga quadriradiata* Cav.), (*Galinsoga ciliata* [Raf.] Blake)

Synonym: Fringed Quickweed, Common Quickweed, Shaggy-soldier

Hairy galinsoga is a summer annual that has hairy stems and leaves. It generally reaches a height of 1 to 2 feet (0.3 to 0.6 m). Leaves are triangular and jagged on the edges, with opposite arrangement on the stem. The plant has white, three-toothed ray flowers around the edge and many yellow disk flowers at the center.

Hairy galinsoga is found throughout the eastern and midwestern states and as far west as California. It is found on disturbed sites in the landscape and is commonly found in gardens. It does not tolerate defoliation and is rarely found in well-maintained turf.

Hemp Sesbania (*Sesbania exaltata* [Raf.] Rydb. ex A. W. Hill)

Synonym: Danglepod

Hemp sesbania is an annual that can reach 6 to 10 feet (1.8 to 3.0 m) in height. It has opposite, pinnately compound leaves with multiple small leaflets. Leaves are smooth above and slightly hairy below. Flowers are yellow and may have purple to reddish spots. It produces a distinct, curved seedpod that may be tipped, with a beak-like structure. The pod gives it the name "danglepod" in some locations.

Hemp sesbania is found in the northeastern states, south to Florida, and west to California. It is usually found in disturbed areas in the landscape and in gardens. It does not tolerate the continuous defoliation of mowing and is generally not found in turf. It can be controlled by persistent cutting and pulling, or by repeat application of broadleaf herbicides.

Henbit (*Lamium amplexicaule* L.)

Synonym: Henbit Deadnettle

Henbit is a winter annual with four-sided stems. Leaves have an opposite arrangement, with upper leaves lacking petioles and being joined closely to the stem. It produces reddish to purple flowers in the spring. Its rounded leaves and square stem are sometimes mistaken for ground ivy. Ground ivy, however, is a perennial that has spreading stems and a strong mint odor.

A similar weed species is purple deadnettle (*Lamium purpureum*). Purple deadnettle is also a winter annual with a square stem. Its leaves are triangular in shape and do not have the deep lobes of henbit. Its upper leaves may be red to purple in color.

Henbit is found throughout the United States. In the Midwest, it generally germinates very late in the season and may not appear before late October or early November. It is one of the few annual broadleaf species to become a problem in fall-seeded cool-season grasses. It may also be found on bare soil and disturbed areas in the landscape.

Lawn Burweed (*Solvia sessilis* Ruiz & Pavon)

Synonym: Field Burweed

Lawn burweed is a spreading winter annual with opposite leaves. Leaves are deeply lobed, with sparse hairs. While flowers are inconspicuous, its seed has a sharp spine that can penetrate the skin.

Lawn burweed is generally limited to the southern states and is usually not found north of the Carolinas. It can be controlled with preemergence herbicides applied before germination in fall or with postemergence broadleaf controls applied while it is actively growing in spring and summer.

Pineapple Weed (*Matricaria discoidea* DC.)

Pineapple weed is a summer annual with a divided fern-like leaf. Its yellow-green flower is located at the end of the stems. When squeezed, the flower produces the aroma of pineapple.

This weed is not very competitive with a well-maintained turf and generally occurs in compacted areas around the edge of the lawn and along sidewalks. Its appearance is a sign that the management of the lawn needs improvement. Proper turf management, including aerification, should be tried first before herbicides are used to remove it.

Prickly Lettuce (*Lactuca serriola* L.)

Synonym: Wild Lettuce and Compass Plant

Prickly lettuce is a winter annual with a deep taproot that may release a milky sap when it is severed. It is a common weed in crop areas, where it grows up to 6 feet (1.8 m) tall. When cut with a mower, it can survive as a rosette. The deeply lobed leaves alternate on the expanded stem. Leaves have rows of spines along the lower side of the mid-vein. It produces yellow flowers in late spring and summer.

Prickly lettuce is found throughout the United States, with the exception of southern Florida. It is commonly found in landscape beds and thin-turf areas. It rarely persists in well-maintained turf with mowing heights of 2 inches (5 cm) or less. It is also called "compass plant" because of its tendency to align its leaves in a north–south position. It is susceptible to most selective broadleaf herbicides.

Prostrate Knotweed (*Polygonum aviculare* L.)

Prostrate knotweed has spreading stems with symmetrical, alternating dark-green leaves. Each leaf has an ocrea (membranous sheath) at the stem attachment. It can form a dense mat that covers the soil surface.

Prostrate knotweed is a summer annual that is found throughout the United States and into Central America. It germinates very early in the spring and can sometimes be found peeking through melting snow. It is generally a sign of soil compaction and is commonly found at the center of compacted sports fields and along sidewalks. At first appearance, it looks like a grass and is often mistaken for emerging crabgrass. It can appear as much as six to eight weeks before crabgrass germination.

Prostrate knotweed can be controlled with preemergence herbicides, but if these materials are applied early enough to control this species, they will lose their effectiveness before the later germinating species such as goosegrass and crabgrass have emerged. While repeat applications of the preemergence herbicides are effective, they are also expensive. Knotweed can be killed with postemergence, selective broadleaf controls. Chemical controls do little good unless the compacted conditions that led to the infestation are changed. Aerification to relieve compaction, followed by sound management techniques to produce a healthy turf, is generally the best method of control.

Prostrate Pigweed (*Amaranthus blitoides* S. Wats.)

Synonyms: Spreading Amaranth, Prostrate Amaranth

Prostrate pigweed is a summer annual with prostrate, spreading, succulent stems that can cover the soil in a thick mat. Leaves are rounded and light green and may have some reddish coloration. Leaves have an alternate arrangement on the stem. The leaves may be slightly prickly to the touch.

Prostrate pigweed is usually found on disturbed sites and bare-soil areas in the landscape. It is occasionally found along the edge of poorly managed turf area. It is easily controlled with postemergence, selective broadleaf herbicides. Proper management of turf is the best way to keep it from becoming established.

Prostrate Spurge (*Euphorbia maculata* L.)

Prostrate spurge is a summer annual with spreading stems that release a milky sap when broken. It has asymmetrical, opposite leaves that are light green and may have a reddish to purple spot at their center. It spreads from a central taproot and may reach 2 feet (0.6 m) or more in diameter.

Prostrate spurge germinates late in the spring and into the summer, usually around four weeks later than crabgrass and two weeks later than goosegrass. Seed produced in late summer can germinate shortly after falling from the plant. Prostrate spurge is one of the most difficult annual weeds to control in the Midwest because of its time of germination. It can be controlled with preemergence herbicides, but its late germination makes that difficult. If the herbicide is applied early enough for crabgrass, it has generally lost its effectiveness before the spurge germinates. Spurge can be controlled with a number of postemergence herbicides, like 2,4-D and MCPP, but it is risky to use these materials in midseason because of damage to trees and shrubs. Spurge generally germinates into open areas in the lawn, and the best way to control it is with proper cultural techniques that prevent it from becoming established. Prostrate spurge can be found in most parts of the United States; however, it is most prevalent in the Midwest and the northeastern states.

Puncture-vine (*Tribulus cistoides* L.)

Synonyms: Goat's Head, Bullhead

Puncture-vine is a summer annual with spreading stems that originate from a taproot. The leaves are pinnately compound with 16 hairy leaflets. Yellow flowers emerge from the leaf axils in late summer to fall. The fruit have stiff spines that can penetrate shoes and bike tires and are a serious problem for dogs.

Puncture-vine is distributed throughout the United States. It is commonly found on disturbed sites and bare-soil areas in the landscape. It does not compete with well-maintained turf, but will often infest thin turf on sandy soils, particularly on steep slopes.

Purslane (*Portulaca oleracea* L.)

Purslane is a summer annual with a reddish, fleshy stem and rounded, smooth leaves that are narrow at the base. It produces yellow flowers in late summer.

Purslane often infests spring-seeded cool-season grasses. It is commonly found in compacted areas along sidewalks and in other areas where the grass is not doing well. It is a common problem in flower beds and disturbed sites in the landscape. It is easy to control with postemergence herbicides and can easily be pulled by hand.

Shepherd's-purse (*Capsella bursa-pastoris* [L.] Medik.)

Shepherd's-purse is a winter annual that produces a rosette of lobed leaves attached to a central tap root. It produces stems a foot (0.3 m) or more in height in the spring, with clusters of white flowers at their tip. The purse carried by ancient shepherds was the testicle sack of a ram, and the two-lobed seed pods attached to short stalks along the stem of the Shepherd's-purse have the appearance of one of these purses.

Shepherd's-purse occurs in most of the United States and is very common in the Midwest and northeastern states. It is generally found in disturbed sites in the landscape and is common in flower beds and gardens. It is rarely found in healthy turf, but may be found in thin areas and bare spots around lawns. It can be controlled with preemergence and postemergence herbicides, but pulling or hoeing is the preferred method of control in many situations.

Sow Thistle (*Sonchus oleraceus* L.)

Sow thistle is an upright-growing annual. Its stem is hollow and releases a milky sap when broken. Lower leaves are stalked, with deep lobes. The upper leaves have no petiole and wrap around the stem. Most of the plant is smooth, without hairs or bristles, unlike other thistles found in the landscape. The flower is yellow and looks somewhat like a dandelion.

Sow thistle is found in all 50 states. It is commonly found in disturbed areas in the landscape and along roadsides. It does not tolerate mowing and is generally not a problem in turf areas.

Virginia Copperleaf (*Acalypha virginica* L.)

Virginia copperleaf is a summer annual that may reach 2 to 3 feet (0.6 to 0.9 m) in height. Virginia copperleaf begins with opposite, green leaves on a hairy stem. As the plant matures, the leaves become alternate on the stem and take on a copper color. It is monoecious, meaning that it has separate male and female flowers on the same plant.

Virginia copperleaf is found in Texas, several of the midwestern states, and all states east of the Mississippi River. It generally appears in disturbed areas and bare-soil areas in the landscape. It may also infest poorly maintained turf.

Yellow Woodsorrel (*Oxalis stricta* L.)

Synonym: Known in the turf industry by its genus name *Oxalis*

Yellow woodsorrel is an annual in the northern states and a perennial in warmer climatic regions. It is easily recognized by its trifoliate (three-part) leaf with distinct heart-shaped leaflets and by its yellow flowers.

Yellow woodsorrel germinates in midsummer in northern regions and often becomes a problem in August in open areas in turf. This species can be controlled with preemergence herbicides, but its late germination makes it difficult to control in this manner. Selective postemergence herbicides will control it; however, postemergence herbicides are usually avoided, to prevent damage to landscape plants during the warmer part of the summer. Proper irrigation, fertilization, and other sound management steps that keep the lawn healthy provide the best means of preventing this weed from becoming a problem. Yellow woodsorrel is distributed through most of the central and eastern states and into the South. It is a common weed problem in gardens and open-soil areas in the landscape, as well as in open spaces in turf.

PERENNIAL BROADLEAVES

Perennial broadleaves are generally controlled with selective postemergence broadleaf weed controls. Preemergence herbicides will usually prevent germination of their seeds, but these plants will survive the winter and do not require germination from seed to continue growth in the next season. Preemergence herbicides will not provide satisfactory control.

Birdsfoot Trefoil (*Lotus corniculatus* L.)

Birdsfoot trefoil is a perennial forage legume that is sometimes used in place of alfalfa on marginal land. It is also widely used as a part of roadside seed mixtures in the

Midwest and eastern United States. It is easily recognized by its attractive yellow flower that has a similar shape to the flower of the garden pea. Its leaves are five parted and have an opposite arrangement on the stem. This species can be highly invasive and is often found in lawns. It is particularly prevalent along the street curb, where it tends to grow down over the edge of the concrete. It can also invade perennial and annual flower beds and is sometimes found in the garden. It is easily controlled in lawns with the standard broadleaf herbicides used for dandelion and clover.

Pennsylvania Bittercress (*Cardamine pensylvanica* Muhl. ex Willd.)

Pennsylvania bittercress can be an annual, biennial, or perennial. Its leaves are elongated and deeply lobed, with the central lobe being broader than the narrow lateral lobes. Leaves have an alternate arrangement on the stem. Its erect stems can be slightly hairy near the base and can grow a foot (0.3 m) or more in height. It has tiny white flowers that appear in spring and produces elongated seed pods in late summer and fall. Flowers have four petals with yellow stamens in the center.

Pennsylvania bittercress is found in much of the eastern United States and south into northern Florida. It can also be found in the Pacific Northwest and parts of Colorado. It is generally found in the landscape in wet areas along wood lots and in disturbed areas. It is also commonly found along streams.

Blackberry (*Rubus* spp.)

Synonym: Bramble

There are several species of blackberry that become weeds in the landscape. They generally form vines with thorny stems that are quite annoying to people and animals. Heavy growth of wild blackberry is often considered to be a good habitat for rats and mice.

Wild blackberry can be found throughout the United States. It is a serious problem weed in much of the Southeast and parts of the northeastern United States. It can spread by seed and by underground runners. It is not persistent in turf areas, but is commonly found along wooded areas and in open sites in the landscape. It is best controlled by cutting the canes back to the ground, followed by treatment with glyphosate to the tips of the severed stems.

Broadleaf Dock (*Rumex obtusifolius* L.)

Synonym: Bitter Dock

Broadleaf dock has wide heart-shaped leaves with smooth margins that originate from a large taproot. Leaves may be 8 inches (20 cm) wide and 10 inches (25 cm) long. Green flowers are produced in summer, at the end of elongated flowering stems. Seed stalks turn brown late in the season and drop shiny brown seeds with toothed edges.

Broadleaf dock can form a rosette of leaves, but is generally a large plant that will not tolerate mowing and is rarely found in turf. It is usually found in disturbed sites, especially along wooded areas in the landscape. It is found throughout the United States. Hoeing is usually effective in eliminating it. With larger infestations, broadleaf herbicides are generally effective.

Broadleaf Plantain (*Plantago major* L.)

Synonym: Common plantain

Broadleaf plantain is a perennial that produces a rosette of broad, rounded leaves that are from 1 to 2 inches (2.5 to 5.0 cm) wide and 3 to 4 inches (7.6 to 10 cm) long. It produces an upright flowering stalk that forms flowers over one-half to three-quarters of the stalk.

Broadleaf plantain leaves lie flat along the ground's surface, and the plant can easily persist under the mowing height of lawns. It is found throughout the United States, with the exception of the far northeastern states. It is quite susceptible to most selective broadleaf controls and is easily controlled. Mechanical means of control, like hoeing, are generally ineffective because broadleaf plantain easily regrows from the severed taproot.

Buckhorn Plantain (*Plantago lanceolata* L.)

Synonym: Narrowleaf plantain

Buckhorn plantain is a persistent perennial that produces a rosette of narrow, lance-shaped leaves. It has a deep taproot. It produces a narrow seed stalk with a tufted flower spike at its tip, unlike broadleaf plantain that has flowers for several inches down the seed stalk.

Buckhorn plantain is often found in older turf areas like cemeteries and parks. It appears to take a long time to become established and is rarely found in turf the first five years after establishment. It is very difficult to kill. Unlike broadleaf plantain that is easily controlled by herbicides, buckhorn plantain will often survive repeated herbicide application. Late-fall applications of postemergence broadleaf controls will usually provide the best control.

Bull Thistle (*Cirsium vulgare* [Savi] Ten.)

Bull thistle forms deeply lobed rosettes with stiff spines on the tips of the lobes. The upper surface of the leaf is covered with hairs. It is a biennial that forms a rosette in the first year and produces a flower stalk in the second year with large lavender flowers. In the first year, it forms a deep taproot. It does not spread by rhizomes, as Canada thistle does.

Bull thistle can be found throughout the United States. It is commonly found in disturbed areas and bare-soil areas in the landscape. Because of its ability to form a rosette, it can become established in lawns, where its stiff spines can be quite painful to those with bare feet. It may prove difficult to control with herbicides because of the hairy leaf that can prevent droplets of herbicide from coming in contact with the leaf surface. Because of its lack of rhizomes, it can be controlled mechanically by persistent defoliation.

Canada Thistle (*Cirsium arvense* [L.] Scop.)

Canada thistle is a problem weed in crops and roadsides, where it can grow to several feet in height. When defoliated by mowing in turf, it forms a rosette with deeply lobed leaves with sharp spines at the tips of the lobes. The leaves are smooth on the upper surface, unlike the bull thistle, which has spines on the upper surface of its leaves. It has a deep root system and spreads by underground rhizomes. If allowed to grow upright, it produces large lavender flowers. It is dioecious, with fertile male flowers on one plant and fertile female flowers on another plant.

Canada thistle, one of many thistles found in lawns, is the most common and grows throughout the United States and Canada. It is hard to kill. Cutting it off at the above-ground tissue will result in other thistle plants emerging from the rhizomes. Standard herbicides like 2,4-D are somewhat weak on thistles. Herbicides containing dicamba, triclopyr, or chlopyralid will generally provide satisfactory control.

Carolina Geranium (*Geranium carolinianum* L.)

Synonym: Wild geranium

Carolina geranium is a winter annual or, in some regions, a biennial. It has hairy stems with pink to red color. Its leaves are hairy, with blunt, deep lobes. It produces purple flowers in spring and five-parted seed capsules that measure up to 1/2 inch (13 mm).

Carolina geranium occurs throughout the United States. It is usually found in open, disturbed sites and may be found in lawns, although it is not as common as dandelion and broadleaf plantain. It is easily controlled by postemergence broadleaf herbicides.

Cat's-ear Dandelion (*Hypochoeris radicata* L.)

Synonym: False Dandelion

Cat's-ear dandelion is a perennial that forms a rosette of lobed, hairy leaves that grow from a deep taproot. In late spring to early summer it produces a yellow dandelion-like flower on narrow flower stalks that can grow from 1 to 2 feet (0.3 to 0.6 m) tall. Stems produce a milky sap when broken.

Cat's-ear dandelion ranges from the northeastern and northwestern states to Florida and the Gulf states. It is commonly found in disturbed sites and open areas in the landscape and may be found in lawns.

Chicory (*Cichorium intybus* L.)

Chicory is an upright-growing perennial that forms erect stems reaching a height of 3 feet (0.9 m) or more. Its upper leaves are alternate and narrow. Lower leaves are larger and deeply lobed, with bases that wrap around the stem. Its flowers are blue with multiple-ray florets with five lobes at their tips.

Chicory is commonly found in the midwestern and southeastern states. It also grows west into Texas and Oklahoma. It is commonly found along roadsides and in disturbed areas in the landscape. It is occasionally found in lawns, but is generally intolerant of mowing.

Common Cinquefoil (*Potentilla simplex* Michx.)

Synonym: Five Fingers

Common cinquefoil is a perennial with upright-growing stems that occurs in open areas, but may creep along the ground's surface in the landscape. Leaves are palmately divided into five toothed leaflets that alternate on the stem. It produces yellow flowers in spring that measure up to 3/4 inch (19 mm) in diameter.

Common cinquefoil is found throughout the northern states, down into the Carolinas, and as far west as Oklahoma. It is generally controlled by cutting or hoeing, but is susceptible to broadleaf weed controls.

Common Speedwell (*Veronica officinalis* L.)

Synonym: Gipsyweed

Common speedwell is a perennial which spreads by prostrate stems that root at the nodes. Its hairy leaves are rounded and opposite on stems. It produces light-blue flowers on flower spikes in late spring to early summer and has the heart-shaped seed capsules typical of the speedwells.

Common speedwell is generally found in the eastern half of the United States, but has also been reported in the northwestern states. Common speedwell can be difficult to control with standard herbicides and may require repeat applications for effective control.

Creeping Beggarweed (*Desmodium incanum* DC.)

Creeping beggarweed has a trifoliate leaf, with hairy leaflets that are rounded at the base and pointed at the tip. It is a perennial that produces a deep taproot and hairy runners, which root at the nodes. It produces pink to red flowers in summer.

Creeping beggarweed is generally found along the Gulf Coast states, particularly Florida and southern Texas. It infests lawns and disturbed areas in the landscape.

Curly Dock (*Rumex crispus* L.)

Synonym: Sour Dock

Curly dock is a rosette-forming perennial that produces a deep taproot. Its wavy leaves have a smooth surface. Leaves are generally a few inches long, but can grow up to a foot (0.3 m) in length. Each leaf is joined to the stem with a membranous sheath. In unmown areas, it will form a tall flowering stock with light-green flowers.

Curly dock can persist in lawns as a rosette below the mowing height for extended periods. It is also found in disturbed areas in the landscape. It can regrow from the taproot several times following removal of leaves, and control by systemic postemergence broadleaf controls is generally recommended.

Daisy Fleabane (*Erigeron strigosus* Muhl. ex Willd.)

Synonym: Rough Fleabane, Prairie Fleabane

Daisy fleabane is an annual that can occasionally live as a biennial. Its elongated leaves have an opposite arrangement on upright stems that generally reach a height of 1 to 2 feet (0.3 to 0.6 m). It produces clusters of attractive flowers in early summer. Individual flowers have multiple white ray florets, with yellow disk florets at the center.

Daisy fleabane is found throughout most of the United States, with the exception of the Southwest. It is generally found along the edge of the landscape and in open, disturbed sites. It does not tolerate defoliation and is usually not found in lawns. It can be controlled by mowing, and it is susceptible to postemergence, selective broadleaf weed controls.

Dalmation Toadflax (*Linaria dalmatica* [L.] P. Mill.)

Synonym: Butter and Egg Snapdragon

Yellow Toadflax (*Linaria vulgaris* P. Mill.)

Dalmation and yellow toadflax were originally introduced from the Mediterranean area as ornamentals, but have escaped as weeds in rangeland and landscape areas. Both are perennials that form thick stems that are woody near the base and smooth above. Dalmation toadflax can grow 3 feet (0.9 m) or more in height, while yellow toadflax is generally less than 3 feet (0.9 m) tall. Dalmation toadflax has waxy, heart-shaped leaves that may also be lanceolate. The leaves of yellow toadflax are narrower and pointed at both ends. Both species have showy bright-yellow flowers with orange centers that grow at the base of the upper leaves.

Dalmation and yellow toadflax can be found in disturbed areas, in cemeteries, and along roadsides. They can be very difficult to control with herbicides. Repeat pulling can be effective on small infestations in the landscape.

Dandelion (*Taraxacum officinale* Weber ex F.H. Wigg.)

Dandelion has deeply lobed leaves that form a rosette, with leaves radiating from a central taproot. Its yellow flowers, appearing in the spring, mature into white, fluffy puff balls with seeds that are dispersed by wind. Dandelion is a highly variable species, and it is common to find plants that have rounded leaves with a few pointed extensions at their base, pointing back into the center of the rosette.

Dandelion is one of the most universally recognized lawn weeds. It is highly competitive with lawn grasses and is often found in the best-managed lawns. Control by pulling or cutting is difficult, and it can regrow several times after defoliation. It is a very easy species to control with selective, systemic herbicides like 2,4-D, MCPP, dicamba, etc., and lawns are easily kept dandelion-free with a single fall application of these materials.

Dollarweed (*Hydrocotyle* spp.)

Synonym: Pennywort

Dollarweed has rounded bright-green shiny leaves with scalloped margins. The petiole attaches to the center of the leaf, giving the leaf an "umbrella-like" appearance. White flowers form "umbrella-shaped" clusters on narrow stalks above the canopy. It spreads by stolons that root from the nodes.

Dollarweed is a perennial that is common to Florida and other southern states, although there are members of this genus found in the northern states. It is generally found in wet areas in the landscape and is quite competitive with lawn grasses.

English Daisy (*Bellis perennis* L.)

Synonym: European Daisy

English daisy has rounded leaves that are slightly toothed and narrow at their bases. Leaves range from smooth to slightly hairy. This perennial was originally introduced as an ornamental and escaped into the landscape. Flowers are daisy-like and grow on narrow flower stalks that are 4 to 5 inches (10 to 13 cm) tall. The outside petals of the flowers are white to pinkish-white, and the center flowers are bright yellow.

English daisy is generally found through the central states, east of the Mississippi. It is commonly found in turf and disturbed areas in the landscape.

Florida Betony (*Stachys floridana* Shuttlew.)

Synonym: Wild Artichoke, Rattlesnake Weed

Florida betony is a rhizomatous perennial with square stems and opposite leaves. Individual leaves are triangular with toothed margins. It is sometimes called rattlesnake weed because of the white tubers it forms underground that look like a rattlesnake's tail. It forms white to pink flowers with purple spots in late spring.

Florida betony is found as far north as North Carolina and Virginia, and south into Texas. Its primary range is Florida. It infests turf areas and open, disturbed areas in the landscape. It has also been observed to move with nursery stock, and new plantings should be screened for this weed within its range of adaptation.

Field Bindweed (*Convolvulus arvensis* L.)

Synonym: Creeping Jenny

Field bindweed is a perennial that can spread readily in turf and landscape areas. It has 1- to 2-inch (2.5- to 5.0-cm)-long pointed leaves with basal lobes that have the appearance of an arrow point. Its funnel-shaped flowers are pink or white. It blooms through summer and into the fall. It spreads by rhizomes and has a very deep root system. It has creeping, above-ground stems that can climb on landscape plants to a height of several feet.

Field bindweed is common in and around lawns in the midwestern United States and into central Canada. It is also called creeping Jenny in much of this region. It is a spreading vine in the morning glory family. It is very hard to control. Herbicide applications will appear to kill it, but it returns in a few weeks. Persistence with repeated applications of herbicides over a two-season period may be required to kill it, and it is still likely to reoccur from shrub and flower-bed areas and from neighboring property.

Field Horsetail (*Equisetum arvense* L.)

Synonyms: Bottlebrush, Jointed Rush, Cat's Tail, Mare's Tail, Puzzleweed

Field horsetail is a perennial with hollow, jointed stems that reach up to 2 feet (0.6 m) tall. It spreads by extensive underground rhizomes that may have storage tubers. It also produces solid stems with distinct spore-bearing cones. It gets its name from sterile stems that produce a whorl of leaf-like branches which have the appearance of a horse's tail or a small pine tree. Fertile stems generally appear early in the spring, and sterile stems appear later in the season.

Field horsetail is found in almost every state, other than Florida and Hawaii. It is commonly found in disturbed sites and open areas along the edge of turf areas. It does not readily tolerate continuous defoliation, but is quite difficult to control once it is established because of its extensive rhizome system.

Greenbriar (*Smilax* spp.)

Synonyms: Bull Briar, Cat Briar, Bridal Creeper

Greenbriar is a perennial that produces a climbing stem with sharp spines. Its leaves are generally alternate and smooth, although there are some variations by species. It produces reproductive bulbs underground that make it very difficult to control. It is generally found in the landscape at the edge of wooded areas, but may also infest disturbed areas and gardens. It can form a thick growth of vines that is nearly impenetrable.

There are several species of *Smilax,* and various types of this difficult-to-control weed are found in most of the continental United States. Persistent cutting can be an effective control, but it will generally regrow from the underground bulbs.

Ground Ivy (*Glechoma hederacea* L.)

Synonym: Creeping Charlie

Ground ivy is a perennial with square, spreading stems which can form a dense mat that may crowd out grass in lawns. It has opposite, rounded leaves that are scalloped along the edges and rough on the upper surface. Purple to blue trumpet-shaped flowers are formed from mid-season to fall.

Ground ivy, or creeping Charlie, is well adapted to shade and was introduced to the United States as a groundcover for shaded areas in the lawn. It spreads readily into full-sun areas. It is known in many parts of the United States as one of the hardest lawn weeds to control. It roots along the length of the spreading stems, and pulling or other forms of mechanical control are nearly impossible. It is very difficult to control with selective, postemergence broadleaf controls. The best time to treat ground ivy with herbicides is in the late fall. Even if satisfactory control is attained, runners quickly reinfest the lawn from fence rows and surrounding lawns, and the problem usually returns within a year of treatment. The only answer to the ground ivy problem is persistence. Repeated application over two or three seasons may be needed to control it.

Hairy Beggarticks (*Bidens pilosa* L.)

Synonym: Common Beggarticks

Hairy beggarticks can be either annual or perennial, depending on location. It has smooth, spreading stems that can root at the nodes. Leaves are compound, with three to nine leaflets that are arranged oppositely on the stem. It produces an attractive flower with white bracts and a yellow center.

Hairy beggarticks is generally restricted to the southern states and is found as far west as California. It is commonly found in the landscape in disturbed areas, but may also be found in open areas along the edge of poorly maintained turf.

Horsenettle (*Solanum carolinense* L.)

Horsenettle is a rhizomatous perennial with prickly stems. It has opposite leaves that range from deeply lobed to wavy margins. It has spines on the veins. Its flowers have five petals and may vary from white to light purple. The fruit is round and smooth, with a light-green to yellow color, and ranges from 1/4 to 1/2 inch (6.4 to 13 mm) in diameter.

Horsenettle is generally a problem in the eastern United States, but may be found in Kansas, Texas, and California. It is often found in flower beds, gardens, and disturbed areas in the landscape. It does not tolerate mowing and can be killed by persistent defoliation. It is also susceptible to most postemergence, broadleaf herbicides.

Japanese Honeysuckle (*Lonicera japonica* Thunb.)

There are several honeysuckles (*Lonicera* spp.) that are considered weeds in various crops. Japanese honeysuckle (*Lonicera japonica*) is the primary species found in turf and landscape areas. It is a perennial woody vine with underground rhizomes. It has elongated, rounded leaves that have an opposite arrangement on the stem and paired, white, tubular-shaped flowers. It produces black berries in late summer to early fall. There are several varieties of *Lonicera japonica* that vary somewhat in flower and berry color and in leaf appearance.

Japanese honeysuckle is found throughout the central and southern states, from California to Pennsylvania. It has also been reported in Wisconsin and into the northeastern states to Maine. It is sometimes used as an ornamental, but readily spreads into disturbed areas and surrounding wooded areas. Japanese honeysuckle is considered a primary noxious weed in several states. It can be controlled by persistent cutting and pulling, although it will readily regrow from rhizomes. It is susceptible to glyphosate and phenoxy herbicides.

Japanese Knotweed (*Polygonum cuspidatum* Sieb. & Zucc.)

Japanese knotweed is an aggressive shrubby plant that spreads by rhizomes and can grow up to 8 feet (2.4 m) high. Its stems are reddish in color. Its leaves are up to 6 inches (15 cm) long, with a rounded base and a pointed tip. It is dioecious, with male and female flowers on separate plants.

Japanese knotweed is found throughout the northeastern and midwestern states and in the Pacific states as far south as California. It is an escaped ornamental that appears as a weed in gardens and disturbed sites in the landscape. It can be difficult to control because of its rhizome system, but persistent cutting and pulling can eliminate it. It is also susceptible to glyphosate and triclopyr.

Kudzu Vine (*Pueraria montana* var. lobata [Willd.] Maesen & S.M. Almeida)

Kudzu vine is a rapidly spreading perennial that can climb to the tops of tall trees. Its stems become woody as it matures. Leaves are compound, with three leaflets that can be as large as 4 inches (10 cm) across. Leaves, which are pubescent on the lower surface, can vary from ovate with a pointed tip to deeply lobed with two to three lobes per leaflet. Leaf arrangement is alternate on the stem. The vines produce purple flowers in late summer that produce a grape-like odor. Seed pods are brown with long tan hairs.

Kudzu was introduced into the United States as a forage and ornamental in the late 1800s. It has become a serious weed problem in warm climatic areas and may cover everything in the landscape, including mature trees and even buildings on abandoned sites. It is generally limited to the southeastern United States and west to Texas and Oklahoma. Fortunately, it is intolerant of frost. It is usually found along roadsides and at the edge of the landscape. Persistent cutting and pulling can prevent its establishment in maintained landscape areas. It is intolerant of mowing and does not become a lawn weed.

Longstalked Phyllanthus (*Phyllanthus tenellus* Roxb.)

Longstalked phyllanthus is an upright-growing perennial that has alternated leaves. Individual leaves have two rows of alternately arranged leaflets. Individual leaflets lack petioles. It produces round green fruit in late summer.

Longstalked phyllanthus is generally restricted to the southwestern states, west to Texas. It is also found in Hawaii. It often occurs in disturbed sites and open areas in the landscape. It is a particular problem in nurseries and may be transported with nursery plants. It can be controlled with preemergence herbicides before germination, or with selective, postemergence broadleaf controls.

Mat Lippia (*Phyla nodiflora* [L.] Greene)

Synonym: Matchweed, Creeping Lip Plant, Turkey Tangle Fogfruit

Mat lippia is a perennial spreading species which forms a dense mat of stems on the soil surface that root at the nodes. Leaves are oblong to slightly rounded and toothed near the tip, with an opposite arrangement on the stem. It forms round clusters of pale pink to purple flowers in summer.

Mat lippia readily tolerates mowing and can be a problem in turf. It has even been proposed as a substitute for grass in some landscape areas. It is particularly well adapted to sandy soils. It is generally limited to the southern states from Florida to California and can be found in Hawaii. It can be controlled with repeat applications of selective broadleaf weed controls.

Milkweed, common (*Asclepias syriaca* L.)

Synonym: Silkweed

Common milkweed is a perennial that grows up to 4 feet (1.2 m) in height. Its leaves are oblong and pubescent, with prominent veins. They have an opposite arrangement on the stem. It produces pink to purple flowers in summer.

Common milkweed is generally found throughout the Midwest and may be found as far south as Oklahoma. It is generally found in disturbed areas in the landscape and is common in road ditches. It is intolerant of mowing and is usually not found in well-maintained lawns.

Mouse-Ear Chickweed (*Cerastium fontanum* ssp. vulgare [Hartman] Greuter & Burdet)

Mouse-ear chickweed is a perennial (winter annual in some areas) that spreads by short runners. Its leaves have an opposite arrangement on the stems and are rounded to oblong and covered with fine hairs above and along veins on the underside. They have the appearance of a mouse's ear. Flowers are white with five deeply lobed petals.

Mouse-ear chickweed can tolerate mowing and often becomes a problem in lawns. It can also be found in golf course greens at mowing heights of 3/16 (4.8 mm) inch and below. It is found throughout the United States. It is difficult to control, primarily because its hairy leaves prevent herbicides from reaching the surface of the leaf. Liquid herbicide applications that contain a surfactant (wetting agent) to improve penetration of the herbicide to the leaf surface will improve control.

Mugwort (*Artemisia vulgaris* L.)

Synonym: Common Wormwood

Mugwort is a perennial that spreads by rhizomes. It has alternate, variegated leaves that are often yellow-green in color. The upper surface of the leaf is usually smooth or slightly hairy, while the lower side of the leaf is distinctly hairy. Leaves are aromatic when crushed. It forms an upright clump of leaves and stems that may reach up to 3 feet (0.9 m) in height. It produces white flowers on dense panicles in summer.

Mugwort is found throughout the northeastern and southeastern states as far south as Florida. It is also common in the midwestern states west to Iowa and Kansas and in the Pacific region, including Washington and Oregon. It is common in gardens and disturbed sites and may occasionally infest lawns. It is easily controlled with standard selective broadleaf weed controls.

Poison Ivy (*Toxicodendron radicans* [L.] Kuntze)

Poison ivy is a perennial vine-forming plant that may spread over the ground, climb tree trunks or other upright structures in the landscape, or form an upright growing bush. Its glossy leaves are composed of three leaflets that may range from smooth to slightly lobed. It may also form a thumb-like notch at the edge of the leaflets. Two of the leaflets are opposite, and the third grows straight from the central leafstalk. It produces small green flowers with five petals on slender branches that emerge from the juncture of the leaf petiole and the branch. It forms waxy berries in the fall.

Poison ivy is very well known for the allergic reaction that it produces in most humans. It is generally found along the edge of the landscape, often in wooded areas. It is found in all 50 states. Avoid contact with stems and leaves. It can be killed with glyphosate and is susceptible to broadleaf herbicides.

Purple Cudweed (*Gamochaeta purpurea* [L.] Cabrera)

Synonym: Catfoot, Rabbit Tobacco

Purple cudweed is a biennial, but also may be a summer or winter annual in northern locations. It can form an upright woody stem that can grow over a foot (0.3 m) in height. It also forms a basal rosette of leaves that can persist in regularly mown turf areas. Leaves are oblong, with wavy edges and blunt tips and may reach 4 inches (10 cm) or more in length. Stems and the underside of leaves are densely hairy.

Purple cudweed is found in most of the United States, but is most commonly found as a landscape weed in the southern states. It is best adapted to sandy soils and dry upland areas. It is often found in disturbed sites and may be a problem in low-maintenance turf. It can often be eliminated through improved turfgrass maintenance and is susceptible to selective broadleaf weed controls.

Prostrate Vervain (*Verbena bracteata* Lag. & Rodr.)

Synonym: Bigbract Verbena

Prostrate vervain can live as an annual, a biennial, and a perennial. It forms multiple, prostrate stems that branch from a central taproot and grow a foot (0.3 m) or more in length. It forms opposite leaves that are up to 3 inches (7.6 cm) long. Lower leaves have multiple lobes, while upper leaves are smaller. Through most of the summer, it produces a terminal spike of purple flowers at the end of the stems.

Prostrate vervain is found throughout most of the Midwest and northeastern states. It is commonly found in disturbed sites and on roadsides. In lawns, it is usually limited to compacted areas along sidewalks and street curbs. It is very drought tolerant and is often noticed during dry summers in non-irrigated turf. It often produces stems that grow down over the curb along streets. It is not very competitive with well-maintained turf. Reducing compaction and improving conditions for growth of turf will usually prevent its establishment. It can be controlled by selective, broadleaf herbicides.

Red Sorrel (*Rumex acetosella* L.)

Synonym: Sheep Sorrel

Red sorrel is a deep-rooted perennial that spreads by rhizomes. It is known for its arrow-or lance-shaped leaves. It produces a deep taproot and rhizomes. It develops reddish flowers on long stalks that protrude above the canopy in late spring through summer. It is dioecious with male and female flowers on separate plants.

Red sorrel is found throughout most of the United States and is common in lawns in the Midwest and northeastern states. It is often associated with acidic soils, but can be found on soils with a pH of 7 or above. In acidic soils, where the turf is thin, it can take over large sections of the lawn.

Virginia Buttonweed (*Diodia virginiana* L.)

Virginia buttonweed is a spreading perennial with opposite leaves on hairy, spreading stems. Leaves are oblong to lance shaped and join directly to the stem. Leaves begin the season as dark green and shiny, but may become yellow in late summer due to the activity of a virus. Hairy fruit are formed in the leaf axil in late summer.

Virginia buttonweed is often found in wet areas. It can persist below the mowing height of lawns and often becomes a problem in turf and disturbed sites in the landscape. It is generally a problem in the southeastern states, but may be found as far north as Missouri and Illinois. It can be controlled by preemergence herbicides applied before germination. Once established in the lawn, it can be controlled with postemergence, selective, broadleaf controls.

Virginia creeper (*Parthenocissus quinquefolia* [L.] Planch)

Virginia creeper is a creeping vine with a woody stem. Its leaves are alternate on the stem. They are palmately compound, with five elliptical leaflets per leaf. Leaves turn dark red in fall. It forms small, inconspicuous green flowers in summer and forms clusters of blue to black berries in late summer and fall.

Virginia creeper is an ornamental that is often used to cover brick walls and the sides of buildings in the landscape. It also becomes a weed and can be found throughout most of the eastern and midwestern states and west to Colorado. It is generally found growing on the edge of the landscape and may climb to the top of tall trees as high as 50 feet (15 m). It can be controlled by persistent cutting and pulling. Treating the severed stem with glyphosate or auxin-like herbicides will help prevent its regrowth.

Western Salsify (*Tragopogon dubius* Scop.)

Synonym: Goat's Beard, Yellow Salsify

Western salsify is an annual or biennial that produces an upright stem with a swelling in the stem just below the flower. It has lanceolate leaves that grow directly around the stem without a petiole. It produces a large yellow flower with long, pointed, green bracts. The flower opens in the morning and is closed by the bracts in midday. When the flower is mature, it forms a dandelion-like puffball of seeds that disperse by wind.

Western salsify is found in all states, with the exception of Hawaii and the far southeastern states. It is commonly found in disturbed areas in the landscape and may become established in gardens and other open-soil sites. It is intolerant of defoliation and is rarely found in lawns, with the exception of very low-maintenance turf.

White Clover (*Trifolium repens* L.)

White clover is recognized by its trifoliate leaf with three rounded leaflets and a white to pink flower that blooms through much of the growing season. Individual leaflets may have a whitish band near the base. It is a perennial that spreads extensively by creeping stems rooting from the nodes.

White clover is quite tolerant of mowing and is one of the most common weeds found in turf. In the 1940s and 1950s, it was often seeded with Kentucky bluegrass in low-maintenance lawns. Because of this practice, white clover is one of the most widely distributed weeds in cool-season lawns in the Midwest and northeastern United States. It can be found in all 50 states in the United States. It is a legume and releases nitrogen that can stimulate surrounding grass. It thrives on moisture and is usually most noticed in wet summers. White clover is somewhat difficult to control by common herbicides, and an application of 2,4-D alone usually will not kill it. MCPP, triclopyr, chlopyralid, and dicamba are effective, as is quinclorac (Drive).

Wild Carrot (*Daucus carota* L.)

Synonym: Queen Anne's Lace

Wild carrot is a biennial that produces a deep, narrow, carrot-like taproot. The plant also has a carrot-like odor when crushed. The first year, it forms a rosette with divided leaves. The second year, a tall flower stalk is formed that may grow 2 feet (0.6 m) or more in height. The flowers form an attractive umbel, with white flowers surrounding a reddish flower at the center, which gives the plant its other common name, Queen Anne's Lace.

Wild carrot is commonly found along roadsides and in open, disturbed sites in the landscape in the midwestern and northeastern states. It also may be found as far south as northern Florida and west into Texas. It is a problem in flower beds and gardens and may occur in low-maintenance turf. Its taproot makes it difficult to control by cutting or pulling, but persistent weeding will kill it. Systemic herbicides will provide good control.

Wild Mustard (*Sinapis arvensis* L.)

Synonym: Charlock Mustard (also listed as *Brassica kaber*)

Wild mustard is a summer or winter annual with branched stems. Its lower leaves have deep lobes, and its upper leaves are narrow and toothed. Leaves are hairy along the veins on the underside of their surface. Its flower is bright yellow, with four petals that grow in clusters at the end of the upright stem.

Wild mustard is a common problem in disturbed sites, in flower beds, and along roadsides. It is found in most of the United States and is a common landscape weed in the Midwest and northeastern states. It can be controlled by cutting and pulling, as well as by both preemergence and postemergence broadleaf herbicides.

Wild Strawberry (*Fragaria virginiana* Duchesne)

Wild strawberry is a perennial that spreads by creeping stolons. It has a trifoliate leaf with toothed margins on the upper part of the leaflet and a hairy petiole. It produces a white flower with five petals and a yellow center. It produces a small, edible strawberry in summer. There is a similar species generally found in the southeastern United States called Indian mock strawberry (*Duchesnea indica*) that has leaflets toothed along the entire margin and produces a yellow flower.

Wild strawberry is found throughout most of the United States. It can tolerate mowing and is often found in lawns. It is also found along the edge of wooded areas, in roadside ditches, and in disturbed sites in the landscape. It can be cut or pulled, and it is susceptible to postemergence broadleaf weed controls.

Wild Violet (*Viola pratincola* Greene)

Wild violet is a perennial that can spread by both stolons and rhizomes. Its leaves are shiny, with a waxy surface. They are heart shaped, with rounded lobes along the margin. It produces an attractive blue to purple flower and can be an ornamental as well as a weed.

Wild violets are found throughout the United States, although the species varies in some locations. It is commonly found throughout the landscape and is often present at the edge of woody areas in lawns. The waxy surface of its leaf makes penetration by herbicides difficult. It will often survive when all other weeds in the lawn have been controlled. Herbicides containing dicamba and triclopyr may provide the best control. Surfactants added to the spray can help with penetration into the leaf.

Yarrow (*Achillea millefolium* L.)

Yarrow is a perennial with finely divided, long, narrow fern-like leaves that are covered with fine hairs. It produces white flat-topped flower clusters at the tips of its branched stems. It spreads by underground rhizomes and can form dense patches in the landscape.

Yarrow is found throughout most of the United States. It is a common lawn weed that can spread extensively below the mowing height. In other areas of the landscape, it can reach a height of a foot or more. It is difficult to control by cutting or pulling because of its rhizome system. Chemical control may require repeat application.

Wintercreeper (*Euonymus fortunei* [Turcz.] Hand.-Mazz.)

Wintercreeper is a perennial, evergreen climbing vine capable of forming a dense mat on the ground's surface or of climbing tall trees. Its leaves are egg-shaped, with a shiny dark-green surface and silvery veins. It produces clusters of small green-white flowers in early summer. It forms a pink to red fruiting capsule that splits open to expose an orange seed in fall.

Wintercreeper is an ornamental that can form attractive beds in the landscape. It easily escapes into surrounding areas, however, and can form dense vine covers in wooded areas along the edge of the landscape. It is found throughout the Midwest and the northeast and south central states. Cutting and pulling can help prevent it from spreading into undesirable areas. Treating severed stems with glyphosate or broadleaf weed controls can help prevent regrowth.

Yellow Rocket (*Barbarea vulgaris* Ait. f.)

Yellow rocket can be a winter annual, biennial, or perennial. It forms a deep taproot. It has shiny green leaves with deep lobes that form a rosette. There are smaller lobes at the base of the leaf and a large terminal lobe at the tip. It can look a little like a dandelion in lawns, but dandelion leaves have pointed lobes that point back to the center of the rosette. Yellow rocket produces small yellow flowers with four petals that grow in clusters at the tips of erect branches.

Yellow rocket is often found in lawns, but is also a common weed in disturbed sites, flower beds, and gardens, where it can reach a height of 2 feet (0.6 m) or more. It can be found throughout most of the United States, but is most commonly found as a lawn weed in the northeastern and midwestern states. It can be cut or pulled, but it may regrow from the taproot. The same herbicides that control dandelions will control yellow rocket in lawns.

Annual Bluegrass (*Poa annua* L.)

Barnyardgrass (*Echinochloa crus-galli* [L.] Beauv.)

Germinating Crabgrass

Large Hairy Crabgrass (*Digitaria sanguinalis* [L.] Scop.)

Smooth Crabgrass (*Digitaria ischaemum* [Schreb.] Schreb. ex Muhl.)

Southern Crabgrass (*Digitaria ciliaris* [Retz.] Koel.)

Tropical Crabgrass (*Digitaria bicornis* [Lam.] Roemer & J.A. Schultes ex Loud.)

Blanket Crabgrass (*Digitaria serotina* [Walt.] Michx.)

Crowfootgrass (*Dactyloctenium aegyptium* [L.] Willd.)

Fall Panicum (*Panicum dichotomiflorum* Michx.)

Yellow Foxtail (*Setaria pumila* [Poir.] Roem. & J.A. Schult.) (left) and **Giant Foxtail** (*Setaria faberi* Herrm.)

Yellow Foxtail (*Setaria pumila* [Poir.] Roem. & J.A. Schult.)

Green Foxtail (*Setaria viridis* [L.] Beauv.)

Goosegrass (*Eleusine indica* [L.] Gaertn.)

Goosgrass (left) and **Crabgrass**

Lovegrass (*Eragrostis pilosa* [L.] Beauv.)

Field Sandbur (*Cenchrus spinifex* Cav.)

Southern Sandbur (*Cenchrus echinatus* L.)

Witchgrass (*Panicum capillare* L.)

Bahiagrass (*Paspalum notatum* Flueggé)

(Courtesy of Dr. David Minner.)

Bermudagrass (*Cynodon dactylon* [L.] Pers.)

Broomsedge (*Andropogon virginicus* L.)

Creeping Bentgrass (*Agrostis stolonifera* L.)

Dallisgrass (*Paspalum dilatatum* Poir.)

Johnsongrass (*Sorghum halepense* [L.] Pers.)

Kikuyugrass (*Pennisetum clandestinum* Hochst. ex Chiov.)

Nimblewill (*Muhlenbergia schreberi* J.F. Gmel.)

Orchardgrass (*Dactylis glomerata* L.)

Perennial Ryegrass (*Lolium perenne* L.)

Quackgrass (*Elymus repens* [L.] Gould)

Quackgrass Rhizome

Rough Bluegrass (*Poa trivialis* L.)

Smooth Bromegrass (*Bromus inermis* Leyss.)

Tall Fescue (*Festuca arundinacea* Schreb.)

Thin Paspalum (*Paspalum setaceum* Michx.)

Timothy (*Phleum pratense* L.)

Smutgrass (*Sporobolus indicus* [L.] R. Br.)

Torpedograss (*Panicum repens* L.)

Tropical Signalgrass (*Urochloa subquadripara* [Trin.] R.D. Webster)

Windmillgrass (*Chloris verticillata* Nutt.)

Zoysiagrass (*Zoysia japonica* [Hack.] Forbes)

Annual Sedge (*Cyperus compressus* L.)

Cylindric Sedge (*Cyperus retrorsus* Chapm.)

False Nutsedge (*Cyperus strigosus* L.)

Green Kyllinga (*Kyllinga brevifolia* Rottb.)

Hurricanegrass (*Fimbristylis cymosa* R. Br.)

Purple Nutsedge (*Cyperus rotundus* L.)

Tufted Kyllinga (*Cyperus tenuifolius* [Stevd.] Dandy)

Yellow Nutsedge (*Cyperus esculentus* L.)

Doveweed (*Murdannia nudiflora* [L.] Brenan)

Spreading Dayflower (*Commelina diffusa* Burm. f.)

Star of Bethlehem (*Ornithogalum umbellatum* L.)

Wild Garlic (*Allium vineale* L.)

Wild Garlic

Wild Onion (*Allium canadense* L.)

Black Medic (*Medicago lupulina* L.)

Carpetweed (*Mollugo verticillata* L.)

Catchweed Bedstraw (*Galium aparine* L.)

Common Chickweed (*Stellaria media* [L.] Vill.)

Common Groundsel (*Senecio vulgaris* L.)

Common Lespedeza (*Kummerowia striata* [Thunb.] Schindler)

Common Mallow (*Malva neglecta* Wallr.)

Common Ragweed (*Ambrosia artemisiifolia* L.)

Corn Speedwell (*Veronica arvensis* L.)

(Courtesy of Gary Fewless, Cofrin Center for Biodiversity, University of Wisconsin, Green Bay, WI.)

Dog Mustard (*Erucastrum gallicum* [Willd.] O. E. Schulz.)

(Courtesy of M.J. Hatfield.)

Field Dodder (*Cuscuta pentagona* Engelm.)

Florida Pusley (*Richardia scabra* L.)

(Courtesy of Kevin Bradley, Division of Plant Sciences, University of Missouri, Columbia, MO.)

Hairy Galinsoga (*Galinsoga quadriradiata* Cav.), (*Galinsoga ciliata* [Raf.] Blake)

Hemp Sesbania (*Sesbania exaltata* [Raf.] Rydb. ex A. W. Hill)

Henbit (*Lamium amplexicaule* L.)

Lawn Burweed (*Solvia sessilis* Ruiz & Pavon)

Pineapple Weed (*Matricaria discoidea* DC.)

Prickly Lettuce (*Lactuca serriola* L.)

Prostrate Knotweed (*Polygonum aviculare* L.)

Prostrate Pigweed (*Amaranthus blitoides* S. Wats.)

Prostrate Spurge (*Euphorbia maculata* L.)

Puncture-vine (*Tribulus cistoides* L.)

Purslane (*Portulaca oleracea* L.)

Shepherd's-purse (*Capsella bursa-pastoris* [L.] Medik.)

Sow Thistle (*Sonchus oleraceus* L.)

Virginia Copperleaf (*Acalypha virginica* L.)

Yellow Woodsorrel (*Oxalis stricta* L.)

Birdsfoot Trefoil (*Lotus corniculatus* L.)

Pennsylvania Bittercress (*Cardamine pensylvanica* Muhl. ex Willd.)

Blackberry (*Rubus* spp.)

Broadleaf Dock (*Rumex obtusifolius* L.)

Broadleaf Plantain (*Plantago major* L.)

Buckhorn Plantain (*Plantago lanceolata* L.)

Broadleaf Plantain (left) and Buckhorn Plantain

Bull Thistle (*Cirsium vulgare* [Savi] Ten.)

Canada Thistle (*Cirsium arvense* [L.] Scop.)

Carolina Geranium (*Geranium carolinianum* L.)

Cat's-ear Dandelion
(*Hypochoeris radicata* L.)

Chicory
(*Cichorium intybus* L.)

Common Cinquefoil (*Potentilla simplex* Michx.)

Common Speedwell (*Veronica officinalis* L.)

Creeping Beggarweed (*Desmodium incanum* DC.)

Curly Dock (*Rumex crispus* L.)

Daisy Fleabane (*Erigeron strigosus* Muhl. ex Willd.)

Dalmation Toadflax
(*Linaria dalmatica* [L.] P. Mill.)

(Courtesy of Neil Harker and Don Dew, Agriculture & Agri-Food Canada, Lacombe Research Centre, Lacombe, Alberta, Canada.)

Yellow Toadflax (*Linaria vulgaris* P. Mill.)

Dandelion (*Taraxacum officinale* Weber ex F.H. Wigg.)

Dollarweed (*Hydrocotyle* spp.)

English Daisy (*Bellis perennis* L.)

Florida Betony (*Stachys floridana* Shuttlew.)

Field Bindweed (*Convolvulus arvensis* L.)

Field Horsetail (*Equisetum arvense* L.)

Greenbriar (*Smilax* spp.)

Ground Ivy (*Glechoma hederacea* L.)

Hairy Beggarticks (*Bidens pilosa* L.)

Horsenettle (*Solanum carolinense* L.)

Japanese Honeysuckle (*Lonicera japonica* Thunb.)

Japanese Knotweed (*Polygonum cuspidatum* Sieb. & Zucc.)

Kudzu Vine (*Pueraria montana* var. lobata [Willd.] Maesen & S.M. Almeida)

Longstalked Phyllanthus (*Phyllanthus tenellus* Roxb.)

Mat Lippia (*Phyla nodiflora* [L.] Greene)

Milkweed, common (*Asclepias syriaca* L.)

Mouse-Ear Chickweed (*Cerastium fontanum* ssp. vulgare [Hartman] Greuter & Burdet)

Mugwort (*Artemisia vulgaris* L.)

Poison Ivy (*Toxicodendron radicans* [L.] Kuntze)

Purple Cudweed (*Gamochaeta purpurea* [L.] Cabrera)

Prostrate Vervain (*Verbena bracteata* Lag. & Rodr.)

Red Sorrel (*Rumex acetosella* L.)

Virginia Buttonweed (*Diodia virginiana* L.)

Virginia creeper (*Parthenocissus quinquefolia* [L.] Planch)

Western Salsify (*Tragopogon dubius* Scop.)

White Clover (*Trifolium repens* L.)

Wild Carrot (*Daucus carota* L.)

Wild Mustard (*Sinapis arvensis* L.)

Wild Strawberry (*Fragaria virginiana* Duchesne)

Wild Violet (*Viola pratincola* Greene)

Yarrow (*Achillea millefolium* L.)

Wintercreeper (*Euonymus fortunei* [Turcz.] Hand.-Mazz.)

Yellow Rocket (*Barbarea vulgaris* Ait. f.)

unit two
Herbicides

CHAPTER 3 **Herbicides in the Plant** CHAPTER 4 **Herbicides in the Environment**
CHAPTER 5 **Herbicide Formulations and Application**

The first turfgrass herbicides date back to the early 1900s when several *inorganic* compounds, including ferrous ($FeSO_4$) and ferric ($Fe_2[SO_4]_3$) sulfates, sodium chlorate ($NaClO_3$), and sodium arsenite ($NaHAsO_2$), were used for selectively controlling some annual grasses and broadleaf weeds. These chemicals could also discolor and burn turfgrasses, especially at high application rates in hot, dry weather. The first of the *organic* herbicides—the dinitrophenols—were discovered in the 1930s in France; however, their utility in turf was limited by their relatively poor selectivity and efficacy. At about the same time, in Germany, the naturally occurring growth-regulating compound indoleacetic acid (IAA) was also discovered, but it was not until World War II began that a synthetic analog of this compound—2,4-dichlorophenoxyacetic acid (2,4-D)—was found to be effective in killing broadleaf weeds in turf at low application rates and with an unusually high degree of selectivity. After the war, the use of 2,4-D expanded rapidly, introducing the modern era of herbicide technology. In the 1950s, other phenoxy herbicides (e.g., silvex, 2,4,5-T), as well as herbicides from other chemical families (e.g., benzoic acids: dicamba and DCPA; dinitroanilines: benefin and trifluralin; triazines: atrazine and simazine), were developed.

Some of the herbicides that were important in turf are no longer used. For example, the arsenates—lead arsenate ($PbHAsO_4$) and tricalcium arsenate ($Ca_3[AsO_4]_2$)—were abandoned because the costs of adhering to more stringent OSHA regulations in the manufacture of these materials were considered too high to warrant the investments required for continued production. All of the mercury fungicides, including phenyl mercuric acetate (PMA), which was also an effective herbicide for controlling crabgrass and other summer annual grasses, were eliminated because of the health hazard associated with their use. Several of the phenoxy herbicides—2,4,5-T and silvex—were terminated because of claims of carcinogenic effects from the use of formulations containing these materials, including the infamous Agent Orange that was used by the U.S. military as a defoliant in Vietnam.

Today, there are many more herbicides from a diverse array of chemical families that provide the turfgrass manager with a powerful arsenal for effectively controlling the vast majority of the weeds that invade turfgrass communities.

HERBICIDE CHEMISTRY AND NOMENCLATURE

Herbicide labels typically list three names: the *chemical* name of the active ingredient, the *common* name of the active ingredient, and the *trade* name of the commercial formulation containing the active ingredient along with the carrier and other chemicals. For example, triclopyr (common name) is 3,5,6-trichloro-2-pyridinyloxyacetic acid (chemical name) and is sold in various formulations with such trade names as Garlon, Remedy, and Turflon.

The chemical name is a detailed description of the chemistry of the compound, listing each chemical component and its position in the compound. One component of triclopyr is the pyridine molecule: a benzene ring with nitrogen substituting for one of the ring's six carbon atoms. Chlorine atoms substitute for hydrogen atoms at three positions on the pyridine ring. As these occur at the #3, #5, and #6 positions (with position #1 being where the nitrogen is located), this compound would be called 3,5, 6-trichloropyridine, were it not for other substitutions. Since an oxyacetic acid group is located at the #2 position along the pyridine ring, however, the compound is called 3,5,6-trichloro-2-pyridinyloxyacetic acid, with the following chemical structure:

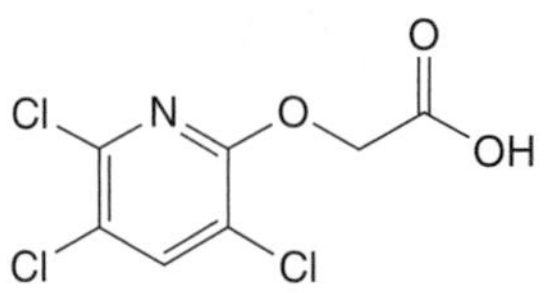

The common name—triclopry—is an abbreviated form of the chemical name, assigned by the American National Standards Institute (ANSI) with input from the Terminology Committee of the Weed Science Society of America. Finally, the trade name is the name given to a commercial formulation of the active ingredient by a manufacturer. When first introduced, a particular herbicide usually has only one trade name; however, as other markets are developed for the herbicide or other manufacturers obtain the rights to formulate and market the herbicide, other trade names may be introduced.

Organic Chemistry of Herbicides

Since most modern herbicides are organic compounds, they are comprised of a carbon (C) skeleton bonded to hydrogen (H) atoms; thus, they are referred to as hydrocarbons. Many herbicides also contain oxygen (O), and some contain nitrogen (N), phosphorus (P), sulfur (S), and various halogens, including chlorine (Cl), fluorine (F), and iodine (I).

Carbon is a relatively small atom with four electrons in its outer energy shell. Since this energy shell can accommodate up to eight electrons, its four electrons can be shared with those of other atoms to form covalent bonds. For example, the single electrons in hydrogen's outer (and only) shell can be shared, forming the following compound:

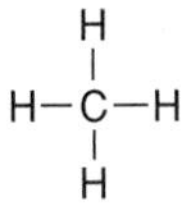

methane

Carbon can also share electrons with other carbon atoms, forming linear chains:

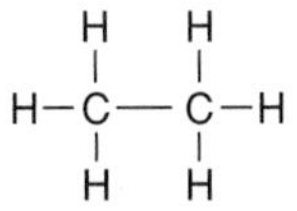

ethane 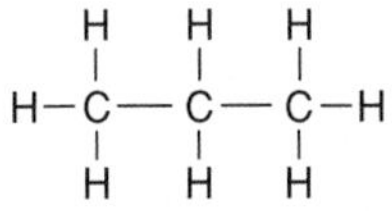   propane 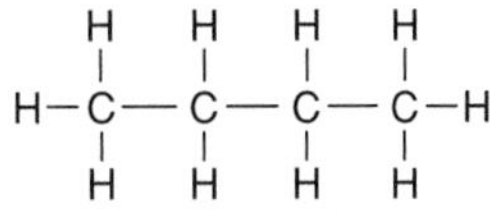   butane

Branched hydrocarbons can also be formed:

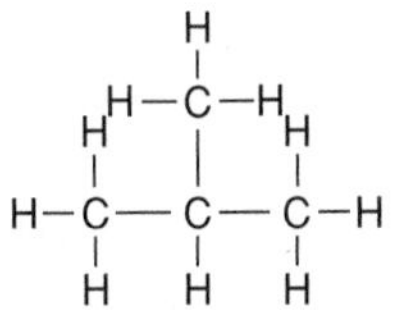

2-methyl propane or isobutane

Carbon atoms can also be organized to form cyclic compounds:

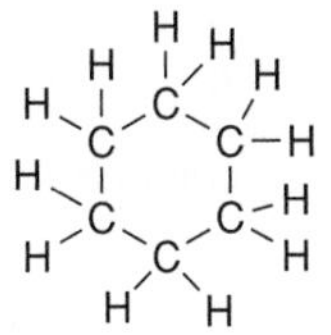

Cyclohexane

If all of the adjacent carbon atoms are connected by single covalent bonds, the compound is said to be saturated; however, if double or triple bonds occur, the compound is unsaturated, as in the following compounds:

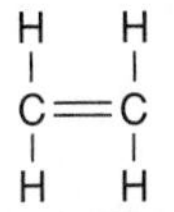

ethylene 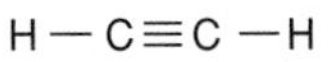 (=)   acetylene

If half of the bonds occurring in six-carbon rings are double, the compound is called benzene:

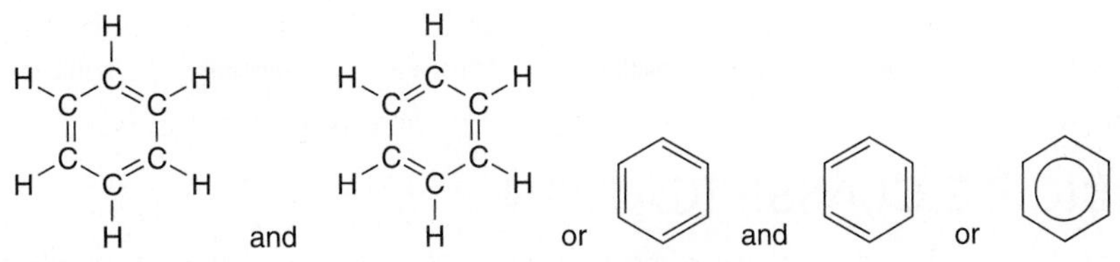

I and II or I and II or "hybrid"

In contrast to the earlier compounds, which are collectively called *aliphatic* (originally meaning *fatty*) compounds, benzene and its derivatives are called *aromatic* (originally meaning *fragrant*) compounds. While benzene is typically illustrated with the structures shown in I or II, which suggest that there are alternating double (shorter, 1.34 A) and single (longer, 1.48 A) bonds connecting adjacent carbons, these illustrations are inaccurate, as all bond lengths are equal (intermediate, 1.39 A), indicating that benzene actually contains six identical bonds that are simply described as *hybrid*, or *benzene*, bonds.

As stated earlier, carbon can bond with other atoms, including oxygen, to form an array of oxygen-containing compounds:

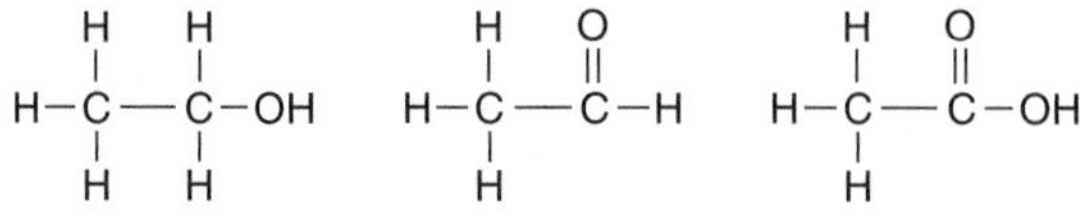

ethanol (alcohol) acetaldehyde (aldehyde) acetic acid (acid)

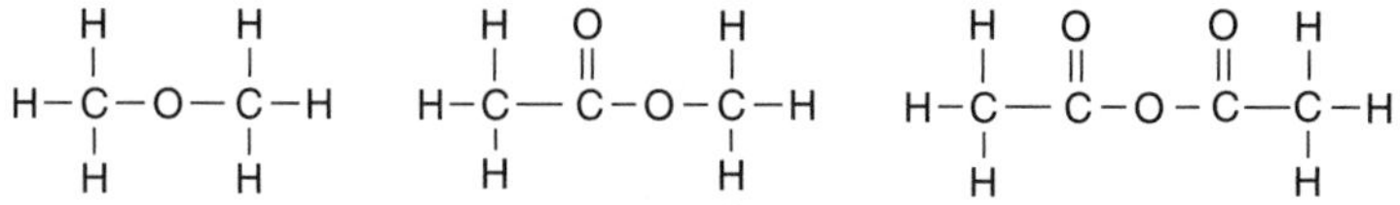

methyl ether (ether) methyl acetate (ester) acetic anhydride (anhydride)

Besides being able to bond with carbon, hydrogen, and oxygen, carbon can bond with other atoms or functional groups as follows:

$-NH_2$ amino; $-NO_3$ nitro; $-SH$ sulfhydryl; $-S-CH_3$ methylthio; $-C\equiv N$ nitrile; $-O-CH_3$ methoxy

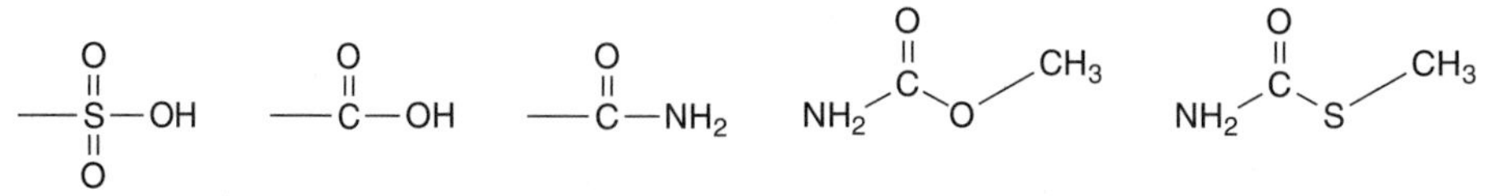

sulfonic carboxyl amide carbamate thiocarbamate

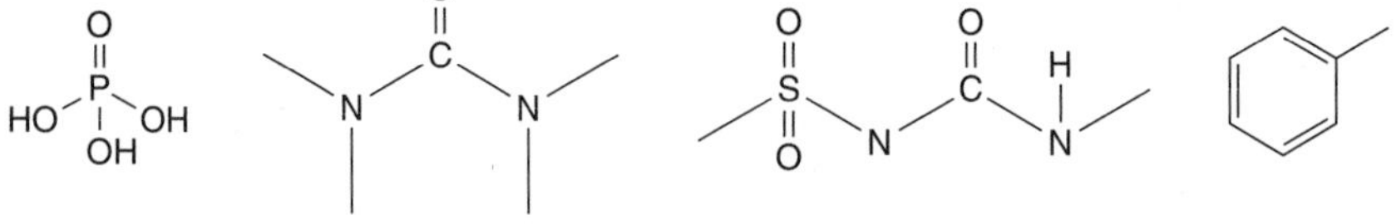

phosphate substituted urea sulfonylurea phenyl

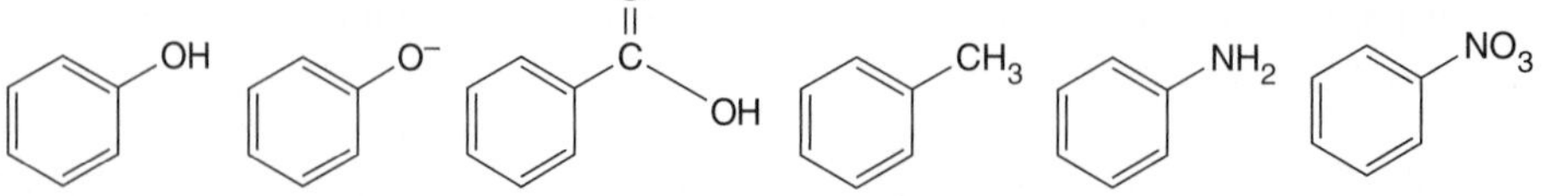

phenol phenoxy benzoic acid toluene aniline nitrobenzene

HERBICIDE CLASSIFICATION

Herbicides are classified on the basis of the timing of their application, site of application, mobility within the plant once absorbed, selectivity in turfgrass communities, mechanism of action, and chemical family.

Timing of Application

Herbicides applied prior to the emergence of target weed species are called *preemergence herbicides.* Once the target weeds have emerged, preemergence herbicides are usually ineffective. The specific date at which a preemergence herbicide should be applied thus depends on the period during which weed seed germination takes place. For example, in temperate climates, crabgrass typically germinates in mid-spring to late spring; a preemergence herbicide should be applied at least several weeks prior to anticipated germination in order to ensure control. The closer to the equator that crabgrass grows, the earlier in the growing season its germination usually occurs, requiring earlier herbicide application. *Postemergence* herbicides are applied after weed emergence; applying them before anticipated or actual emergence usually will not result in control. Many annual grasses and broadleaf weeds can be controlled with either preemergence or postemergence herbicides. Perennial weeds are usually treated with postemergence herbicides.

Site of Application

Herbicides are applied to the foliage, where they are absorbed, or to the soil underlying the grass shoots, where they are absorbed by the roots or immature organs of germinating seeds. Postemergence herbicides are usually *foliar applied,* while preemergence herbicides are *soil applied.* This distinction is of practical importance, since a foliar-applied herbicide that is washed off the foliage by irrigation or rainfall shortly after application may not be effective. Conversely, a soil-applied herbicide that is retained on the foliage for an extended period may break down before reaching the soil (or thatch) and hence not be effective. Granular formulations of soil-applied herbicides may offer some advantage over sprayable formulations in that they more easily bypass the foliage en route to the soil, especially when the foliage is dry. Where foliar-applied herbicides are formulated as granules, they should be applied to wet leaves. Wetting the leaves increases adherence of the granules to the foliage, which aids in absorption.

Mobility within the Plant

Postemergence herbicides include contacts and systemics. *Contact* herbicides enter and are active in destroying those portions of the weed plant with which they come in contact. For annual weeds, contact herbicides may be quite effective. Perennial weeds often recover following treatment with a contact herbicide, because of new growth from belowground regenerative organs. *Systemic* herbicides are translocated within the plant following absorption and are therefore more effective than contact herbicides for controlling perennial weeds.

Selectivity

Most herbicides used for controlling annual grasses and broadleaf weeds are *selective.* When applied in accordance with directions on the herbicide container, they control target weeds without seriously injuring desired turfgrasses. Currently, there are few herbicides with adequate selectivity for controlling most perennial grasses; therefore, *nonselective* herbicides must be used to control these weeds. Since nonselective herbicides will kill or injure all plants, they should be applied only to the target weed, with a directed spray to minimize damage to the turf. On small sites, nonselective herbicides can be applied with a soft brush or sponge to limit their contact with plants within the area of application and therefore minimize injury to nontarget plants.

HERBICIDE REGULATION

The authority to regulate pesticide development and use in the United States is provided by two laws: the Federal Insecticide, Fungicide, and Rodenticide Act (FIFRA) and portions of the Federal Food, Drug, and Cosmetic Act (FFDCA). FIFRA is administered by the U.S. Environmental Protection Agency (EPA) and attempts to balance the benefits of pesticide use with concerns over public health and environmental impact (Radosevich et al., 1997; Monaco et al., 2002). FIFRA covers the registration and cancellation of pesticides, creates a toxicity-based pesticide classification system, and allows states to regulate pesticides when addressing emergencies (Section 18) and special local needs (Section 24c). FFDCA is administered by the U.S. Food and Drug Administration (FDA) and regulates the establishment of pesticide tolerances in food, feed, fiber, and water. In 1996, both FIFRA and FFDCA were amended by the Food Quality Protection Act (FQPA), which established additional criteria for pesticide registration, covering exposure to infants and children and risks posed by pesticides with similar modes of action. FQPA requires sufficient evidence that a pesticide poses a *reasonable certainty of no harm* before it can be registered. It replaced the *Delaney Clause*, which prohibited the registration of any pesticide that caused cancer in test animals at *any test rate.*

Pesticide Toxicity

Pesticide toxicity is determined by standardized procedures performed with test animals to establish a pesticide's likely effects on humans. A *toxicity category* is assigned to each pesticide according to the hazard indicator level, as shown in the following table from Maddy et al., 1989:

Hazard Indicators	Toxicity Category I	Toxicity Category II	Toxicity Category III	Toxicity Category IV
Oral LD_{50}	<50 mg/kg	50 to 500 mg/kg	500 to 5,000 mg/kg	>5,000 mg/kg
Inhalation LC_{50}	<0.2 mg/L	0.2 to 2 mg/L	2 to 20 mg/L	>20 mg/L
Dermal LD_{50}	200 mg/kg	200 to 2,000 mg/kg	2,000 to 20,000 mg/kg	>20,000 mg/kg
Eye Effects	Corrosive corneal capacity, not reversible within 7 days	Corneal capacity; reversible within 7 days; irritation persisting 7 days	No corneal capacity; irritation reversible within 7 days	No irritation
Skin Effects	Corrosive	Severe irritation at 72 hours	Moderate irritation at 72 hours	Mild or slight irritation at 72 hours

The toxicity categories I through IV correspond to the signal words on the pesticide label; the signal word *danger* is required for pesticides meeting any of the criteria in category I, *warning* for category II, and *caution* for categories III and IV. The LD_{50} is the lethal dose required to kill 50 percent of a test population of animals and is expressed in mg of pesticide per kilogram of body weight (mg/kg or parts per million, ppm). The LC_{50} is the lethal concentration required to kill 50 percent of a test population of animals and is expressed in milligrams of pesticide inhaled as a mist or dust per liter (L) of air (mg/L or ppm). Eye and skin effects refer to a single dose applied directly to the eye or skin, respectively. In contrast to other pesticides, the vast majority of herbicides have relatively low toxicity in higher animals and carries the signal word *caution* on the label. This reflects the fact that, at the molecular level, most herbicides work at sites of action (Table 3.1) that are specific to plants (and perhaps microorganisms), but not to higher animals.

Table 3.1
LISTING OF HERBICIDES BY MECHANISM OF ACTION, SITE OF ACTION, AND CHEMICAL FAMILY

Mechanism of Action	Site of Action	Chemical Family	Common Names
Plant Growth Regulators (PGR)	IAA-like	Phenoxy Carboxylics	2,4-D, MCPA Dichloroprop Mecoprop
		Benzoics	Dicamba
		Pyridine Carboxylics	Clopyralid Fluroxypyr Triclopyr
		Quinoline Carboxylics	Quinclorac
	Gibberellic Acid Biosynthesis	Pyrimidines	Flurprimidol Fenarimol
		Triazoles	Paclobutrazol
		Acylcyclohexanediones	Trinexapac-ethyl
Amino Acid Biosynthesis Inhibitors	EPSP Enzyme	Glycines	Glyphosate
	ALS Enzyme	Sulfonylureas	Chlorsulfuron Foramsulfuron Halosulfuron Metsulfuron Rimsulfuron Sulfometuron Sulfosulfuron Trifloxysulfuron
		Imidazolinones	Imazaquin
		Pyrimidinyloxybenzoates	Bispyribac-sodium
Carotenoid Biosynthesis Inhibitors	PDS Enzyme	Pyridazinones	Norflurazon
	HPPD Enzyme	Callistemones	Mesotrione
Lipid Biosynthesis Inhibitors	ACCase Enzyme	Aryloxyphenoxy-propionates	Diclofop-methyl Fenoxyprop-ethyl Fluazifop-butyl
		Cyclohexanediones	Clethodim Sethoxydim
	Not ACCase Enzyme	Benzofuranes	Ethofumesate
Respiration Inhibitors	Uncouplers of Oxidative Phosphorylation	Organic Arsenicals	CAMA, DSMA MSMA
Cell Membrane Disruptors	Photosystem II (PSII)	Triazines	Atrazine Simazine
		Triazinones	Metribuzin
		Benzonitriles	Bromoxynil Dichlobenill
		Benzothiadiazoles	Bentazon
	Photosystem I (PSI)	Bipyridyliums	Paraquat Diquat
	Protox (PPO) Enzyme	Diphenylethers	Oxyfluorfen
		Triazolinones	Carfentrazone Sulfentrazone
		Oxadiazoles	Oxadiazon
	Glutamine Synthetase (GS)	Phosphinic Acids	Glufosinate

(continued)

Table 3.1
LISTING OF HERBICIDES BY MECHANISM OF ACTION, SITE OF ACTION, AND CHEMICAL FAMILY *(CONTINUED)*

Mechanism of Action	Site of Action	Chemical Family	Common Names
	Unknown	Fatty Acids	Pelargonic Acid
Cell Growth Disruptors and Inhibitors	Mitotic Disruptors	Dinitroanilines	Benefin Oryzalin Pendimethalin Prodiamine Trifluralin
		Pyridines	Dithiopyr
		Benzamides*	Pronamide
		Phthalic Acids	DCPA
		Carbamates	Asulam
	Shoot and/or Root Inhibitors	Chloroacetamides*	Metolachlor
		Acetamides*	Napropamide
		Phosphorodithioates	Bensulide
		Phenylureas	Siduron
Cell Wall Formation Inhibitors	Cellulose Biosynthesis	Benzonitriles	Dichlobenil
		Benzamides*	Isoxaben
		Quinoline carboxylics	Quinclorac
Soil Sterilant		Dithiocarbamate	Metham
		Thiadiazine	Dazomet

*Collectively called the acid amide herbicides.

It is important to recognize that the toxicity of a substance is directly related to its dosage. For example, common table salt has an LD_{50} of 3,000 mg/kg. While small quantities are common in human diets, ingesting 0.52 lb (235 g) would be sufficient to reach the LD_{50} for a 175-lb person. The LD_{50} of aspirin is 1,500 mg/kg; ingesting just 0.26 lb (118 g) would be sufficient for this person. And caffeine is even more lethal, with an LD_{50} of 192 mg/kg, requiring only about 0.034 lb (15 g). Therefore, showing that a particular substance can cause adverse effects, or even be lethal, if administered in a dose sufficient to cause the effects simply ignores the dose/toxicity relationship and could be used to indict virtually any substance.

Another effect that is sometimes confused with pesticide toxicity is *multiple chemical sensitivity* (MCS). This is the response that some people have when they are exposed to the odors associated with common petrochemical-based substances, including perfume, fresh paint, and pesticides. Disorders commonly seen in MCS patients include headache (often, migraine), chronic fatigue, musculoskeletal aching, chronic respiratory inflammation, attention deficit, and hyperactivity in young children (Ziem and McTamney, 1997). In a telephone survey conducted by Caress and Steinemann (2004), the authors found that 11.2 percent of respondents reported an unusually severe hypersensitivity to chemical products and 2.5 percent reported that they had been medically diagnosed with MCS. Local registries of MCS patients have been created for use by lawn-care companies to enable the lawn-care personnel to inform MCS patients of spraying operations so that those sensitive to offending substances can temporarily isolate themselves from exposure to pesticide formulations applied in their immediate vicinity.

Herbicide Documents

Herbicide documents include the herbicide label and the Material Safety Data Sheet (MSDS). The label affixed to a herbicide container is considered a legal document that permits the manufacturer to distribute and sell the product. It specifies how the material should be used to ensure safety and effectiveness. Any use of the product in a manner that is inconsistent with the instructions provided on the label is a violation of federal law.

Other information that must be provided on the label includes the following:

- Product trade name
- Manufacturer of the product
- Net weight of contents
- EPA registration number
- Factory registration number
- Name and percentage of active ingredients
- Percentage of inert ingredients
- Use classification (general or restricted)
- Signal word (*danger*, *warning*, or *caution*)

MSDSs should be provided to all purchasers of herbicides and other pesticides and should include information on

- Product safety
- Physical and chemical properties
- Health and physical hazards
- Primary routes of chemical entry
- Exposure limits
- Precautions for safety in use
- Emergency first aid
- Responsible party contacts

Other documents of importance include those providing information on *Restricted Use* pesticides. This classification restricts a product to use by a certified pesticide applicator and restricts the use of a product to be under the direct supervision of a certified applicator. Updated Restricted Use Products (RUP) Reports and related documents can be obtained at the following website: http://www.epa.gov/pesticides/regulating/restricted.htm.

3

Herbicides in the Plant

In order for an herbicide to be effective, it must first be absorbed by the plant. Some herbicides must then be translocated to a site of action where toxicity is expressed through the disruption of some process vital to plant growth. While in the plant, an herbicide may be metabolized to CO_2, or to some intermediary metabolite with altered biological properties. Because of differences in the ways specific plants respond to the presence of herbicides through these processes, some plant species may be more tolerant than others, resulting in some measure of selectivity.

HERBICIDE ABSORPTION

Effective use of herbicides in turf requires sufficient herbicide absorption to disrupt some vital process in target plants, causing their death or disappearance from the turfgrass community. Absorption occurs through root and shoot tissues, with soil-applied herbicides being absorbed primarily by roots and foliar-applied herbicides by shoots. An understanding of absorption is aided by an understanding of the terms *symplast* and *apoplast.* The symplast includes all living cells minus their respective cell walls. The apoplast is the nonliving portions of the plant that surround the symplast, often interposed between the symplast and the external environment. The apoplast is made up of the cell walls and intercellular spaces, the xylem portion of the vascular system, and the cuticle covering the leaf surfaces. Herbicides that enter the plant do so by penetrating the apoplast in order to get to the symplast.

Foliar Absorption

Foliar absorption is the uptake of an herbicide principally through the leaves, but possibly through stems as well. Foliar absorption of herbicides involves retention of spray droplets applied to leaf surfaces, penetration of the leaf surface by the herbicide, and movement of the herbicide into the cytoplasm of plant cells. Factors influencing leaf absorption are (1) surface tension of the spray solution, (2) wettability and permeability of the leaf surface as influenced by the nature of the cuticle and trichomes (hairs) present, and (3) surface area and orientation of leaves available for intercepting and retaining incoming spray droplets.

Surface Tension Most herbicide applications are made with water as the carrier, and the high surface tension of water—due to hydrogen bonding between water molecules—restricts the coverage of leaf surfaces by water droplets. Incorporation of a surfactant into the spray solution reduces surface tension, causing the droplets to spread out and cover more leaf surface area.

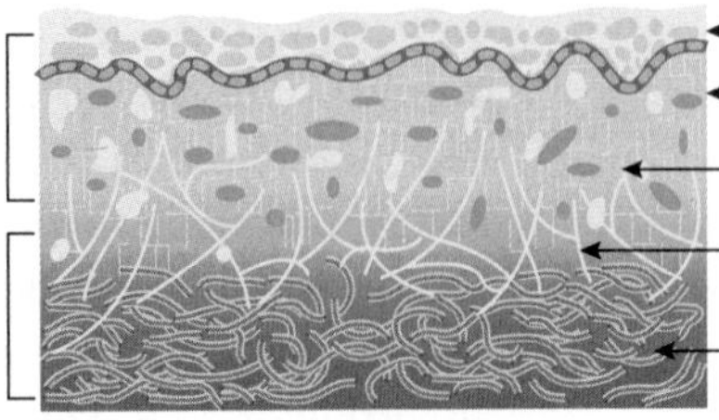

Wettability Wettability is influenced by the thickness, composition, and surface topography of the cuticle, a coating on leaf surfaces that is synthesized and secreted by the leaf epidermis and serves as a protective barrier to water loss and dehydration (Figure 3.1). The thickness of the cuticle ranges from 0.1 to 10 μm, reflecting plant species, leaf age (older leaves generally have thicker cuticles than younger leaves), and environmental conditions (thicker cuticles typically develop in nonshaded locations and in response to environmental stresses). Because cuticle thickness is inversely related to herbicide absorption, thicker cuticles are often associated with reduced herbicide efficacy. The composition of the cuticle—including wax, cutin, and pectin—influences its permeability to foliar-applied herbicides. The waxy portion is made up of long-chain hydrocarbons. It is divided between the epicuticular wax, which covers the cutin and forms the surface of the cuticle, and the intracuticular wax, which is embedded within the cutin. The cuticular waxes are nonpolar and hydrophobic, and form a water-repellent barrier between the epidermal cells of the leaf and the atmosphere surrounding it. The type of epicuticular wax can dramatically affect the wettability of the leaf surface; waxes occurring on monocot plants tend to have a crystalline structure and are highly resistant to wetting, while those occurring on dicot plants tend to be amorphous (i.e., lacking a crystalline structure) and less resistant to wetting. In some cases, herbicide selectivity may be a reflection of differences in absorption due to the type and amount of epicuticular wax present. Cuticular waxes generally resist penetration by water-soluble (polar) herbicide formulations, but are more easily penetrated by oil-soluble (nonpolar) formulations. The bulk of the cuticle's volume is cutin, which consists of 16-carbon and 18-carbon fatty acids. It forms a porous matrix in which some of the pores are filled with intracuticular (embedded) wax. Pectin, a polysaccharide, occurs in the form of strands at the base of the cuticle, attaching themselves to the epidermal cell walls (Monaco et al., 2002).

Cuticle
Cell wall
Epicuticular wax
Embedded wax
Cutin matrix
Pectin strands

Figure 3.1
Illustration of the cuticle, showing the distribution of epicuticular and embedded wax above and within the cutin, and pectin strands connecting the cutin and the cell walls of the leaf epidermis.

The lipophilic (wax-loving) and hydrophilic (water-loving) properties of herbicides are important in understanding how they are absorbed across cuticles. Lipophilic herbicide formulations, including oil-soluble, water-emulsifiable, and water-dispersible formulations, readily traverse the cuticular barrier by simple diffusion through the waxy constituents, especially the more permeable amorphous waxes. Water-soluble herbicide formulations, including salts and soluble powders, diffuse more slowly due to the presence of epicuticular waxes at the surface of the cuticle. As a consequence, less herbicide is absorbed and efficacy is correspondingly reduced. Once across this barrier, however, subsequent diffusion through the more hydrophilic cutin matrix and pectin strands is relatively rapid. Where cracks occur in the cuticle layer, due to traffic, mowing, insects, and other environmental factors, the absorption of water-soluble herbicides may be substantially increased.

The permeability of the cuticle is also influenced by its hydration level. When the plant absorbs sufficient water to become fully turgid, the cutin imbibes water and expands, the intracuticular wax within the pores is moved farther apart, and the cutin's water-permeability is increased. When the plant is under moisture stress, the cutin shrinks, the intracuticular wax is pulled closer together, and the cutin's water permeability is reduced. Thus, turgid plants absorb water-soluble herbicides more readily than moisture-stressed plants, resulting in greater efficacy.

Trichomes are hairs emerging from epidermal cells that, like the cuticle, reduce evaporative water loss from leaf surfaces. Where they are densely packed, trichomes may enable the leaf to resist wetting by preventing direct contact between herbicide droplets and the leaf surface. On the other hand, loosely arranged trichomes may increase wetting and absorption of herbicides by enhancing the retention of herbicide droplets onto leaf surfaces. And because the cuticle in the immediate vicinity of the trichomes tends to be shallower, the trichomes may mark sites where herbicides are preferentially absorbed.

Penetration through stomata—small openings in the leaf epidermis that are surrounded by guard cells—is believed to be of little importance in the absorption of foliar-applied herbicides. This may reflect the fact that, in many dicot species, stomata occur primarily on the underside (abaxial side) of the leaves, while foliar sprays are typically intercepted by the upper (adaxial) side of the leaves. Thus, herbicides must penetrate the cuticle of these species in order to enter the symplast and effectively control them.

Leaf Surface Area and Orientation Larger-leaved plants can intercept and absorb more foliar-applied herbicide than smaller plants can. Also, plants in which the leaves are more horizontal in their orientation can intercept, retain, and absorb more herbicide than plants with a more vertical orientation. Therefore, relatively broad, horizontally oriented dicot plants (i.e., broadleaf weeds) will be more likely to succumb to foliar-applied herbicides than will the relatively narrow, vertically oriented monocot plants, including many turfgrass species, within a mixed plant community.

Root Absorption

Soil-applied herbicides—as well as some foliar-applied herbicides that reach the soil surface directly or as a result of the washing action of rainfall or irrigation—enter the plant primarily through the roots. Many preemergence herbicides are absorbed by embryonic roots as they emerge from germinating seeds. Root absorption can be quite rapid, reflecting the absence of a significant wax layer or cuticle that might otherwise restrict uptake. Entry into the plant occurs through comigration with water absorbed primarily by root hairs. Diffusion of water and dissolved substances, including herbicides, can occur apoplastically through the intercellular space and pores within cell walls of the root's epidermal and cortex cells. Once the endodermis is reached, however, further apoplastic movement is blocked by a band of suberin, called the Casparian strip, located in the radial walls of these cells. As a consequence, movement across the endodermis must occur through the membranes (plasmalemma) of this single layer of cells; these membranes enclose the central core of the root (stele) containing the vascular tissues (xylem, phloem) and serve the purpose of moving water and materials throughout the plant. The plasmalemma and other cell membranes are lipid bilayers with proteins embedded in them. Most herbicides cross the plasmalemma by simple diffusion. This is called *passive* absorption. Nutrient ions and some ionized herbicides require some type of transport mechanism to get across this barrier. Since metabolic energy is employed to activate the process, its mechanism in this case is called *active* absorption (Monaco et al., 2002).

For those herbicides crossing the membranes by diffusion, the rate of movement reflects the herbicide concentration difference on either side of the membrane; that is, the herbicide moves from a region of higher concentration to one of lower concentration. When the concentration is the same on both sides of the membrane, movement stops; however, subsequent movement of the herbicide into and through xylem and phloem ducts reduces the herbicide concentration within the stele, causing a concentration gradient to develop that promotes additional herbicide movement across the membrane. An additional factor influencing the diffusion rate is the partition coefficient (solubility) of the herbicide within the plasmalemma. Lipophilic (oil-soluble) herbicides move more freely across the plasmalemma, which is also lipophilic, than do hydrophilic (water-soluble) herbicides. Phenylureas, sulfonylureas, imidazolinones, and triazines are root-absorbed herbicides that readily move across the endodermis and upward through the plant.

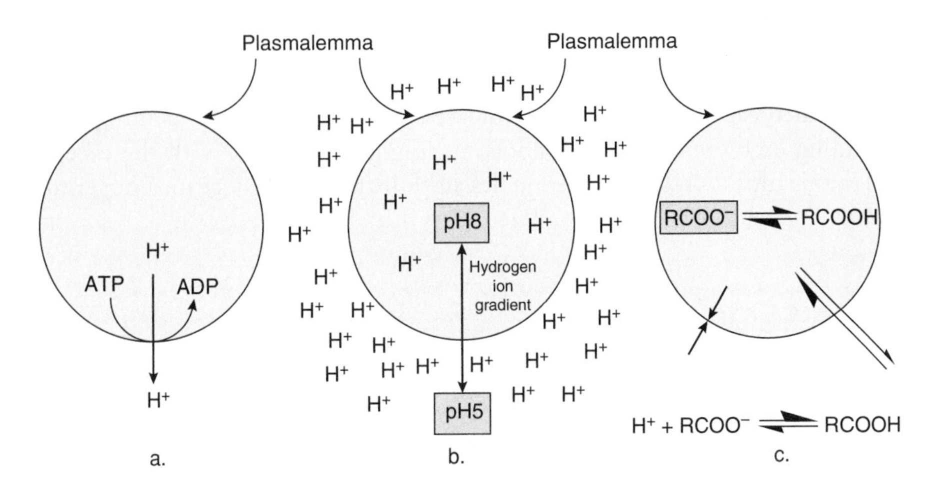

Figure 3.2
Illustration of the movement of a hydrophilic herbicide (RCOOH) across the plasmalemma against a concentration gradient due to the driving force of an ATP-generated proton pump.

Some hydrophilic herbicides containing ionizable groups (—COOH, —OH) can move across the plasmalemma and other cell membranes against a concentration gradient. The additional driving force employed for this type of movement is an ATP-generated hydrogen ion (H^+) or proton (since a hydrogen ion is made up of a single proton, with no neutron or electron present) pump that propels hydrogen ions across the plasmalemma from the inside to the outside of the cell (Figure 3.2). As a consequence of the reduced concentration of hydrogen ions, the pH inside the cell increases, while the pH of the environment surrounding the cell decreases, reflecting the increased concentration of hydrogen ions. This difference in pH and associated charges creates a proton (H^+) gradient that serves as the driving force for moving ionized herbicides ($RCOO^-$) into the cell. The ionized (more hydrophilic) herbicide is transformed into a protonated (more lipophilic) herbicide (RCOOH) that more easily traverses the plasmalemma. Once inside the cell, the herbicide readily loses its proton, transforming back into the ionized form ($RCOO^-$), and there it accumulates due to the relative difficulty of diffusing out of the cell in this form. This phenomenon is sometimes referred to as *ion trapping* (Ross and Lembi, 1999). Examples of ionizable herbicides are bentazon, 2,4-D, and sethoxydim.

Another form of active absorption is facilitated by carriers. The only herbicides known to move across the plasmalemma through a carrier-mediated process are glyphosate, which moves across the plasmalemma by using a phosphate carrier; 2,4-D and some related phenoxy herbicides, which move across by using the auxin carrier; and paraquat, which moves by means of a protein carrier.

Shoot Absorption from Soil

Some soil-applied herbicides may be absorbed by emerging seedling shoots from the soil solution. The cuticle on seedling plants is poorly developed, with little or no wax accumulation, making it more easily penetrated by herbicides. Furthermore, there is no Casparian strip to block movement within the shoot. Shoot entry is largely by diffusion of herbicides dissolved in the soil solution or, for highly volatile herbicides, in the vapor form within soil pores. DCPA, metolachlor, and trifluralin (as well as other dinitroaniline herbicides) are examples of shoot-absorbed preemergence herbicides.

Seed Absorption from the Soil

Soil-applied herbicides may be adsorbed onto the surfaces of seeds present in the soil, perhaps remaining there until conditions favorable for germination occur. Adsorbed herbicides, or herbicides dissolved in the soil solution, may be absorbed by the seeds as they embibe water shortly before germination. Some volatile herbicides or soil fumigants may be absorbed by dry seeds immediately after application. And, as explained earlier, soil-applied herbicides may be absorbed by seedling structures, including embryonic roots or shoots, during germination. With the exception of soil fumigants, soil-applied herbicides have little or no effect on nongerminating seeds.

TRANSLOCATION

Once an herbicide has entered a plant, either through the shoots or the roots, in order to be effective it must move to the site of action to cause some adverse effect on plant growth. With some herbicides, very little movement is required for this purpose, as the site of action is very close to the site of absorption. With others, considerable movement within the plant may be necessary for effective control. With respect to their mobility within plants, herbicides may be divided among four major groups: (1) herbicides with limited or no mobility, called *contact* herbicides when applied to the foliage; (2) xylem-mobile herbicides, with their movement largely restricted to the apoplast; (3) phloem-mobile herbicides, with their movement largely restricted to the symplast; and (4) ambimobile herbicides, with movement throughout the plant, occurring symplastically within the phloem and apoplastically within the xylem.

Limited-Movement Herbicides

Foliar-applied herbicides that kill plant tissue quickly following application have very limited movement within plants; these are the *contact* herbicides. Classical examples of contact herbicides are the bipyridyliums: paraquat and diquat, along with glufosinate and bromoxynil. These herbicides cause membrane destruction and necrosis of shoot tissues, often within hours of application. Since they do not move very much within the plant, complete coverage of leaf surfaces is usually required for effective control of target plants. Some soil-applied herbicides move only a short distance into embryonic roots (and, in some cases, shoots) before effectively inhibiting cell division and growth. Examples include several preemergence herbicides: trifluralin and other dinitroanilines, bensulide, and DCPA.

Xylem-Mobile Herbicides

The xylem is a continuous series of nonliving (apoplastic) cells extending from the roots to the shoots of plants. Its main function is to provide the plant with water along with the dissolved nutrients absorbed with the water from the soil by the roots. Herbicide residues that are in the soil and that enter the plant through the roots may move upward in the xylem via mass flow with the transpiration stream. With the evaporation of water from the surfaces of mesophyll cells in plant leaves and its loss to the atmosphere (transpiration), the water potential inside the leaves is reduced, causing a water-potential gradient that pulls water and dissolved materials, including many herbicides, up the plant through the xylem. Highly xylem-mobile herbicides include bentazon, metolachlor, napropamide, and norflurazon, along with the phenylureas and triazines.

While many herbicides are xylem mobile, some are not. This lack of xylem mobility of some herbicides may be explained by adsorption onto cell walls or other cellular components, compartmentalization within cellular components (e.g., vacuoles or plastids), conjugation to cellular substrates that are not xylem mobile, and degradation to inactive forms.

Phloem-Mobile Herbicides

The phloem is composed of a continuous series of living (symplastic) cells extending from the shoots to the roots of plants. Its main function is to provide all living cells with photosynthates (principally, sucrose) produced in the photosynthesizing cell chloroplasts of the fully formed leaves of the plant. Photosynthates move into the phloem against a concentration gradient because the concentration of photosynthates is substantially higher in the sieve elements of the phloem than in the mesophyll cells where they are produced. This movement is called *phloem loading*, and since it works against a concentration gradient, it is an active transport process requiring metabolic energy from ATP. The ATP-generated hydrogen ion (H^+) pump propels hydrogen ions across the plasmalemma from the inside to the outside of the phloem cells, establishing a relatively high H^+ concentration there. This is similar to the process for ion trapping, described earlier under the heading *Root Absorption*, in that sucrose is cotransported with an H^+ ion into the phloem cells, through a specialized carrier protein within the plasmalemma called a sucrose-H^+ symporter. These photosynthates are subsequently transported to other portions of the plant where they are unloaded, also by an active transport process, for utilization or storage.

Since the movement of foliar-applied herbicides from the leaves to stems and roots follows the same path as that of photosynthates, factors that optimize net photosynthesis—including high light intensity, adequate soil moisture, and optimum temperatures—can maximize herbicide movement within the plant. Also, since transport is via living phloem tissues, care must be taken to avoid killing leaf tissues by excessive herbicide application rates; otherwise, the efficacy of the herbicide, especially against perennial plants, can be dramatically reduced. For this reason, several applications of an herbicide or of a combination of herbicides at relatively low rates often provide better results than a single larger dose that kills too rapidly. Because contact herbicides, such as paraquat and diquat, destroy the living tissues shortly after being absorbed by the foliage, they are not phloem mobile, as they destroy the system that would otherwise transport them.

Some other herbicides are also limited in their phloem mobility, but for different reasons. For example, atrazine can diffuse into the phloem, but tends to diffuse back out, easily entering the xylem, which is located in close proximity to the phloem. Since movement within the xylem tends to be much faster than in the phloem (because of rapid transpirational water movement), the net movement of herbicides that freely move between the apoplast (xylem) and symplast (phloem), including the triazines, is in the direction of the transpiration stream within the xylem. Thus, the triazines tend to be mostly xylem mobile.

Herbicides that translocate primarily or substantially within the phloem do so by two possible mechanisms. The first is related to the chemical structure of the herbicide molecule. Most herbicides that are primarily phloem mobile have ionizable groups (usually carboxyl groups), which exist in both ionized (COO^-) and protonated (COOH) forms. These forms are in an equilibrium ($COO^- + H^+ \leftrightarrow COOH$) in which the favored form is dependent upon the pH of the surrounding medium. With pHs of 5 and 8 outside of and within the phloem, respectively (established by the phloem-loading process described earlier), there is a tendency for some herbicides to become protonated when outside the phloem due to the relatively high H^+ concentration there, and ionized after moving inside of the phloem due to the relatively low H^+ there, resulting in ion trapping due to the low permeability of the plasmalemma to the ionized form of the herbicide. The tendency of herbicide molecules to gain or lose a proton is also a function of their respective ionization constants (K_a), calculated from the hydrogen ion concentration times the concentration of the ionized herbicide divided by the concentration of the protonated herbicide: $K_a = [H^+][COO^-][COOH]$.

Compounds with higher K_a values are stronger acids. Because these values can vary by several factors of 10, ionization constants are usually expressed in logarithmic form ($-\log_{10}[K_a]$) as pK_a. When the pH is equal to the pK_a, there will be an equal distribution of molecules in the ionized and protonated forms. When the pH is higher than the pK_a, it is more likely that the molecule will lose a proton so that more of it is in the ionized form; conversely, when the pH is lower than the pK_a, it is more likely that the molecule will gain a proton so that more of it is in the protonated form. This tendency to gain or lose a proton determines the hydrophilic-lipophilic balance that exists for each herbicide, and thus the ease with which it moves into the phloem and the likelihood that it will remain there. Therefore, herbicides that tend to be predominantly phloem mobile are those with relatively high pK_a values; that is, they are stronger acids with a higher likelihood of losing a proton once they have gained entrance into the phloem.

A second factor influencing phloem mobility is the membrane partition coefficient of herbicides. The partition coefficient is a measure of the solubility of a compound in oil, divided by its solubility in water; thus, the higher the partition coefficient, the more likely the compound is to dissolve into the plasmalemma and diffuse through it to enter the phloem. Compounds with low partition coefficients will not permeate the plasmalemma very readily, while those with high partition coefficients will readily move in, as well as out, of the phloem; however, herbicides with intermediate partition coefficients can move into the phloem and remain there in quantities sufficient to allow for significant translocation via the phloem.

An herbicide that is considered highly phloem mobile is glyphosate; however, it can also be ambimobile in some plant species.

Ambimobile Herbicides

These herbicides are highly mobile within the plant, continuously circulating through the phloem and xylem and causing cell disfunction at many locations. When foliar applied, they can move to and within the phloem, often in conjunction with photosynthates. When soil applied, they can move to and within the xylem, usually in water moving along the transpiration stream. As discussed in the previous section, the capacity of these herbicides to maintain significant concentrations in both phloem and xylem tissues probably reflects their specific partition coefficients and pK_a values. Highly ambimobile herbicides include asulam, along with the benzoics, cyclohexanediones, imidazolinones, organic arsenicals, pyridine-carboxylics, and sulfonylureas.

METABOLISM

Once an herbicide is absorbed by plants, it is susceptible to metabolism and an associated loss of biological activity. Therefore, the faster an herbicide is metabolized, the less likely it is that it will translocate to the site of action and cause cellular dysfunction and death.

Overall metabolism of herbicides in plants can be divided into four phases:

- Phase I—structural change from a variety of reactions, including oxidation, reduction, or hydrolysis.
- Phase II—conjugation to cell constituents, including glucose, glutathione, and amino acids.
- Phase III—transport across cell membranes into the vacuole or cell wall.
- Phase IV—transformation into insoluble or bound residues.

Phase-I reactions often add hydroxyl (—OH), amino (—NH_2), sulfhydryl (—SH), or carboxyl (—COOH) groups that usually change phytotoxicity and predispose the compound to further metabolism. Oxidation reactions sometimes employ mixed-function oxidase (MFO) enzymes that bind molecular oxygen (O_2), incorporating one of its atoms into the herbicide and reducing the other to form water (H_2O). An example is the hydroxylation of bentazon:

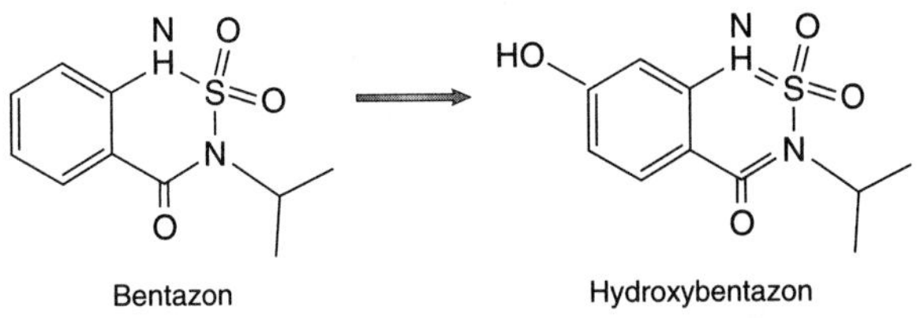

Bentazon → Hydroxybentazon

Similar aromatic hydroxylations occur with 2,4-D and chlorsulfuron. Other types of oxidation processes include dealkylation: the removal of alkyl groups of varying sizes and types, such as methyl (—CH_3), carboxyl (—COOH), ethyl (—CH_2CH_3), acetyl (—CH_3COOH), and propyl (—$CH_2CH_2CH_3$); deamination: the removal of amino groups (—NH_2); and sulfoxidation: the oxidation of sulfur molecules (—SCH_3 to $SOCH_3$).

Reduction is another form of phase-I metabolism. For example, following the dealkylation of the amino group of trifluralin, both of the nitro (—NO_2) groups are reduced to amino (—NH_2) groups, through a process called *nitroreduction.* Nonenzymatic photochemical reduction, called *photoreduction,* occurs with paraquat when it is exposed to ultraviolet light.

Another way herbicides undergo phase-I metabolism is by hydrolysis, which is the splitting of a molecule through the addition of water (H_2O). The hydrolysis of simazine results in the displacement of the chlorine (Cl) atom by a hydroxyl (OH) group, resulting in the production of nontoxic hydroxysimazine:

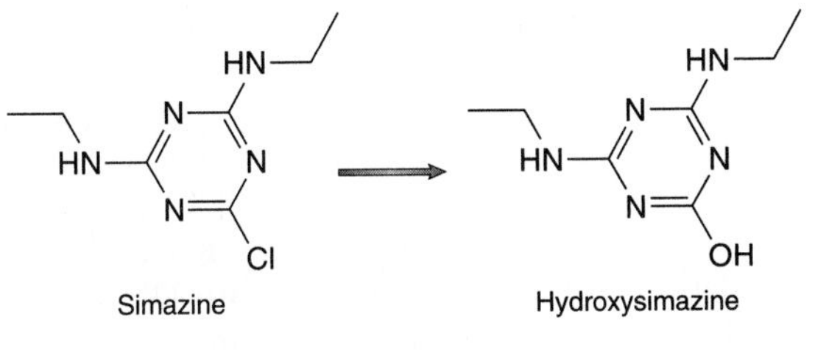

Simazine → Hydroxysimazine

Phase-II metabolism involves conjugation with natural plant constituents and results in the synthesis of metabolites of higher molecular weight, but with reduced or no phytotoxicity. Conjugation can also result in higher water solubility and reduced mobility within the plant. Herbicides and herbicide metabolites can be conjugated to sugars, amino acids, proteins, and lignins. After an herbicide has been hydroxylated (i.e., by the addition or substitution of a hydroxyl [—OH] group) through phase-I oxidation or hydrolysis, the metabolite frequently will conjugate with glucose or another sugar through a process called *glycosidation.* Examples of herbicides that can conjugate with glucose include hydroxylated forms of 2,4-D, bentazon, and chlorsulfuron. Some herbicides or herbicide metabolites conjugate with glutathione, a tripeptide composed of three amino acids: glutamate, cysteine, and glycine. These include the chloroacetamides, aryloxyphenoxy propionates, and triazines, as well as the sulfoxide metabolite of metribuzin.

Phase-III metabolism usually involves the transport and storage of the conjugated metabolite to cell walls or vacuoles, where they can no longer have any impact on the living portions of the cell. Phase-IV metabolism involves the further processing of the conjugated metabolite to yield insoluble or bound residues. This compartmentalization and detoxification of herbicide metabolites is important, as plants have no mechanism for excreting these materials (Radosevich et al., 1997).

SELECTIVITY

Some plants are easily killed with herbicide doses that have little effect on other plants. This characteristic is called *selectivity.* Selectivity is a reflection of many interrelated physical, chemical, and biological factors influencing the response of different plants to specific herbicides.

Physical Factors Influencing Herbicide Selectivity

Herbicide application rate and retention are physical factors influencing herbicide selectivity. Within the proper application rate range, susceptible species are killed and tolerant species are not. When herbicides are applied at rates below this range, efficacy is reduced; above it, selectivity is reduced. Variations in leaf shape, orientation, and surface characteristics account for substantial differences in the retention of foliar-applied herbicides. The relatively large, horizontally oriented leaves of dicot species are more likely to intercept and retain herbicide spray droplets than are the narrower, more vertically oriented leaves of many monocots, including turfgrasses. Also, leaves with a moderate degree of surface roughness or hairiness usually retain spray droplets better than do leaf surfaces that are smooth and either densely hairy or without hairs. Herbicides, such as ethofumesate, that reduce the formation of cuticular waxes on leaf surfaces can dramatically change the efficacy and selectivity of other foliar-applied herbicides. Sometimes, greater selectivity is obtained with soil-applied preemergence herbicides by bypassing the foliage and thus avoiding foliar uptake and, in some cases, phytotoxicity. The selectivity of dichlobenil in ornamental planting beds is due to its physical placement. In treated soils, germinating weeds are killed while established deep-rooted ornamental plants are tolerant, as dichlobenil does not readily leach in the soil.

Chemical Factors Influencing Herbicide Selectivity

The specific properties of a particular herbicide—phytotoxicity, solubility, volatility, and absorption—can be substantially altered by changes in the configuration of the molecule or the formulation. For example, the phytotoxicity of 2,4-D can be reduced or eliminated by a shift in the positions of the chlorine (Cl) atoms (e.g., #2 and #4) on the benzene ring. Substituting a methyl (CH_3) group for the chlorine in the #2 position converts 2,4-D to MCPA, which has higher efficacy on selected broadleaf weeds. Substitution of another methyl group for hydrogen (H) at the α-carbon converts the acetic acid aliphatic group to propionic acid at the #1 position on the benzene ring, changing MCPA to MCPP (mecoprop), which is less phytotoxic to bentgrass and more efficacious on clovers and other broadleaf weeds. Conversion of the acid form of these herbicides to a salt increases their solubility in water, while conversion to an ester increases their solubility in oil. The salt forms are nonvolatile, but are limited in their foliar absorption, while the ester forms are volatile, but more readily absorbed by the foliage.

Biological Factors Influencing Herbicide Selectivity

The biological factors affecting herbicide selectivity include absorption, translocation, and metabolism of the herbicide by plants. Differential absorption is most important as a selectivity factor among plants at different stages of maturity. Because the thickness and composition of cuticles change as a plant matures, seedling plants are much more susceptible to foliar-applied herbicides than mature plants are. With soil-applied herbicides, however, differential absorption is unlikely to be very important as a selectivity factor, regardless of plant age. Following absorption, herbicides may be differentially translocated within plant species. For example, 2,4-D and other phenoxy herbicides translocate more slowly in grasses than in susceptible broadleaf weeds, which accounts in part for the selectivity of these herbicides.

Differential metabolism within plants is the major reason for selectivity among species. For example, the removal of the chlorine atom from atrazine, resulting in its conversion to hydroxyl atrazine, accounts for the majority of detoxification processes in tolerant (most warm-season or C_4) turfgrasses. Likewise, the selectivity of other triazines, as well as bentazon, ethofumesate, imazaquin, and the sulfonylurea herbicides, is due to their rapid metabolism to nontoxic derivatives in tolerant turfgrasses. The tolerance of selected turfgrasses to the cyclohexadiones and aryloxyphenoxypropionates, however, is due to the presence of a tolerant form of ACCase enzyme in these species.

MODE, MECHANISM, AND SITE OF ACTION

The *mode of action* is defined as the entire chain of events that occurs within the plant from first contact by an herbicide to its final effect. This is distinguished from the *mechanism of action*, which is the series of biophysical (e.g., inhibiting electron flow, binding to a protein, disrupting cell division) or biochemical (e.g., inhibiting a specific enzyme) processes that express the herbicide's ultimate effect on plants. The *site of action* is the specific site in plant cells where these processes (e.g., photosynthesis in chloroplasts, respiration in mitochondria, mitosis in nuclei) take place. While some herbicides appear to work via several mechanisms of action, one of them is usually considered the primary mechanism of action. Herbicides and herbicide families are currently organized by their respective mechanisms and sites of action (Table 3.1).

Plant Growth Regulators

Plant growth regulators include auxinic and non-auxinic subgroups. Members of the auxinic subgroup include the phenoxy-carboxylics (2,4-D, MCPA, mecoprop, and dichloroprop), benzoics (dicamba), pyridine-carboxylics (clopyralid, fluroxypyr, triclopyr), and quinoline-carboxylics (quinclorac).

Auxinics Auxinic growth regulators mimic the effects of natural plant auxins, such as indol-3-acetic acid (IAA). Auxins regulate cell growth and development, and their activity within the plant is under direct metabolic control. Auxinic growth-regulating herbicides are not under metabolic control, and they cause abnormal growth in susceptible plants. Because grasses can rapidly inactivate these herbicides by conjugation, they are generally tolerant while many broadleaf (dicot) plants, which cannot, are relatively intolerant. Auxinic growth regulators are believed to have multiple sites of action at which they disrupt the balance of hormones and alter the synthesis of proteins and nucleic acids. These effects lead to various plant-growth abnormalities, especially on new tissues. Broadleaf (dicot) species exhibit stem and petiole twisting (called *epinasty*), leaf malformations (e.g., parallel venation, crinkling, leaf strapping, and cupping), stem callus formation, and stunted root growth. Reproduction is also affected, as evidenced by sterile florets and nonviable seeds.

Results from recent research with quinclorac suggest that the activity of the ACC synthase (1-aminocyclopropane-1-carboxylic acid synthase) enzyme is stimulated, resulting in the production of 1-aminocyclopropane-1-carboxylic acid, which is

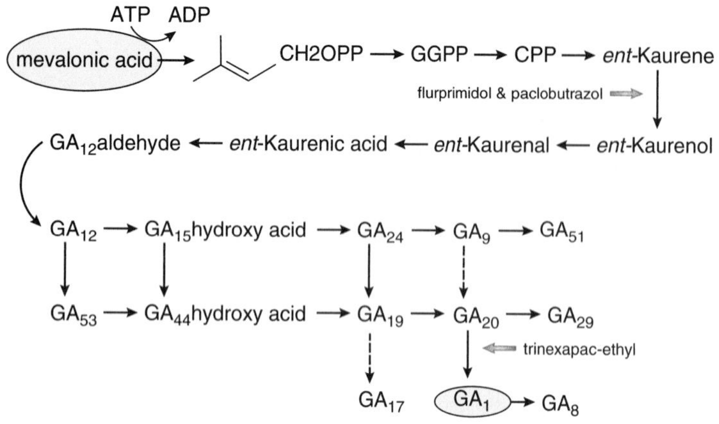

Figure 3.3
Illustration of the gibberellic acid biosynthesis pathway in which the ent-kaurene *oxidase enzyme that catalyzes the conversion of* ent-kaurene *to* ent-kaurenoic *is inhibited by flurprimidol and paclobutrazol, and the 3 β-hydroxylase enzyme that catalyzes the generation of GA_1 from GA_{20} is inhibited by trinexapac-ethyl.*

oxidized by ACC oxidase to form ethylene (CH_2CH_2) and cyanide (HCN) (Grossmann, 1998 and 2003). Ethylene causes epinastic growth and tissue swelling and triggers the biosynthesis of abscisic acid (ABA). Elevated levels of ABA induce stomatal closure, limiting photosynthetic activity and biomass production, and stimulate the overproduction of oxygen radicals, causing cell membrane disruption and cell necrosis. Also, cyanide (HCN), if not detoxified (usually by conjugation with alanine, an amino acid), can be directly phytotoxic to plant cells. Quinclorac is unique among the growth-regulating herbicides in that, in addition to its auxinic mechanism of action in broadleaf weeds, it also inhibits cellulose biosynthesis in annual grass weeds.

Non-auxinics. Members of the non-auxinic subgroup of plant growth regulators include the triazoles (paclobutrazol), pyrimidines (flurprimidol, fenarimol), and acylcyclohexanediones (trinexapac-ethyl). The triazoles and pyrimidines are root-absorbed, xylem-translocated plant-growth regulators that inhibit gibberellic acid (GA) biosynthesis in plants by blocking the *ent*-kaurene oxidase enzyme that catalyzes the conversion of *ent*-kaurene to *ent*-kaurenoic acid, as shown in Figure 3.3 (Rademacher, 2000).

As the conversion of *ent*-kaurene to *ent*-kaurenoic acid occurs relatively early in the GA biosynthesis pathway, its blockage results in a reduction in the concentrations of all GAs occurring after the formation of *ent*-kaurenoic acid, including GA_{12}, GA_{15}, GA_{24}, GA_9, GA_{53}, GA_{44}, GA_{12}, GA_{19}, GA_{20}, and finally, the biologically active GA_1.

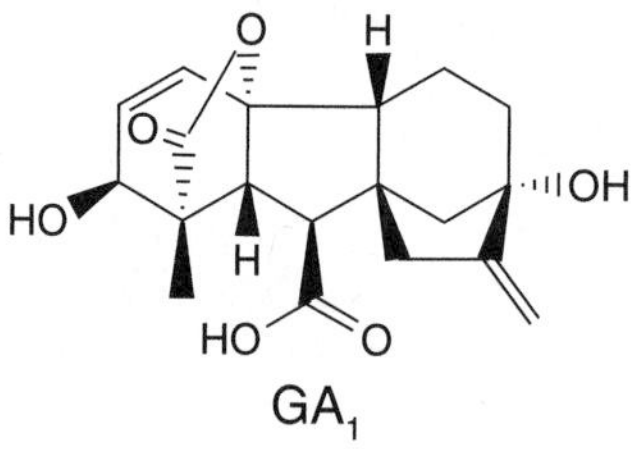

GA_1

The triazoles also influence plant growth by altering the concentrations of other phytohormones, including cytokinins, abscisic acid (ABA), and ethylene. The cytokinin concentration is increased by the influence of triazoles on the isoprenoid pathway from which cytokinins are derived. The concentration of ABA is increased by the inhibiting effect of triazoles on the conversion of ABA to the physiologically inactive phaseic acid. And the concentration of ethylene is decreased by the inhibiting effect of triazoles on ACCase, the enzyme involved in the conversion of the ethylene precursor 1-amino-cyclopropane-1-carboxylic acid (ACC) to ethylene.

The effect of triazoles on treated plants include a reduction in leaf and stem length; darker green leaves; increases in leaf width, leaf thickness, epicuticular wax on leaf surfaces, shoot density, root thickness, and root–shoot ratio; and a retardation of senescence development. The magnitude of these effects, however, varies with different triazoles and plant species. While many of these effects may be directly related to the influence of triazoles on GA biosynthesis and activity, some of them may be due to their influence on other phytohormones. For example, the darker green color is usually associated with larger chloroplasts in triazole-treated plants, perhaps reflecting the increased cytokinin concentration, as cytokinins are known to be involved in chloroplast development. An increase in epicuticular wax on leaf surfaces probably reflects an influence of triazoles on lipid metabolism and the accumulation of fatty acids. The retardation of senescence is probably due to the reduction in ethylene levels. The increased tolerance of some triazole-treated plants to drought stress may be due to the effects of elevated ABA levels on stomatal closure, which increases diffusive resistance and water conservation. Finally, increases in the levels of both antioxidants (e.g., tocopherol and ascorbic acid) and antioxidant enzymes (e.g., SOD and peroxidases) in treated plants increase their ability to scavenge the free radicals produced in response to environmental stresses. Subsequent application of GA to triazole-treated plants causes a reversal of these effects.

The triazole and pyrimidine chemical families include several commercial fungicides, which are called the sterol demethylation inhibitors (DMIs) because they inhibit sterol 14α-demethylase, an enzyme that catalyzes the C-14 demethylation reaction in the biosynthesis of ergosterol, the predominant component of fungal cell membranes. Myclobutanil, propiconazol, and triadimefon are triazole fungicides, but they also have plant-growth regulating properties. Paclobutrazol—also a triazole—is a plant-growth regulator, but it also has fungitoxic properties. Fenarimol is a pyrimidine fungicide, but it has also been used for preemergence and early postemergence control of annual bluegrass in overseeded bermudagrass turfs. Flurprimidol—also a pyrimidine—is a plant-growth regulator, but it has fungitoxic properties as well. Therefore, combinations of triazole or pyrimidine fungicides and plant-growth regulators can have significant additive effects, in that higher than anticipated levels of disease control and plant-growth regulation may result from their applications.

The acylcyclohexanediones are foliar-absorbed phloem-translocated plant-growth regulators that block the 3 β-hydroxylase enzyme that catalyzes the generation of GA_1—the biologically active form of GA—from GA_{20}, its immediate precursor in the GA biosynthesis pathway (Figure 3.3). Acylcyclohexanediones also affect plant growth as well as the concentrations of the other phytohormones, in a fashion similar to that of the triazoles. Because this inhibition occurs near the end of the GA biosynthetic pathway, an accumulation of GA_{20} and perhaps other precursors is often associated with a growth surge in treated turfgrasses once the inhibitory effects have worn off. The acylcyclohexanediones do not have fungitoxic properties, however; nor do they differentially suppress annual bluegrass in mixed communities with other cool-season turfgrasses to the extent that the triazoles and pyrimidines do.

Phenoxy-Carboxylic Acids The phenoxy-carboxylics—commonly referred to as the phenoxys—include 2,4-D and various chemical analogues, including MCPA, dichlorprop, and MCPP (mecoprop):

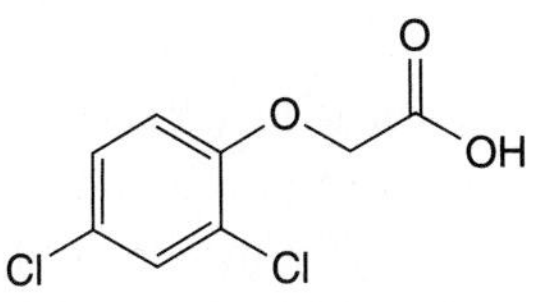

2,4-D

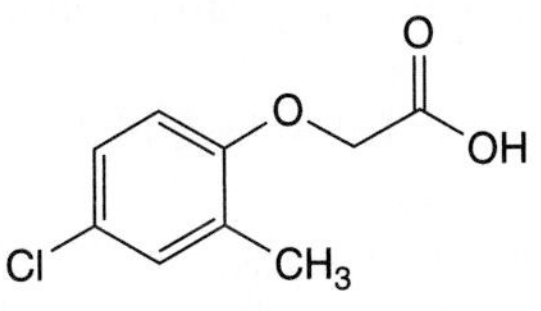

MCPA

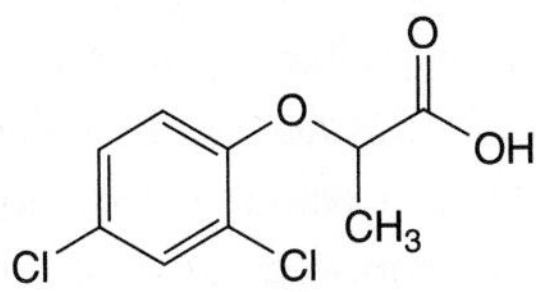

Dichlorprop

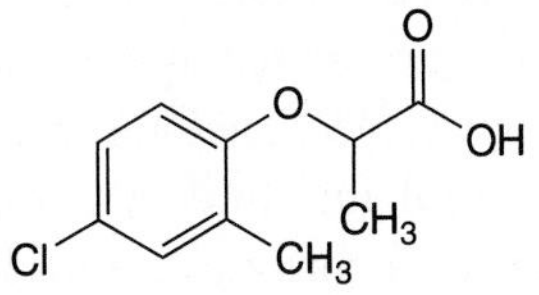

MCPP (mecoprop)

The molecular structure of the phenoxy-carboxylics includes a benzene ring, an oxygen atom substituted for one of the hydrogens bonded to the ring, a carboxyl (COOH) group bonded indirectly to the oxygen atom by an aliphatic side chain of one or more carbon atoms, and at least two other substituents on the ring. The differences among these chemical analogues are in the sizes of the aliphatic side chains and in the substituents at the #2 position on the benzene ring. In 2,4-D and MCPA, the aliphatic side chain is a two-carbon acetic acid, while in dichlorprop and MCPP, it is a three-carbon propionic acid. Furthermore, in 2,4-D and dichlorprop, there is a chlorine (Cl) at the #2 position on the benzene ring, while in MCPA and MCPP, there is a methyl group (CH_3) at this location. These variations account for the differences in phytotoxicity among target weed species, as well as for other properties. (See Table 4.1.)

MCPA is (4-chloro-2-methylphenoxy)acetic acid. 2,4-D is (2,4-dichlorophenoxy)acetic acid. Dichlorprop, or 2,4-DP, is 2-(2,4-dichlorophenoxy)propanoic acid. Mecoprop, or MCPP, is 2-(4-chloro-2-methylphenoxy)propanoic acid. Salt formulations have generally lower vapor pressures (indicating a relatively low volatility hazard) and higher water solubility, while ester formulations have generally higher vapor pressures (indicating a moderate to high volatility hazard, depending on the particular ester) and much lower water solubility.

Salts (and water) are produced by reacting the parent acid with a base:

$$RCOOH + NaOH \longrightarrow RCOONa + H_2O$$

Salt forms include the sodium (Na^+) salt (as just shown), as well as the ammonium (NH_4^+), dimethylamine ($[CH_3]_2NH^+$), isopropylamine ($[CH_3]_2CHNH_2^+$), and triethanolamine ($[CH_2CH_2OH]_3N^+$) salts. These salt concentrates (SC) dissolve readily in water as the anion (phenoxy acetate or propionate) and the cation (Na) dissociate, forming true solutions.

$$\underset{\text{salt}}{RCOONa} \longrightarrow \underset{\text{anion}}{RCOO^-} + \underset{\text{cation}}{Na^+}$$

In hard water, the calcium (Ca^{2+}) and magnesium (Mg^{2+}) ions present could replace the other cations (e.g., Na^+, NH_4^+, etc.) to form insoluble salts, resulting in the formation of precipitates at the bottom of the spray tank or in suspension, reducing the herbicide concentration in solution and clogging the screens and orifices of the spray equipment. The amine salts are less affected by hard water than are the sodium and ammonium salts. As insoluble salts of the herbicide can also occur in aqueous mixtures with liquid fertilizers, tests should be conducted that involve mixing representative amounts of the herbicide and fertilizer in a bottle of water, shaking, and observing for the formation of precipitates. Esters (and water) are produced by reacting the parent acid with an alcohol:

$$RCOOH + CH_3CH_2OH \longrightarrow RCOOCH_2CH_3 + H_2O$$

Ester forms include the ethyl (CH_2CH_3) ester (as just shown), along with the isopropyl ($CH[CH_3]_2$), butoxyethyl ($CH_2CH_2O[CH_2]_3CH_3$), and isooctyl ($CH_2[CH_2]_4CH[CH_3]_2$) esters. These are water insoluble, but dissolve readily in oil. They are typically formulated as emulsifiable concentrates (EC) that form emulsions when added to water. Because esters are nonionic and do not dissociate in water, they do not react with calcium and magnesium in hard water to form precipitates. The short-chain esters (i.e., four carbons or less) of phenoxy-carboxylic acid herbicides are highly volatile and thus constitute a serious hazard when used in the vicinity of susceptible plants in the landscape. The long-chain esters are less volatile, but can still be hazardous to use in close proximity to susceptible plants. Care must be exercised to minimize vapor drift with ester forms and spray drift with both salt and ester forms of these herbicides. Carrots, cotton, grapes, tobacco, and tomatoes are especially sensitive to the phenoxy-carboxylic herbicides.

The salts of the phenoxy-carboxylic herbicides may be readily washed from plant leaf surfaces by water; however, if they remain in contact with the leaves for at least six hours following application, sufficient absorption should have taken place to preclude a significant loss in phytotoxicity. The ester forms of these herbicides are more rapidly absorbed by plant leaves and are often used with greater effectiveness, especially in summer when the cuticles on some target plants have become dehydrated, making these plants more difficult to control.

Some granular (G) formulations are also used, primarily with fertilizers to facilitate their coordinated application. Best results are obtained when the foliage is moist, which promotes adherence of the granules to the leaves and thus enhances foliar absorption. Slowly available nitrogen carriers are therefore used to minimize the potential for foliar burning. The phenoxy-carboxylic acid herbicides, alone or in various combinations, are used for controlling broadleaf weeds in turf.

Benzoic Acids The benzoic acid most often used in turf is dicamba:

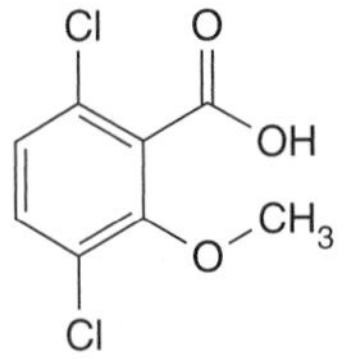

Cl O OH O CH₃ Cl

Dicamba

The molecular structure of a benzoic acid herbicide includes a benzene ring, with a carboxyl group (COOH) bonded to one of the ring carbons. Dicamba also has three other ring substituents: a methyl (CH_3) group bonded indirectly to the ring through an oxygen, and two chlorine atoms. Compared with the phenoxy-carboxylics, dicamba is more mobile within the vascular systems of treated plants.

Dicamba is 3,-6-dichloro-2-methoxybenzoic acid. While it is also more mobile in soils, leaching is not a serious problem due to its rapid microbial degradation. It is typically formulated as salt concentrates (SC) or granules (G), usually in mixtures with one or more phenoxy-carboxylic herbicides and in granular formulations with fertilizer nutrients. Dicamba, usually in combination with 2,4-D and/or other phenoxy-carboxylic herbicides, is used for controlling broadleaves weeds in turf.

Pyridine-Carboxylic Acids The pyridine-carboxylics include clopyralid, fluroxypyr, and triclopyr:

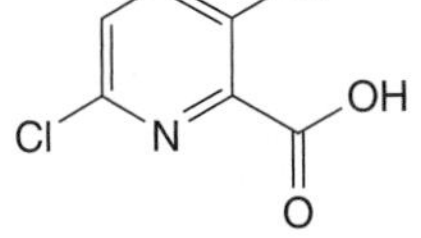

Clopyralid

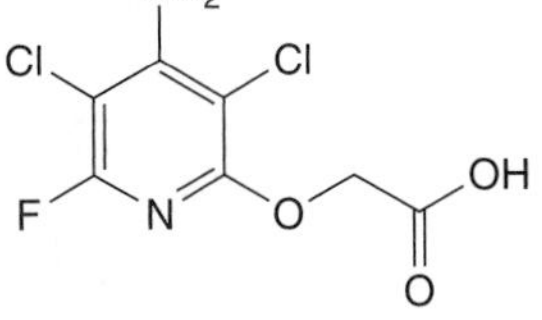

Fluroxypyr

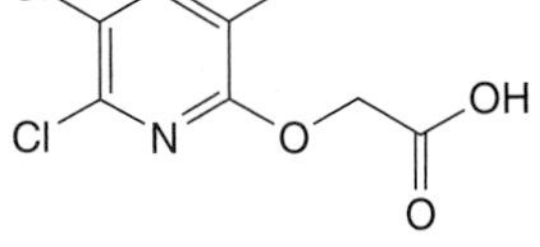

Triclopyr

The molecular structure of the pyridine-carboxylics includes a pyridine ring (in which one of the carbons in the aromatic ring is replaced by nitrogen). While each has an aliphatic group attached to a carbon adjacent to the ring nitrogen, the aliphatic group in clopyralid is a carboxyl (COOH) group, while fluroxypyr and triclopyr have acetic acids (CH_3COOH) bonded indirectly to the ring through an oxygen. Other substituents include two chlorines (Cl) in clopyralid; two chlorines, one fluorine (F), and one amino group (—NH_2) in fluroxypyr; and three chlorines in triclopyr. These variations account for differences in phytotoxicity among target weed species and soil persistence and mobility, as well as other properties (e.g., vapor pressure, water solubility, soil half-life, and LD_{50}).

Clopyralid is 3,6-dichloro-2-pyridinecarboxylic acid. Fluroxypyr is 1-methylheptyl (4-amino-3,5-dichloro-6-fluro-2-pyridyloxyl)acetic acid. Triclopyr is ([3,5,6-trichloro-2-pyridinyl]oxy)acetic acid. The pyridine-carboxylics are highly mobile within the vascular systems of treated plants. They are also highly mobile in the soil; their persistence within the soil, while appreciably longer that that of the phenoxy-carboxylic and benzoic acid herbicides, is far less than that of another pyridine-carboxylic herbicide—picloram, which is not used in turf. Clopyralid and triclopyr are formulated as salt concentrates (SC); fluroxypyr ester is formulated as an emulsifiable concentrate (EC). Clopyralid (Lontrel), fluroxypyr (Spotlight), and triclopyr (Turflon), as well as the combination of triclopyr and clopyralid (Confront), are primarily used for controlling broadleaf weeds in turf.

Quinoline-Carboxylic Acids The quinoline-carboxylic acid herbicide used in turf is quinclorac:

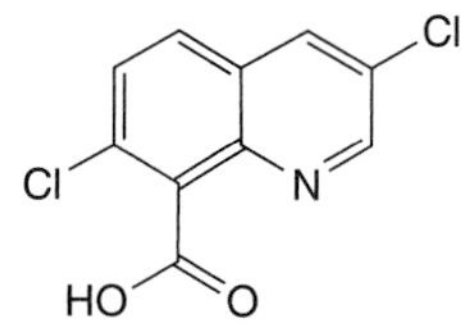

Quinclorac

The molecular structure of a quinoline-carboxylic acid herbicide includes benzene and pyridine rings fused together, called a quinoline group, with a carboxyl group (COOH) attached directly to the benzene. Quinclorac also has two chlorine (Cl) substituents.

Quinclorac is 3,7-dichloro-8-quinolinecarboxylic acid. It is highly mobile in treated plants, moving in both the phloem and xylem, but not very mobile in the soil. It is usually formulated as a wettable powder (WP), water-dispersible granule (WDG), or dry flowable (DF). Quinclorac (Drive) is used primarily for selective control of crabgrass and an array of broadleaf weeds in cool-season and some warm-season turfgrass communities.

Pyrimidines The pyrimidines include flurprimidol and fenarimol:

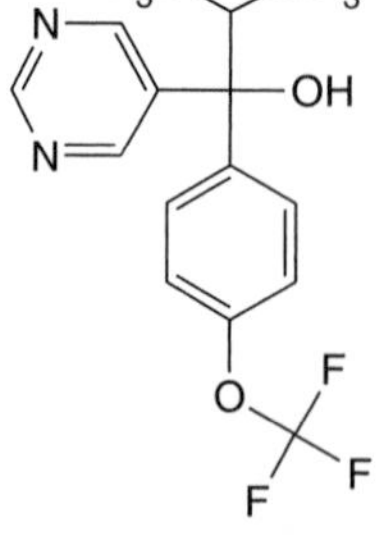

Flurprimidol

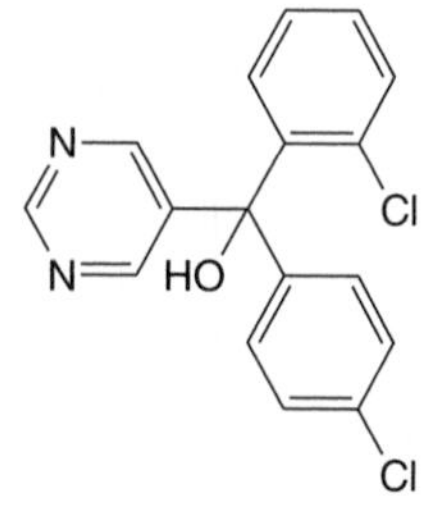

Fenarimol

While the molecular structure of pyridine is an aromatic ring in which one of the carbons has been replaced by nitrogen, a pyrimidine has an aromatic ring in which two of the carbons have been replaced by nitrogen. In flurprimidol, an aliphatic group ($COHCH_2[CH_3]_2$) connects pyrimidine to a benzene ring in which a trifluoromethoxy group (OCF_3) is one of the substituents. In fenarimol, a COH group connects pyrimidine to two benzene rings, both of which have a chlorine (Cl) substituent positioned on the ring.

Flurprimidol is α-(1-methylethyl)-α-[4-(trifluoro-methoxy) phenyl] 5-pyrimidine-methanol. Fenarimol is α-(2-chlorophenyl)-α-(4-chlorophenyl)-5-pyrimidinemethanol. While neither is formally considered an herbicide, both are used to selectively suppress or control annual bluegrass in turfgrass communities. Fenarimol is also used as a systemic fungicide for controlling an array of turfgrass diseases. While flurprimidol does provide some disease suppression, it is not commercially used for this purpose. Both flurprimidol and fenarimol are usually formulated as wettable powders (WP); fenarimol is also formulated as an emulsifiable concentrate (EC).

Triazoles The triazole currently used for growth regulation in turf is paclobutrazol:

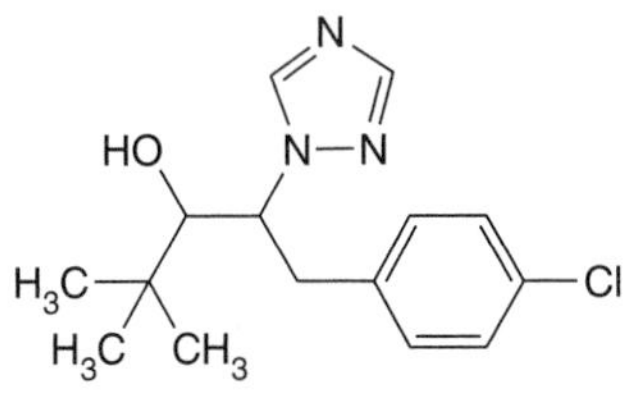

Paclobutrazol

The triazoles have a five-member azole ring in which three of the carbons have been replaced by nitrogen (making it a *tri*azole). In paclobutrazol, one of the nitrogens in the triazole ring is linked by a dimethyl pentanol ($CH_2CHCHOHC[CH_3]_3$) group to a benzene ring with a single chlorine substituent. Paclobutrazol is structurally similar to myclobutanil, propiconazol, and triadimefon—three triazole systemic fungicides used for controlling an array of turfgrass diseases.

Paclobutrazol is (2*RS*,3*RS*)-1-(4chlorophenyl)-4,4dimethyl-2-1,2,4-triazol-1-yl penta-n-3-ol. Because paclobutrazol also has fungitoxic properties, combinations of triazole fungicides with paclobutrazol will have additive effects with respect to disease control efficacy and plant-growth regulation. It is typically formulated as a salt concentrate (SC). Paclobutrazol (Trimmit) is primarily used for growth regulation of cool-season and warm-season (bermudagrass, St. Augustinegrass) turfgrasses and for selective suppression of annual bluegrass in creeping bentgrass turf.

Acylcyclohexanediones The only acylcyclohexanedione is trinexapac-ethyl:

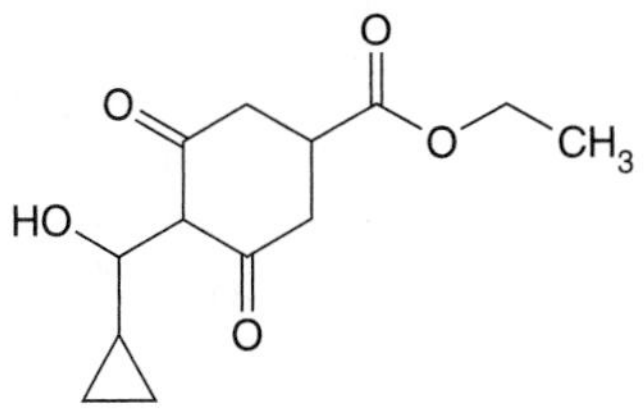

Trinexapac-ethyl

The acylcyclohexanediones contain a six-carbon non-aromatic ring with two doubly bonded oxygen substituents. In trinexapac-ethyl, the ring substituents also include a cyclopropyl-hydroxy-methylene ($CH_2CH_2CHCHOH$) group and a carboxylic acid ethyl ether ($COOCH_2CH_3$) group.

Trinexapac-ethyl is ethyl 4-cyclopropyl[hydroxy]methylene-3,5-dioxocyclohexane-carboxylate. The acylcyclohexanediones—unlike the pyrimidine and triazole

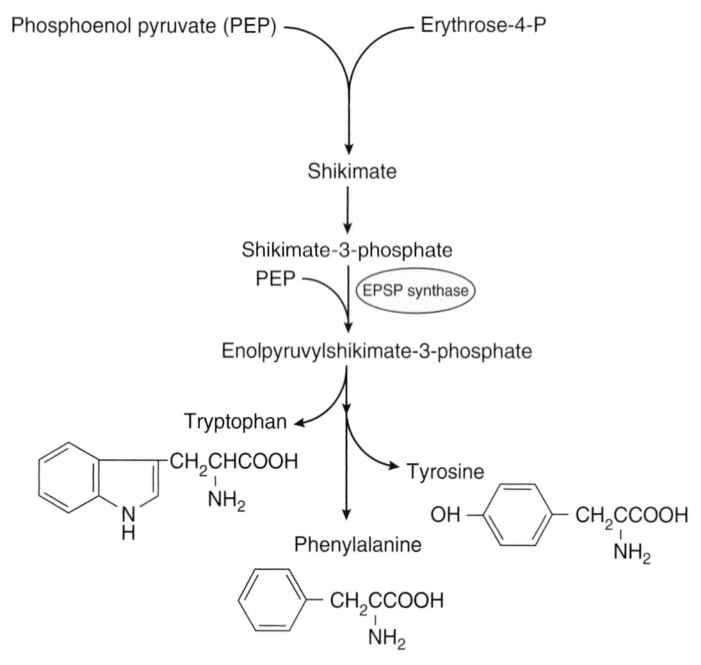

Figure 3.4
Illustration of the shikimic acid biosynthesis pathway by which the aromatic amino acids—tryptophan, tyrosine, and phenylalanine—are produced.

plant-growth regulators—do not have fungitoxic properties. Trinexapac-ethyl is formulated as an emulsifiable concentrate (EC) or a wettable powder packaged in a water-soluble bag (WSB). Trinexapac-ethyl (Primo) is widely used for turfgrass growth regulation on greens and fairways.

Amino Acid Biosynthesis Inhibitors

The amino acid biosynthesis inhibitors include the glycines (glyphosate), imidazolinones (imazaquin), pyrimidinyloxybenzoates (bispyribac-sodium), and sulfonylureas (chlorsulfuron, foramsulfuron, halosulfuron, metsulfuron, rimsulfuron, sulfometuron, sulfosulfuron, and trifloxysulfuron). The amino acid biosynthetic pathway inhibited by these herbicides occurs within the chloroplasts of sensitive plants.

Glyphosate is the only herbicide that prevents the production of the aromatic amino acids—phenylalanine, tryptophan, and tyrosine—by blocking the conversion of phosphoenolpyruvate (PEP) and shikimate-3-phosphate (SP) to enolpyruvylshikimate-3-phosphate (EPSP); it does this by inhibiting the EPSP synthase (EPSPS) enzyme (Figure 3.4). As this enzyme is not found in animals, glyphosate has low mammalian toxicity. These amino acids are essential in plants as precursors for cell wall formation and the production of phytohormones and other plant constituents. It has been estimated that more than 20 percent of the carbon fixed by leaves through photosynthesis passes through the aromatic amino acid pathway, and that up to 30 percent of the plant's dry weight is aromatic molecules derived from this pathway. These effects probably account for the relatively slow death (i.e., plants usually begin to die 5 to 10 days after treatment) of glyphosate-treated plants.

The imidazolinones, pyrimidinyloxybenzoates, and sulfonylureas prevent the production of the branch-chain aliphatic amino acids—isoleucine, leucine, and valine—by inhibiting the acetolactate synthase (ALS) enzyme (Figure 3.5). In the production of leucine and valine, ALS catalyzes the conversion of pyruvate to α-acetolactate. In the production of isoleucine, ALS catalyzes the conversion of pyruvate and α-ketobutyrate to α-keto-α-hydroxy butyrate. While the depletion of these amino acids would surely limit plant growth, it is not known whether this is the sole cause of death from ALS-inhibiting herbicides. Large accumulations of α-aminobutyrate in some treated plants led some investigators to conclude that this might actually be the toxic agent; however, this conclusion was later disproved. More recently, investigators have hypothesized that the inhibition of the ALS enzyme might generate oxygen radicals that, in sufficient concentrations, could be involved in the mechanism of action of ALS-inhibiting herbicides, but this has yet to be confirmed (Monaco et al., 2002). As the branch-chain aliphatic amino acid biosynthesis pathway does not exist in mammals, including humans, the ALS-inhibiting herbicides have very low mammalian toxicity.

Figure 3.5
Illustration of the branch-chain amino acid pathway.

Glycines The glycines include glyphosate:

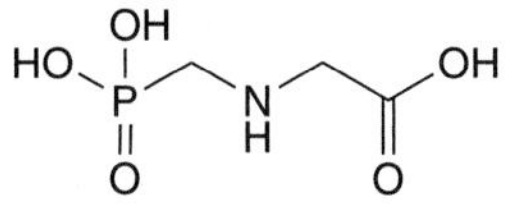

Glyphosate

Glycine (NH_2CH_2COOH) is composed of a methylene group ($—CH_2$-) connecting amino (NH_2) and carboxyl (COOH) groups. Glyphosate is a derivative with a phosphonomethyl group ($PO_3H_2CH_2$) attached to glycine's nitrogen. Glyphosate is N-[phosphonomethyl]-glycine. Glyphosate is a foliar-applied nonselective herbicide that is readily absorbed and translocated throughout the plant. Because it is rapidly adsorbed onto soil surfaces, it has no residual activity in the soil. It is formulated as a salt concentrate (SC). Glyphosate (Roundup, Touchdown) is a nonselective herbicide used for general vegetation control or for controlling winter weeds in dormant bermudagrass and bahiagrass turfs.

Imidazolinones The imidazolinones include imazaquin:

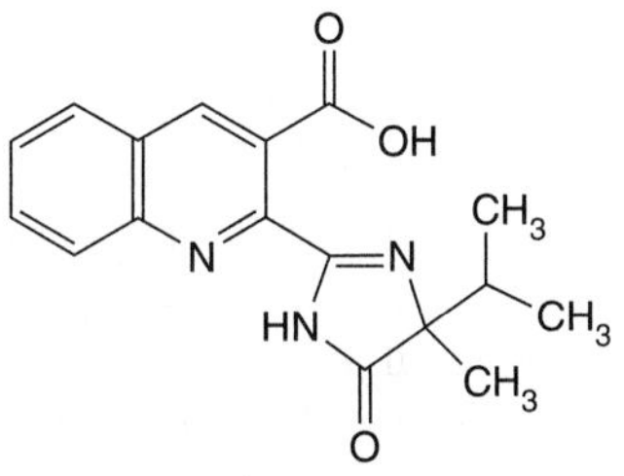

Imazaquin

The imidazolinones have a five-member azole ring in which two nonadjacent carbons have been replaced by nitrogen (making it an *imid*azole); the ring also has an oxygen doubly bonded to one of the ring's carbons. Imazaquin also has a methyl group (CH_3), a propyl group ($[CH_3]_2CH$), and a quinoline group (i.e., benzene and pyridine rings fused together) attached to its imidazole ring by carbon-to-carbon (C—C) bonds; in addition, there is a carboxyl group (COOH) as a substituent on the pyridine portion of the quinoline. Imazaquin is 2-[4,5-dihydro-4-methyl-4-(1-methylethyl)- 5-oxo-1H-imidazol-2-yl]-3-quinoline carboxylic acid. It is formulated as a dispersible granule (DG). Imazaquin (Image) is used on warm-season turfgrasses and in ornamental planting beds, primarily for controlling sedges and broadleaf weeds.

Pyrimidyloxybenzoates The pyrimidyloxybenzoates include bispyribac-sodium:

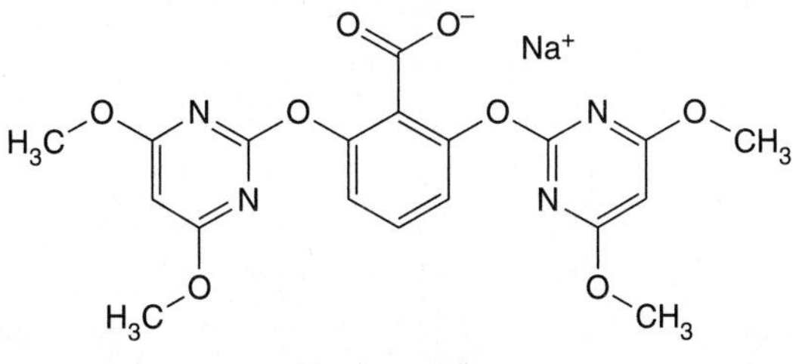

bispyribac-sodium

The pyrimidinyloxybenzoates have two primary components: a benzoate (i.e., benzene ring with a carboxylate [COO^-] group as a substituent) and at least one pyrimidine (aromatic ring in which two of the carbons were replaced by nitrogen) connected to the benzoate via an oxygen bridge (called an ether linkage).

Bispyribac-sodium has two pyrimidines, each with two methoxy groups (OCH_3) as ring substituents.

Bispyribac-sodium is sodium 2,6-bis([4,6-dimethoxypyrimidin-2-yl]oxy) benzoate. It is usually formulated as a salt concentrate (SC). Bispyribac-sodium (Velocity) is used for selectively controlling annual bluegrass and rough bluegrass, as well as several broadleaf species in creeping bentgrass and perennial ryegrass fairways and tees.

Sulfonylureas The sulfonylureas include chlorsulfuron, foramsulfuron, halosulfuron, metsulfuron, rimsulfuron, sulfometuron, sulfosulfuron, and trifloxysulfuron:

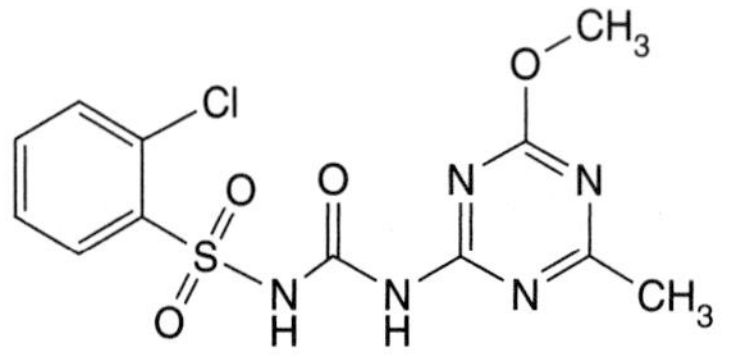

Chlorsulfuron

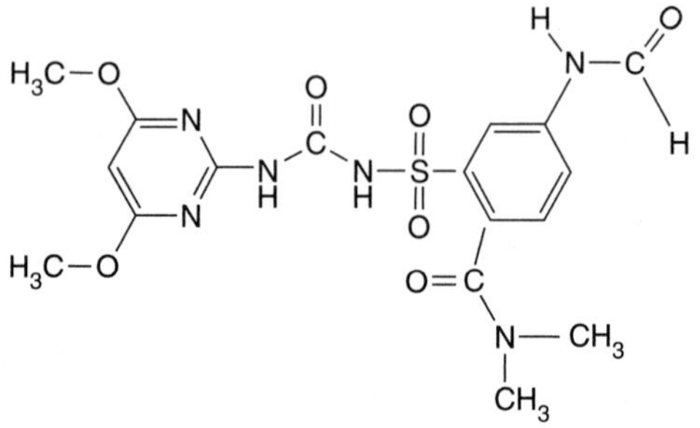

Foramsulfuron

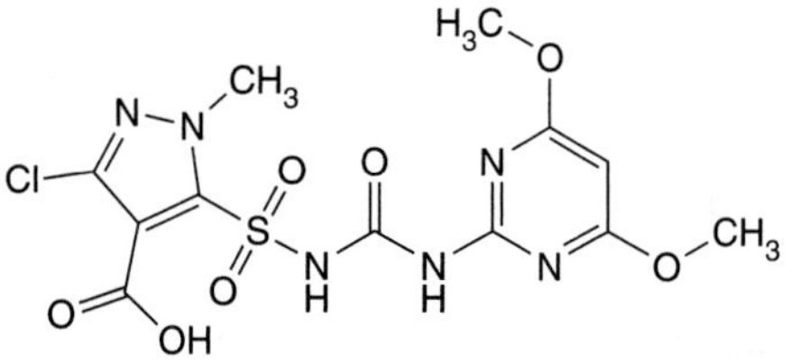

Halosulfuron

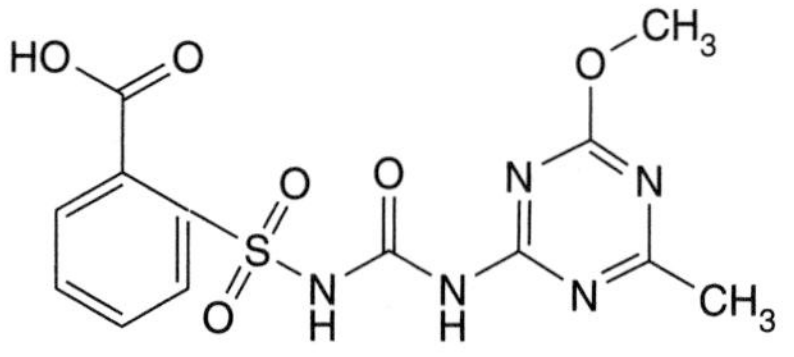

Metsulfuron

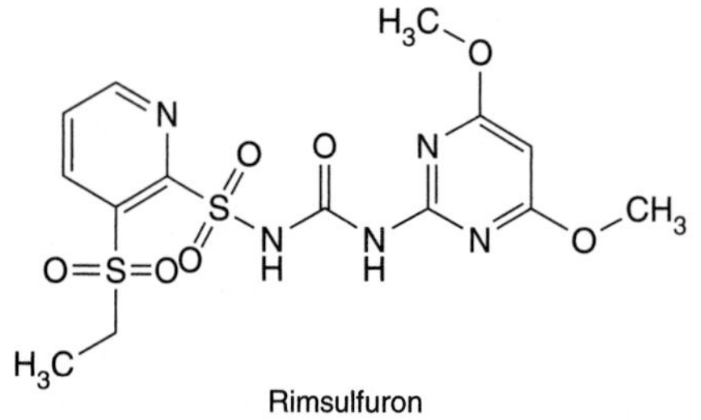

Rimsulfuron

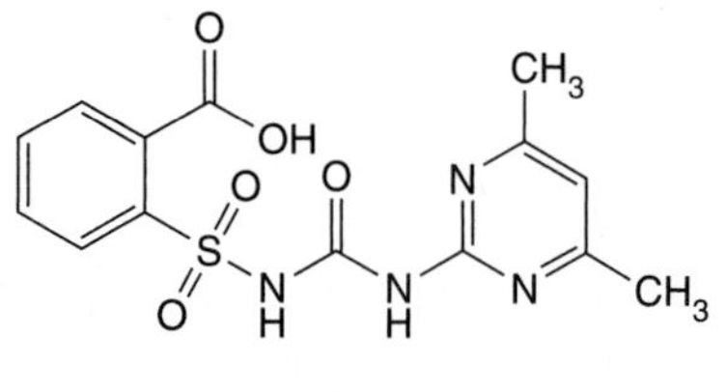

Sulfometuron

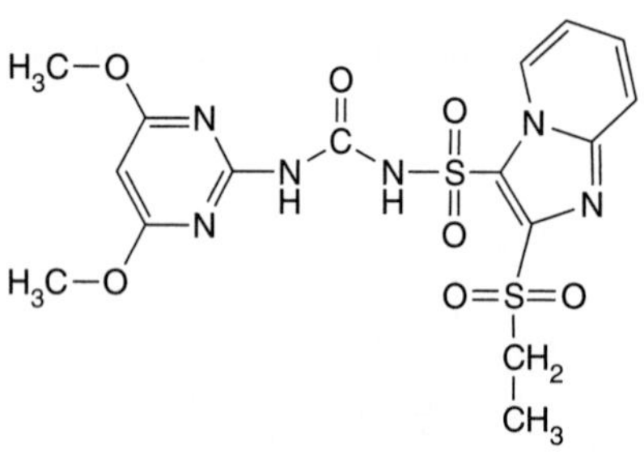

Sulfosulfuron

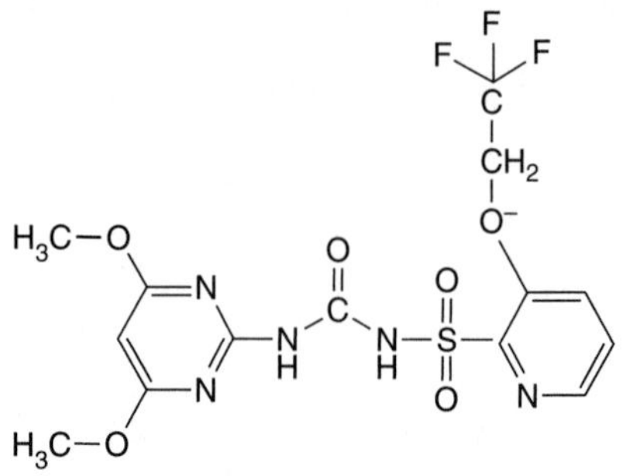

Trifloxysulfuron

The sulfonylureas have two common components: a urea group (—NHCONH—) at the center of the compound and a sulfonyl group (SO_2) connecting the sulfur to a urea nitrogen and to an adjacent aromatic or aliphatic group.

Chlorsulfuron has bonded to the sulfur a chlorobenzene (benzene ring with a chlorine [Cl] substituent), and bonded to the nitrogen is a triazine ring (in which three of the aromatic ring's carbons have been replaced by nitrogen) with two ring substituents: a methyl group (CH_3) and a methoxy group (OCH_3). Chlorsulfuron is 2-chloro-N-[(4-methoxy-6-methyl-1,3,5-triazin-2-yl) aminocarbonyl]-benzenesulfonamide. It is formulated as a dispersible granule (DG) and dry flowable (DF). Chlorsulfuron (TFC, Corsair) is used for spot treating clumps of tall fescue and perennial ryegrass in turf.

Foramsulfuron has bonded to the sulfur a benzene (with two substituents: a foramido [NHCOH] group and a dimethylcarbamoyl [$CON(CH_3)_2$] group), and bonded to the nitrogen is a pyrimidine (aromatic ring in which two of the carbons were replaced by nitrogen) with two methoxy (OCH_3) groups as ring substituents. Foramsulfuron is 1-(4,6-dimethoxypyrimidin-2-yl)-3-[2-(dimethylcarbamoyl)-5-formamidophenylsulfonyl]-urea. It is formulated as a suspension concentrate. Foramsulfuron (Revolver) is primarily used for controlling annual bluegrass, rough bluegrass, and ryegrasses from bermudagrass and other warm-season turfs.

Halosulfuron has bonded to the sulfur a pyrazole (five-member azole in which two adjacent carbons have been replaced by nitrogen, making it a *pyr*azole) group with three substituents: carboxyl [COOH], methyl [CH_3], and chlorine [Cl] substituents); and bonded to the nitrogen is pyrimidine (aromatic ring in which two of the carbons were replaced by nitrogen) with two methoxy (OCH_3) groups as ring substituents. Halosulfuron is methyl 5-[[(4,6-dimethoxy-2-pyrimidinyl)amino] carbonylamino-sulfonyl]-3-chloro-1-methyl-H-pyrazole 4-carboxylate. It is formulated as a wettable powder (WP). Halosulfuron (Sledgehammer) is primarily used for controlling annual and perennial sedges in turf.

Metsulfuron has bonded to the sulfur a benzoic acid (benzene with carboxyl group as a substituent), and bonded to the nitrogen is a triazine ring (in which three of the aromatic ring's carbons have been replaced by nitrogen) with two ring substituents: a methyl group (CH_3) and a methoxy group (OCH_3). Metsulfuron is 2-[[[[(4-methoxy-6-methyl-1,3,5-triazin-2-yl) amino] carbonyl]-amino] sulfonyl]-benzoate. It is formulated as a dry flowable (DF). Metsulfuron (Manor, Blade, Escort) is primarily used for controlling bahiagrass, ryegrass, and numerous broadleaf weeds in warm-season turfgrass communities; it is also used to hasten the spring transition from ryegrass to bermudagrass.

Rimsulfuron has bonded to the sulfur a pyridine ring (in which one of the carbons in the aromatic ring is replaced by nitrogen) with a sulfonyl group (SO_2) connecting the ring to a methyl group (CH_3), and bonded to the nitrogen is pyrimidine (aromatic ring in which two of the carbons were replaced by nitrogen) with two methoxy (OCH_3) groups as ring substituents. Rimsulfuron is N-((4,6-dimethyoxypriminidin-2-yl)aminocarbonyl)-3-(ethylsulfonyl)-2-pyridinesulfonamide. It is formulated as a dispersible granule (DG). Rimsulfuron (TranXit) is primarily used for controlling annual bluegrass, rough bluegrass, and ryegrasses in bermudagrass and other warm-season turfgrasses.

Sulfometuron has bonded to the sulfur a benzoic acid (benzene with carboxyl group as a substituent), and bonded to the nitrogen is pyrimidine (aromatic ring in which two of the carbons were replaced by nitrogen) with two methyl (CH_3) groups as ring substituents. Sulfometuron is 2-[[[[(4,6-dimethyl-pyrimidinyl)amino]carbonyl]amino]sulfonyl]benzoic acid. It is formulated as a dispersible granule (DG). Sulfometuron (Oust) is primarily used on warm-season utility turfs for controlling bahiagrass and tall fescues.

Sulfosulfuron has bonded to the sulfur an imidazole (a five-member azole ring in which two nonadjacent carbons have been replaced by nitrogen, making it an *imid*azole) fused with a pyridine ring (in which one of the carbons in the aromatic ring is replaced by nitrogen) and bonded to an ethyl group (CH_2CH_3) via a sulfonyl group (SO_2), and bonded to the nitrogen is pyrimidine (aromatic ring in which two of the carbons were replaced by nitrogen) with two methoxy (OCH_3) groups as ring substituents. Sulfosulfuron is 1-(4,6-dimethoxypyrimidin-2-yl)-3-(2-ethylsulfonylimidazo[1,2-a]pyridin-3-yl)sulfonylurea. It is formulated as a dry flowable (DF) or water dispersible granule (WDG). Sulfosulfuron (Certainty) is primarily used for controlling sedges, grasses, and broadleaf weeds in warm-season turfgrass communities (except bermudagrass greens). It can also be used for controlling rough bluegrass in creeping bentgrass fairways and tees, and or controlling quackgrass in kentucky bluegrass turf.

Trifloxysulfuron has bonded to the sulfur a pyridine ring (in which one of the carbons in the aromatic ring is replaced by nitrogen) with a trifluoroethoxy group (OCH_2CF_3) as a ring substituent, and bonded to the nitrogen is pyrimidine (aromatic ring in which two of the carbons were replaced by nitrogen) with two methoxy (OCH_3) groups as ring substituents. Trifloxysulfuron is *N*-[[(4,6-dimethoxy-2-pyrimidinyl)amino]carbonyl]-3-(2,2,2-trifluoroethoxy)-2-pyridinesulfonamide. It is formulated as a dry flowable (DF) or water-dispersible granule (WG). Trifloxysulfuron (Monument) is primarily used for controlling sedges, grasses, and broadleaf weeds in bermudagrass and zoysiagrass turfs. It can also be used for removing overseeded ryegrasses and rough bluegrass to aid spring transition of bermudagrass.

Carotenoid Pigment Biosynthesis Inhibitors

The carotenoid pigment biosynthesis inhibitors include two herbicides used for weed control in turf: norflurazon, a member of the pyridazinone chemical family, and mesotrione, a member of the callistemones family. Pyridazinone herbicides block the conversion of phytoene to phytofluene, an essential step in the carotenoid pigment biosynthesis pathway, by inhibiting the PDS (phytoene desaturase) enzyme (Figure 3.6). As a consequence, the formation of lycopene and other carotenoid pigments is prevented. Callistemones herbicides block the conversion of 4-hydroxyphenylpyruvate to homogentisate by inhibiting the enzyme 4-hydroxyphenylpyruvate dioxygenase (HPPD), resulting in the depletion of plastoquinones (Figure 3.7). As these are needed as cofactors for the proper functioning of the phytoene desaturase enzyme,

Carotenoid Biosynthesis
(in chloroplast)

phytoene
phytoene desaturase (PDS)
phytofluene
zeta-carotene
neurosporene
lycopene
cyclization
α-carotene
β-carotene
lutein
zeaxanthin

Figure 3.6 *Illustration of the carotenoid pigment biosynthesis pathway.*

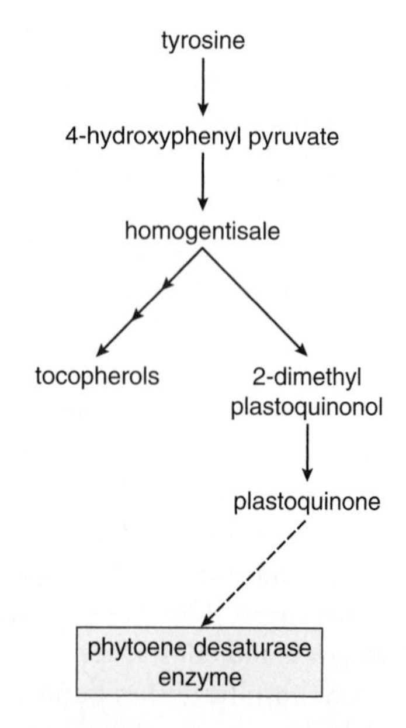

Figure 3.7 *Illustration of the plastoquinone biosynthesis pathway.*

carotenoid pigment production is inhibited. The carotenoid pigments protect chlorophyll from photooxidation. In their absence, chlorophyll is destroyed in the new growth following treatment. The bleached white appearance of affected tissues signals the loss of photosynthetic activity, resulting in a cessation of growth and eventual death. Plant tissues formed before treatment with pigment biosynthesis inhibitors do not show bleaching symptoms; however, with the turnover of carotenoid pigments, these tissues will eventually succumb as well.

Supplying treated plants with homogentisic acid can prevent toxicity from developing from either pyridazinone or callistemones herbicides.

While the carotenoid biosynthesis pathway is specific to plants, the HPPD enzyme functions in the catabolism of tyrosine in mammalian systems (Shaner, 2003). As tyrosine is converted to 4-hydroxyphenylpyruvate by the HPPD enzyme, inhibition of this conversion results in the accumulation of tyrosine in the blood. While this accumulation can lead to the development of corneal lesions in rats, the effect appears to be specific to this mammalian species. In humans, there are no apparent effects from accumulated tyrosine (from HPPD inhibitors), as they are readily excreted.

Pyridazinones The pyridazinones include norflurazon:

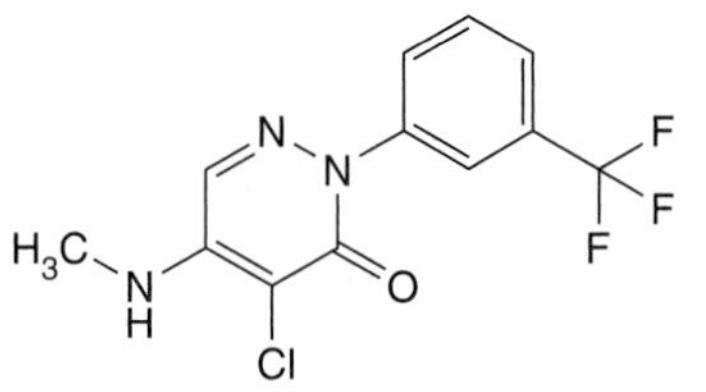

Norflurazon

This chemical class contains a pyridazinone core composed of an aromatic ring in which two adjacent carbons are replaced by nitrogen and an oxygen atom is doubly bonded to a carbon immediately adjacent to one of the nitrogens. Norflurazon has bonded to one of the ring nitrogens a benzene ring with a trifluoromethyl group (CF_3) as a substituent and a methyl amino group ($NHCH_3$) bonded to a carbon in the pyridazinone core. Norflurazon is 4-chloro-5-(methylamino)-2-(3-[trifluoromethyl]phenyl)-3(2H)-pyridazinone. It is formulated as a wettable powder (WP). Norflurazon (Predict) is primarily used for controlling annual grasses and some broadleaf weeds in woody, ornamental planting beds.

Callistemones The callistemones (also called benzoylcyclohexanediones) include mesotrione:

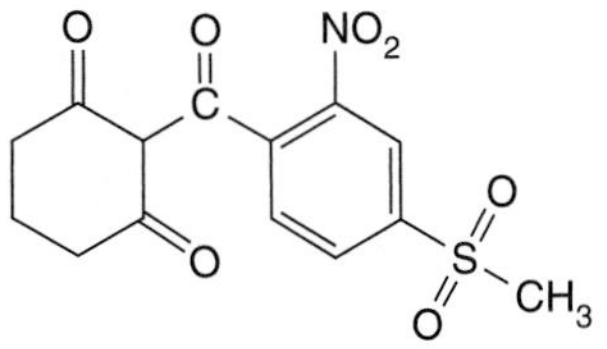

Mesotrione

This chemical class contains a cyclohexane ring with two doubly bonded oxygen atoms at the #1 and #3 positions, making it a cyclohexanedione. This is connected via a carbonyl (C=O) group to a nitrobenzene, with a methyl sulfonly (CH_3SO_2—) group at the #4 position. Mesotrione is 2-[4-(methylsulfonyl)-2-nitrobenzoyl]-1,3-cyclohexanedione. It is formulated as a salt concentrate (SC). Mesotrione (Tenacity) is primarily used for controlling crabgrass in turf. It also shows promise for controlling bentgrass in Kentucky bluegrass turf.

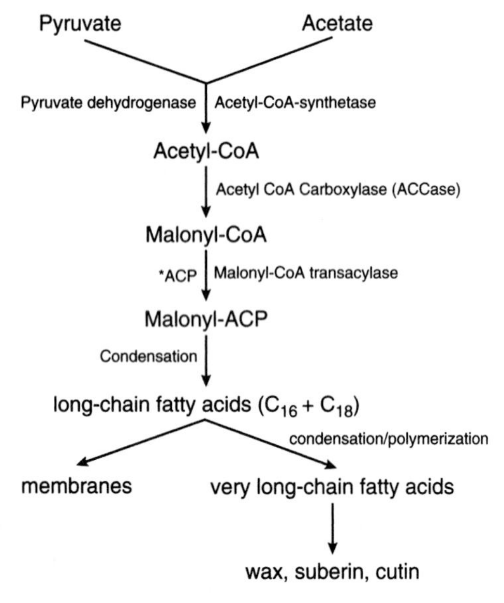

Figure 3.8
Illustration of the fatty acid biosynthesis pathway. Pyruvate and acetate are sources of acetyl-CoA, which is converted to malonyl-CoA by the acetyl-CoA (ACCase) enzyme. The addition of an acyl carrier protein (ACP) transforms malonyl-CoA to malonyl ACP, which can undergo successive condensations to form the long-chain fatty acids used to construct membranes. Further condensations or polymerization of the long-chain fatty acids results in the formation of very long-chain fatty acids that are used to form cuticular waxes and cutin or suberin on root cells.

Lipid Biosynthesis Inhibitors

The lipid biosynthesis inhibitors include the aryloxyphenoxy-propionates (diclofop-methyl, fenoxyprop-ethyl, and fluazifop-butyl) and the cyclohexanediones (clethodim and sethoxydim). Sometimes, the benzofuranes (ethofumesate) are included as well. The aryloxyphenoxy-propionate and cyclohexanediones herbicides block the conversion of acetyl-CoA to malonyl-CoA, an essential step in the lipid biosynthesis pathway, by inhibiting the ACCase (acetyl-CoA carboxylase) enzyme (Figure 3.8). As a consequence, the lack of lipids leads to a loss of membrane integrity in plant cells, arrested growth, and eventual death. This effect is specific to grasses, as broadleaf (dicot) plants have a different form of the enzyme in their chloroplasts and thus are largely resistant to these herbicides. The benzofurane herbicides do not inhibit the ACCase enzyme; however, there is a decrease in epicuticular wax formation on treated plants, suggesting that fatty acid synthesis is in some way inhibited.

While the ACCase enzyme is involved in fatty acid biosynthesis in mammals, none of the mammalian forms of this enzyme are inhibited by the aryloxyphenoxypropionate or cyclohexanedione herbicides.

Aryloxyphenoxy-Propionates The aryloxyphenoxy-propionates include diclofop-methyl, fenoxyprop-ethyl, and fluazifop-butyl:

Diclofop-methyl

Fenoxyprop-ethyl

Fluazifop-butyl

The common structural features of this chemical family include an esterified derivative (e.g., methyl- ethyl- and butyl-) of propionic acid (CH_3CH_2COOH) derivative, a phenoxy group (e.g., benzene with an oxygen substituent), and an "aryloxy" group composed of an aromatic structure indirectly bonded to the phenoxy group by an oxygen atom. The aryloxy group in diclofop-methyl is a dichlorophenoxy group. In fenoxyprop-ethyl, it is a chlorobenzoxazole group composed of a chlorobenzene (benzene with one chlorine substituent) fused to an oxazole (five-member heterocyclic ring with one oxygen, one nitrogen, and three carbons forming the ring) indirectly bonded to the phenoxy group by an oxygen atom. And in fluazifop-butyl, it is a trifluoromethyl pyridine (an aromatic ring in which one of the carbons is replaced by nitrogen, and with a trifluoromethyl [CF_3] substituent) indirectly bonded to the phenoxy group by an oxygen atom.

Diclofop-methyl is (+/−)-2-[4-(2,4-dichlorophenoxy)phenoxy]propionic acid. It is formulated as an emulsifiable concentrate (EC). Fenoxyprop-ethyl is (+/−)-ethyl 2-4-((6-chloro-2-benzoxazolyloxy)-phenoxy) propionic acid. It is formulated as an emulsifiable concentrate (EC). Fluazifop-butyl is (*R*)-2-[4-[[5-(trifluoromethyl)-2-pyridinyl]oxy]phenoxy] propionic acid. It is formulated as an emulsifiable concentrate (EC). Diclofop-methyl (Illoxan) is primarily used for goosegrass and ryegrass control in bermudagrass turf. Fenoxyprop-ethyl (Acclaim) is primarily used for controlling crabgrass and other summer annual grasses in cool-season and warm-season turfgrass communities. Fluazifop-butyl (Fusilade) is primarily used for controlling annual and perennial grasses in tall fescue and zoysiagrass turfgrass communities and in ornamental planting beds.

Cyclohexanediones The cyclohexanediones include clethodim and sethoxydim:

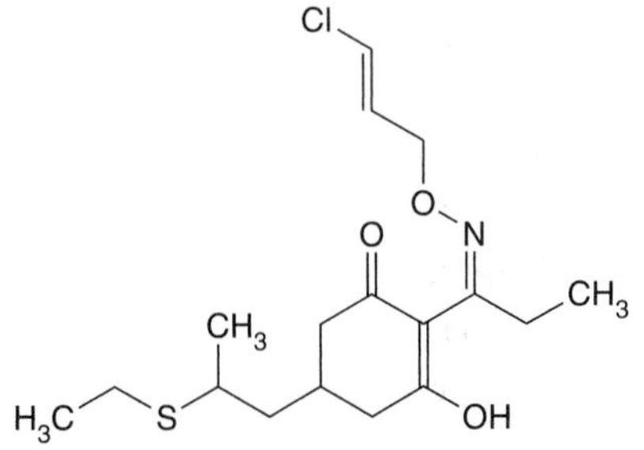

Clethodim

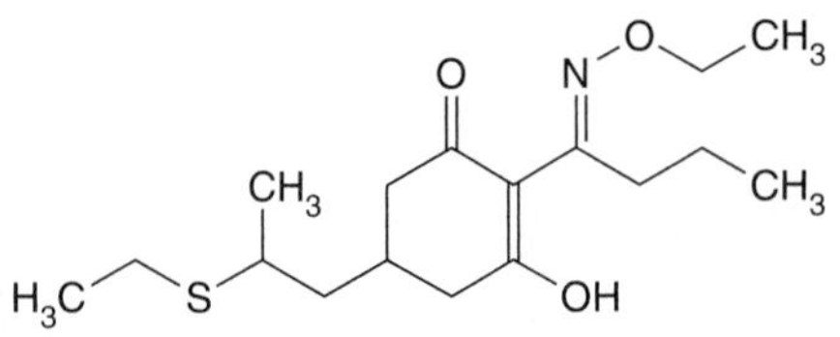

Sethoxydim

The common structural features of this chemical family include a cyclohexene (six-member ring containing one double bond between adjacent carbon atoms) with four substituents:

- #1 position: a doubly bonded oxygen.
- #2 position: an aliphatic group with an oxyimino (O—N=) link to another aliphatic group.
- #3 position: an hydroxy group (OH).
- #5 position: a propyl group (CH_3CHCH_2) bonded to an ethyl-thio group (CH_3CH_2S).

In clethodim, the aliphatic group bonded directly to the cyclohexene is a propyl group (CH_3CH_2C) that is in turn bonded to a chloropropene group ($ClCH=CHCH_2$) through the oxyimino (O—N=) link. In sethoxydim, the aliphatic group bonded directly to the cyclohexene is a butyl group (CH3CH_2CH_2C) that is in turn bonded to an ethane group (CH_3CH_2) through the oxyimino (O—N=) link.

Clethodim is E,E-(±)-2-[1-[[3-chloro-2-propenyl)oxy]-imino]propyl]-5-[2-(ethylthio)-propyl]-3-hydroxy-2-cyclohexen-1-one. It is formulated as an emulsifiable concentrate (EC). Sethoxydim is ±2[1-ethoxyimino)butyl]-5-(ethylthio)propyl]-3-hydroxy-2-cyclohexen-1-one. It is formulated as an emulsifiable concentrate (EC). Clethodim (Envoy) and sethoxydim (Poast) are primarily used for controlling annual and perennial grasses on ornamental planting beds.

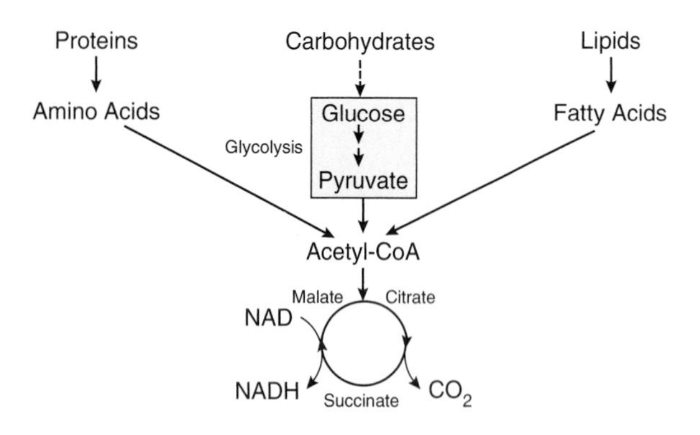

Figure 3.9
Illustration of the processes involved in respiration, including the breakdown of proteins, carbohydrates (including the glycolytic transformation of glucose to pyruvate), and lipids into acetyl-CoA, and the production of NADH from NAD in the Krebs cycle.

Benzofuranes The benzofuranes include ethofumesate:

Ethofumesate

The benzofuranes have a benzene ring fused with a furan (five-member heterocyclic compound in which an oxygen has been substituted for one of the carbons). The benzene ring has one substituent, which is a methylsulfonate group (CH_3SO_3), while the furan ring has three substituents: an ethoxy group (CH_3CH_2O) at the #2 position and two methyl groups (CH_3) at the #3 position.

Ethofumesate is (±)2-ethoxy-2,3-dihydro-3,3-dimethyl-5-benzofuranyl methane- sulphonate. It is formulated as an emulsifiable concentrate (EC). Ethofumesate (Prograss) is primarily used for controlling annual bluegrass in ryegrass, or in dormant bermudagrass overseeded with ryegrass. It can also be used on other cool-season and some warm-season (St. Augustinegrass) turfs.

Respiration Inhibitors

The herbicides that are believed to work, at least in part, as respiration inhibitors are the organic arsenicals: CAMA, DSMA, and MSMA. This action may reflect their role as uncouplers of oxidative phosphorylation. In respiration, enzymes extract two electrons and two protons from glucose and an array of metabolites, transferring both electrons and one proton to the coenzyme nicotinamide adenine dinucleotide (NAD^+) to form NADH, with the other proton released into the surrounding solution (Figure 3.9). The two electrons carried by NADH are passed along the electron transport chain—a series of alternately oxidized and reduced molecules embedded within the inner membrane of mitochondria—to an oxygen (O_2) receptor at the end of the chain (Figure 3.10). The oxygen then joins with hydrogen ions to form water (H_2O). As the electrons move down the chain, they release energy in a stepwise fashion. Much of this energy is used to pump protons from the interior of the mitochondria to the intermembrane space between the inner and outer mitochondrial membranes. This results in a proton gradient, as the pH within the mitochondrion is 8.5 while that of the intermembrane space is 7.0, reflecting the higher H^+ concentration there. This proton gradient constitutes the potential energy for oxidative phosphorylation—the process by which adenosine-triphosphate (ATP) is synthesized from adenosine-diphosphate (ADP) and phosphate. Because the inner mitochondrial membrane is impervious to protons, they can flow back into the mitochondrion only through protein channels within the membrane that are part of the ATP synthase enzyme complex. The protons move along the proton gradient (i.e., from higher to lower H^+ concentrations) through the protein channels, releasing energy that is used by ATP synthase to produce ATP. The energy stored in ATP is then used to drive all of the cell's biosynthesis, mechanical, and

Figure 3.10
Illustration of the electron transport chain through which the NADH generated from the Krebs cycle transfers its reducing power in a series of redox (reduction and oxidation) reactions to form water from oxygen and protons. This is coupled to the creation of high-energy ATP from ADP and inorganic phosphate in oxidative phosphorylation.

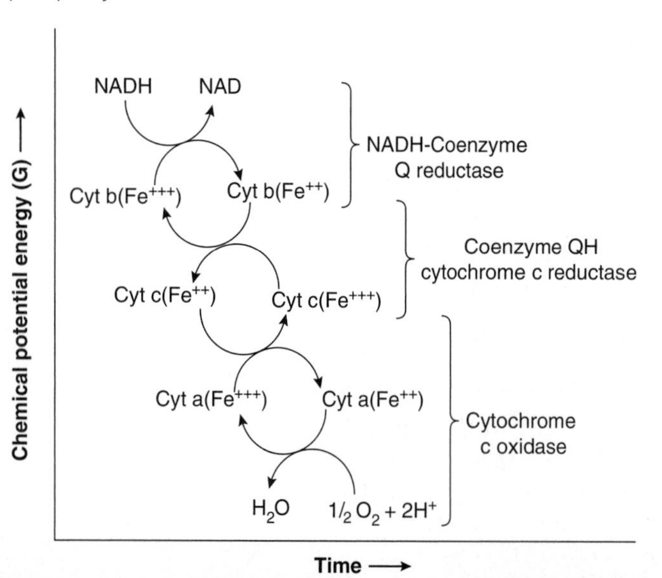

transport activities. To the extent that the organic arsenicals uncouple oxidative phosphorylation, this probably occurs where arsenate substitutes for phosphate in oxidative phosphorylation, forming ADP-arsenate, which is easily hydrolyzed; thus, energy from the electron transport chain merely generates heat instead of being captured in the formation of ATP.

In addition to uncoupling oxidative phosphorylation, other possible effects of arsenate on respiration may include binding with sulfhydryl enzymes on cell membranes, altering membrane integrity, and destroying the proton gradient so that there is no energy conservation and no ATP synthesis. Also, arsenate can block the conversion of fumarate to malate—a step in the Krebs cycle through which Acetyl-CoA from the oxidation of glucose is further oxidized, generating ATP directly, as well as indirectly through the formation of NADH—by inhibiting the furarase enzyme.

The organic arsenicals generally are considered to have low mammalian toxicity; however, as the same sites of action for these herbicides exist in both plant and animal species, arsenates can pose a significant hazard to humans, especially if converted to the inorganic form prior to or following ingestion.

Organic Arsenicals The organic arsenicals include calcium methanearsonate (CAMA), disodium methanearsonate (DSMA), monosodium methanearsonate (MSMA):

CAMA: $CH_3-As(=O)(O^-)-O^-$ Ca^{2+}

DSMA: $H_3C-As(=O)(O^-)-O^-$ Na^+ Na^+

MSMA: $HO-As(=O)(O^-)-CH_3$ Na^+

The common feature of the organic arsenicals is the pentavalent arsenic with one doubly bonded and two singly bonded oxygens. In CAMA, there is a divalent calcium (Ca^{2+}) cation balancing two negatively charged oxygens. In DSMA, there are two monovalent sodium (Na^+) cations balancing two negatively charged oxygens. And in MSMA, there is one monovalent sodium cation balancing a negatively charged oxygen. They are formulated as salt concentrates (SC). The organic arsenicals are primarily used alone or in combinations with metribuzin (Sencor) for controlling crabgrass, goosegrass, and dallisgrass in warm-season turfgrass communities.

Cell Membrane Disruptors

Since the herbicides that disrupt cell membranes as their primary mechanism have different sites of action within the chloroplasts of plant cells, they are usually grouped by their respective sites of action.

Cell membrane disruptors operating in the Photosystem II reaction center as their site of action include the triazines (atrazine, simazine), triazinones (metribuzin), benzonitriles (bromoxynil, dichlobenil), and benzothiadiazoles (bentazon). The Photosystem II (PS II) reaction center is a pigment–protein complex where electron transport is initiated by solar energy harvested with chlorophyll and carotenoid pigments to split water molecules (i.e., photolysis), generating reducing power (i.e., excited electrons, protons [H^+ ions]) and molecular oxygen (O_2). The electrons and protons are passed along the electron transport chain—a series of alternately reduced and oxidized molecules embedded within the thylakoid membranes of chloroplasts—to the Photosystem I reaction center, where it reduces nicotinamide adenine dinucleotide phosphate [$NADP^+$] to NADPH for later use

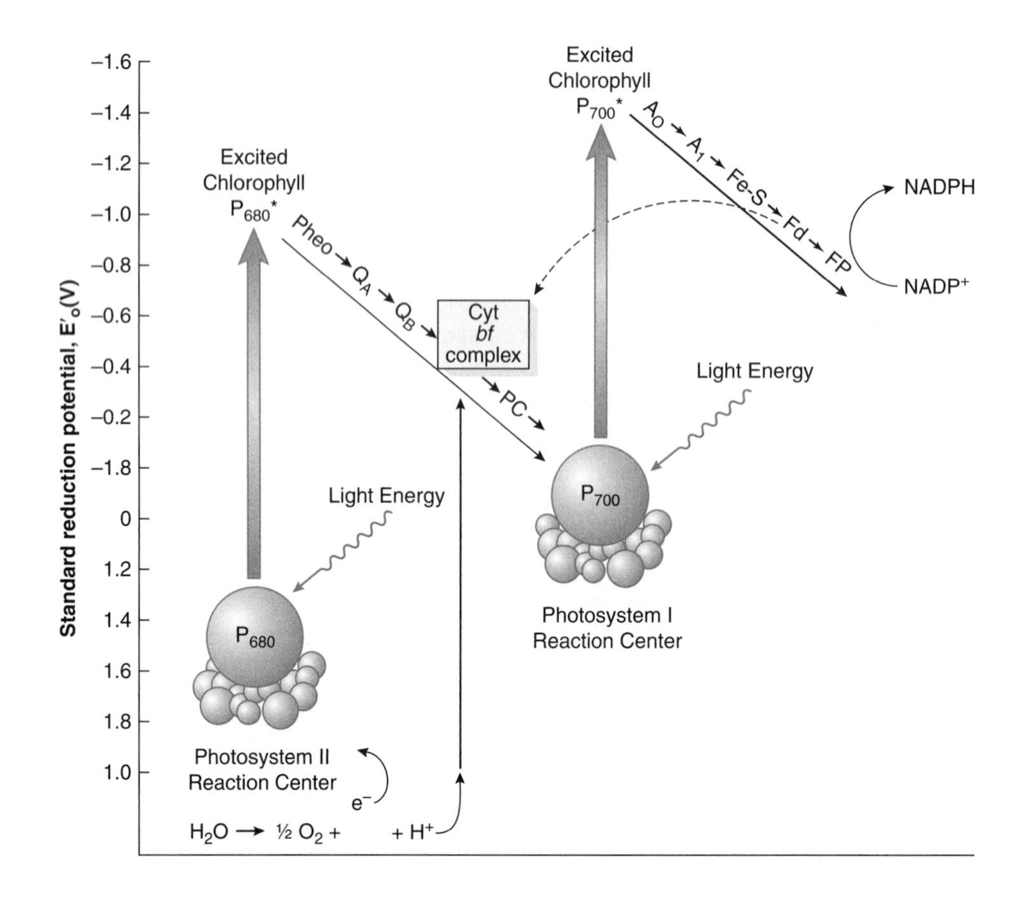

Figure 3.11
Illustration of the Z scheme encompassing Photosystems I (right) and II (left).

in the light-independent conversion of carbon dioxide [CO_2] to glucose [$C_6H_{12}O_6$] and other carbohydrates (Figure 3.11).

As the electrons move down the chain, they release energy in a stepwise fashion, generating ATP from ADP and phosphate through a process called photophosphorylation. Actually, the energy is used to pump protons from the stroma of the chloroplasts across the thylakoid membrane and into the center of the thylakoid, called the lumen, where they accumulate and lower the pH. This results in a proton gradient, constituting the potential energy for photophosphorylation. Because the thylakoid membrane is impervious to protons, they can flow back into the stroma only through protein channels within the membrane that are part of the ATP synthase enzyme complex. The protons move along the proton gradient (i.e., from higher to lower H^+ concentrations) through the protein channels, releasing energy that is used by ATP synthase to produce ATP.

The herbicides operating in the Photosystem II reaction center bind to the Q_B protein in the normal electron transport sequence, blocking electron transport to the plastoquinone (PQ) pool so that no reducing power is generated in photosynthesis. Earlier, it was believed that plants treated with PS II-inhibiting herbicides died from carbohydrate starvation. In fact, they die from cell membrane disruption caused by lipid peroxidation. As stated earlier, when chlorophyll molecules are excited by light energy, the excitation energy is used for photolysis; however, when electron flow is blocked, this energy is used instead to form oxygen radicals (e.g., singlet oxygen [O_2*] and hydrogen peroxide [H_2O_2]), which are highly reactive forms of oxygen that cause lipid peroxidation, destroying membrane integrity and leading to the loss of cellular compartmentalization and, eventually, cell death. Initial chlorosis of herbicide-treated plants reflects the photo-oxidation of chlorophyll and carotenoids (whose role is to protect chlorophyll from photo-oxidation) from the excess energy transferred to these pigments.

As the Photosystem II site of action does not exist in mammals, including humans, the PSII-inhibiting herbicides have very low mammalian toxicity.

Cell membrane disruptors operating in the Photosystem I reaction center as their site of action include the bipyridyliums paraquat and diquat. The Photosystem I (PS I) reaction center is a pigment–protein complex where light energy is absorbed and electrons transmitted from the PS II reaction center are passed along an electron transport chain to eventually reduce $NADP^+$ and form (with H^+) NADPH (for later use in the light-independent conversion of carbon dioxide [CO_2] to glucose [$C_6H_{12}O_6$] and other carbohydrates). The herbicides operating in the Photosystem I reaction center inhibit electron transport from plastocyanin (a small water-soluble copper-containing protein) to ferrodoxin (a small water-soluble iron-sulfur protein), so that no reducing power is generated in photosynthesis. The bipyridylium herbicides do this by accepting electrons directly and, as a result, becoming free radicals (instead of passing the electrons on in the electron transport chain) and by initiating a series of reactions leading to lipid peroxidation, cell membrane disruption, and death. The bipyridylium free radicals do not directly cause membrane damage; rather, they interact with water and oxygen to form superoxide ($O_2{}^{*-}$), which reacts with the superoxide dismutase (SOD) enzyme to form hydrogen peroxide (H_2O_2), which in turn forms hydroxy radicals (OH^*)—a highly potent biological oxidant that quickly causes lipid peroxidation. The herbicides are then converted back to the parent compounds—a process called auto-oxidation—to undergo successive cycles of electron acceptance and hydroxy radical formation.

Since the bipyridilium herbicides can also accept electrons from the electron transport chain in the mitochondria, forming oxygen radicals that oxidize membranes, they are extremely toxic to mammals, including humans. In animals, these herbicides appear to target the lungs, where they accumulate in the alveolar epithelium (membranes surrounding the air sacs) and generate oxygen radicals that cause necrosis of these cells (Shaner, 2003).

Cell membrane disruptors operating in the chlorophyll biosynthesis pathway as their site of action include the diphenyl ethers (oxyfluorfen), triazolinones (carfentrazone, sulfentrazone), and oxadiazoles (oxadiazon). The chlorophyll biosynthesis pathway occurs within the chloroplasts and consists of more than a dozen steps, beginning with glutamate, an amino acid, and ending with chlorophyll a. Herbicides operating in the chlorophyll biosynthesis pathway block the conversion of protoporphyrinogen IX to protoporphyrine IX—two intermediaries in the pathway—by inhibiting the protoporphyrinogen oxidase (PROTOX) enzyme; thus, these herbicides are often referred to as PROTOX inhibitors (Figure 3.12). As protoporphyrinogen molecules accumulate after treatment with PROTOX-inhibiting herbicides, they diffuse from the chloroplasts into the cytoplasm, where they are oxidized to protoporphyrine molecules. Protoporphyrine readily interacts with oxygen and light to form singlet oxygen ($O_2{}^*$), which initiates a series of reactions leading to lipid peroxidation, cell membrane disruption, and death.

In mammals, porphyrin biosynthesis is critical for the production of heme, which are iron-containing molecules that bind with proteins as a cofactor to form haemoproteins (e.g., haemoglobin and myoglobin) that function as oxygen carriers. The effects of PPO enzyme inhibition in animals are highly species-dependent, as mice can be seriously affected at dosages that have little or no effect on rabbits; however, the mammalian toxicity of these herbicides appears to be minimal at the rates they are used (Shaner, 2003).

Cell membrane disruptors operating in a nitrogen assimilation pathway as their site of action include the phosphinic acids (glufosinate). There are several

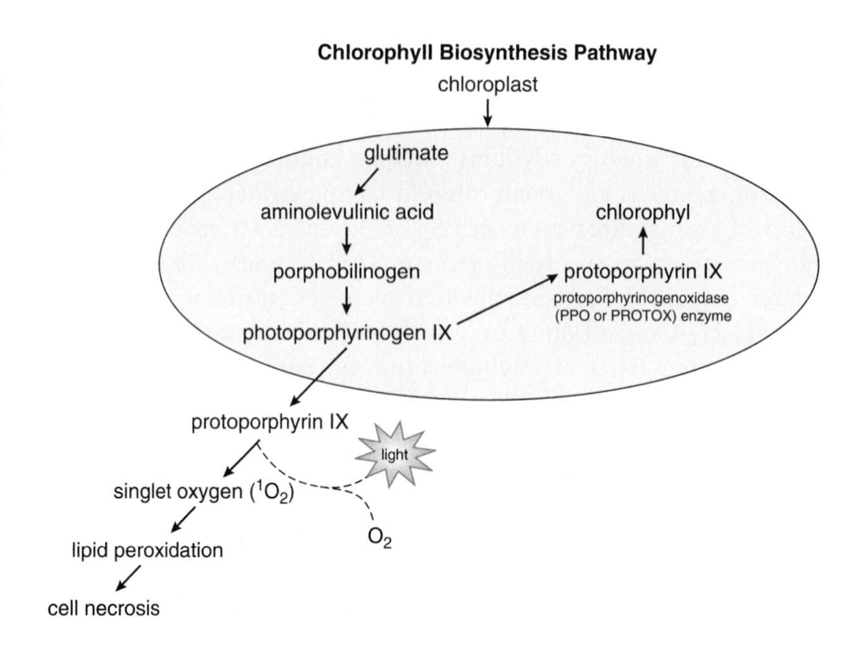

Figure 3.12
Illustration of the effects of a protoporphyrinogen oxidase (PROTOX) enzyme inhibition on singlet oxygen formation, resulting in lipid peroxidation and cell-membrane disruption.

nitrogen assimilation pathways through which ammonium ions (NH_4^+) are assimilated into amino acids. One pathway involves the conversion of glutamate and ammonium ions to glutamine by the glutamine synthetase (GS) enzyme (Figure 3.13). Glufosinate blocks this conversion by inhibiting the GS enzyme, resulting in increased levels of potentially phytotoxic ammonia and decreased levels of glutamine, as well as several other amino acids—glutamate, aspartate, asparagine, alanine, and serine—which are dependent on the presence of glutamine for their synthesis. Following glufosinate treatment, the decrease in amino acids not only influences the subsequent formation of proteins, but causes other reactions that are dependent upon amino acids as carriers of amino groups (—NH_2). One such reaction is the transamination of glyoxalate (HCOCOOH) to form glycine (NH_2CH_2COOH) in the photorespiration cycle. With the absence of amino donors in glufosinate-treated plants, glyoxalate concentrations accumulate to levels that inhibit RUBP carboxylase (also known as Rubisco)—the key enzyme for carbon fixation in the Calvin cycle. Inhibition of Rubisco leads to an inhibition of the light reactions of photosynthesis. As stated earlier, when chlorophyll molecules are excited by light energy, the excitation energy is used for photolysis; however, when electron flow is blocked, this energy is used instead to form oxygen radicals (e.g., singlet oxygen [O_2*], hydrogen peroxide [H_2O_2]), which are highly reactive forms of oxygen that cause lipid peroxidation, destroying membrane integrity and leading to the loss of cellular compartmentalization and, eventually, cell death.

In mammals, the GS enzyme regulates ammonia levels in the muscles, liver, and other organs; it also recycles glutamate—a neurotransmitter—in the brain (Shaner, 2003). In cases where people have ingested large quantities of glufosinate in suicide attempts, some memory loss occurred, suggesting that high concentrations of the herbicide can have toxic effects in the brain as a result of GS inhibition; however, these effects were relatively short term, as glufosinate is rapidly excreted from the system. In these suicide attempts, much of the toxicity was related to the surfactant included in the formulation rather than to the active ingredient.

Figure 3.13
Illustration of inhibition of glutamine synthetase by glufosinate, blocking the conversion of glutamate to glutamine.

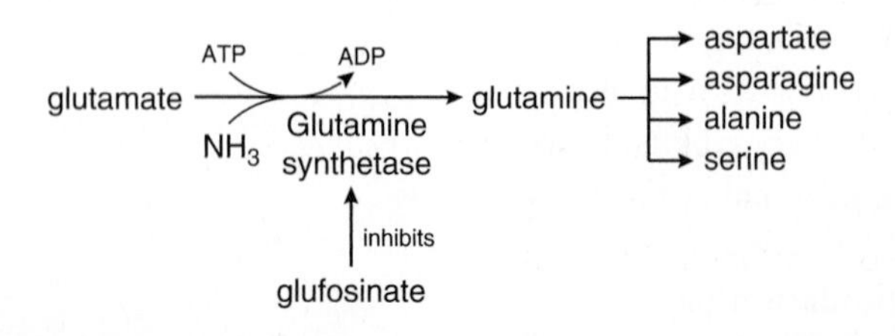

The site of action of one cell membrane disruptor—pelargonic acid—is currently unknown. Pelargonic acid and some other fatty acids are presumed to be membrane disruptors because of substantial ion leakage that occurs from their use.

Triazines The triazines include atrazine and simazine:

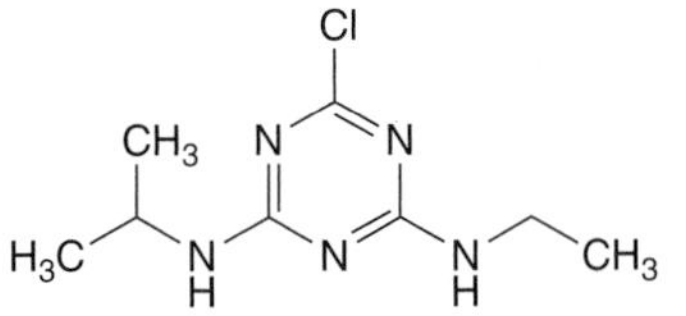

Atrazine

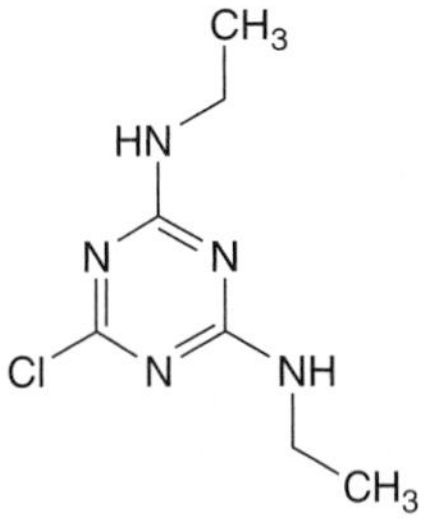

Simazine

The molecular structure of the triazines includes a triazine ring (in which three of the aromatic ring's carbons have been replaced by nitrogen) and two substituted amino groups as ring substituents. One of the amino groups in atrazine is bonded to a propyl group (CH_2CHCH_3), while the other is bonded to an ethyl group (CH_3CH_2); in simazine, both amino groups are bonded to ethyl groups. Both atrazine and simazine have a chlorine (Cl) atom as an additional ring substituent.

Atrazine is 2-choloro-4-(ethylamino)-6-(isopropylamino)-s-triazine. It is formulations are water dispersible granules (WG) and liquids (L). Simazine is 2-chloro-4,6-bis(ethylamino)-s-triazine. It is formulated as granules (G), wettable powders (WP), water dispersible granules (WDG), and dry flowables (DF). Atrazine and simazine are available under many trade names and are primarily used for controlling annual weeds in centipedegrass, St. Augustinegrass, and zoysiagrass turfs. Simazine is also used for weed control in woody ornamental planting beds.

Triazinones The triazinones include metribuzin:

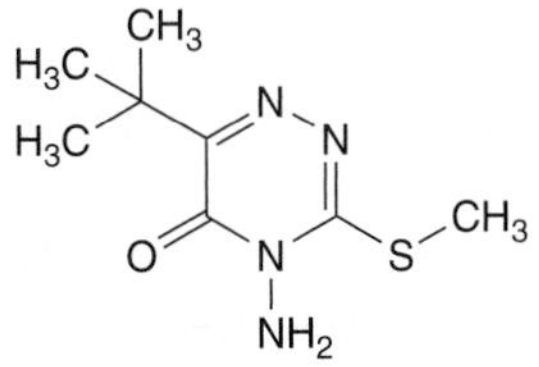

Metribuzin

The molecular structure of the triazinones includes a triazine ring (in which three of the aromatic ring's carbons have been replaced by nitrogen) and to which an oxygen atom has been doubly bonded to one of the ring's carbons. In metribuzin, the other ring substituents include a methyl thio group (CH_3S) and a butyl group ($[CH_3]_3C$) bonded to ring carbons and an amino group (NH_2) bonded to a ring nitrogen.

Metribuzin is 4-amino-6-tert-butyl-3-(methylthio)-as-triazin-5(4H)-one. It is formulated as a dry flowable (DF), water dispersible granule (WG), solution concentrate (SC), and water dispersible liquid (L). Metribuzin (Sencor) is primarily used alone or in combination with organic arsenical herbicides for controlling goosegrass and annual broadleaf weeds in bermudagrass turf.

Benzonitriles The benzonitrile include bromoxynil:

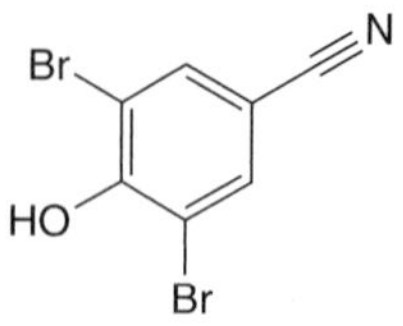

Bromoxynil

The molecular structure of the benzonitriles include bromoxynil: with a nitrile group (CN) as a ring substituent. Bromoxynil also has two bromine atoms (Br) and a hydroxy group (OH) as ring substituents.

Bromoxynil is 3,5-dibromo-4-hydroxybenzonitrile. It is formulated as an emulsifiable concentrate (EC). Bromoxynil (Buctril) is primarily used for controlling broadleaf weeds in seedling turfgrass stands.

Benzothiadiazoles The benzothiadiazoles include bentazon:

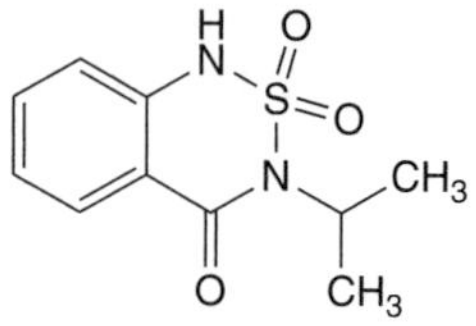

Bentazon

The molecular structure of the benzothiadiazoles includes a benzene ring fused to a six-member heterocyclic ring in which two of the ring's carbons have been replaced by nitrogens and one by sulfur. Diathiazole ring substituents include two oxygens doubly bonded to the ring sulfur and a propyl group ($[CH_3]_2CH$) bonded to a ring nitrogen.

Bentazon is 3-(1-methylethyl)-(1*H*)-2,1,3-benzothiadiazin-4(3*H*)-one 2,2-dioxide. It is formulated as a water dispersible liquid (L). Bentazon (Basagran) is primarily used for controlling sedges in turf.

Bipyridyliums The bipyridyliums include paraquat and diquat:

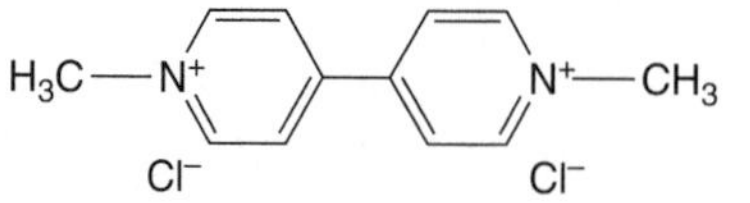

Paraquat

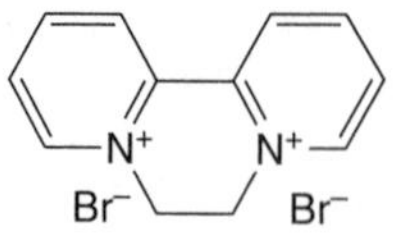

Diquat

The chemical structure of the bipyridyliums includes two pyridine rings fused together by a C—C bond to form a bipyridinium ion. Paraquat has methyl groups (CH_3) bonded to the ring nitrogens, and chlorine anions (Cl^-) are use to form a salt with its bipyridinium ion. In diquat, the nitrogens in the pyridine rings are connected via an ethylene (—CH_2CH_2—) group, and bromine anions (Br^-) are use to form a salt with its bipyridinium ion.

Paraquat is 1,1′=dimethyl-4,4′-bipryidinium ion. It is formulated as a liquid water soluble (L) in the dichloride form. Diquat is 6,7-dihydropyrido(1,2-α:2′,1′-c) pyrazinediium ion. It is formulated as a liquid water soluble (SL) in the dichloride form. Paraquat (Gramoxone) and diquat (Reward) are non-residual herbicides used for general vegetation control under trees and for turf renovation.

Diphenyl Ethers The diphenyl ethers include oxyfluorfen:

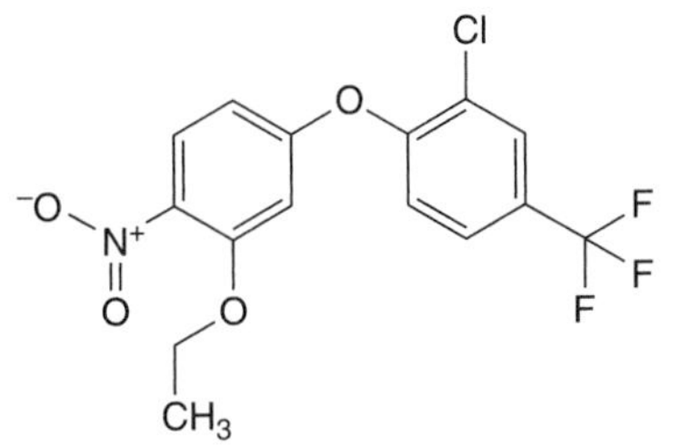

Oxyfluorfen

The chemical structure of the diphenyl ethers includes two benzene rings connected via an oxygen atom. Oxyfluorfen has a nitro group (NO_2) and a methoxy group (OCH_3) as substituents on one of the rings, and a trifluoromethyl group (CF_3) and a chlorine atom (Cl) as substituents on the other ring.

Oxyfluorfen is 2-chloro-1(3-ethoxy-4-nitrophenoxy)-4-(trifluoromethyl)benzene. It is formulated as an emulsifiable concentrate (EC). Oxyfluorfen (Goal) is primarily used for controlling weeds in ornamental planting beds.

Triazolinones The triazolinones include carfentrazone and sulfentrazone:

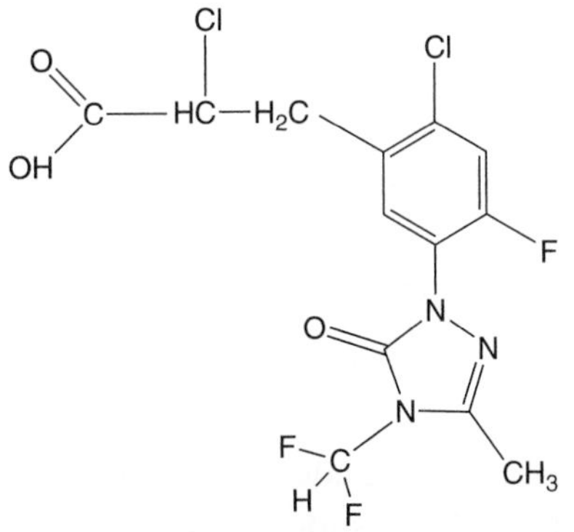

Carfentrazone

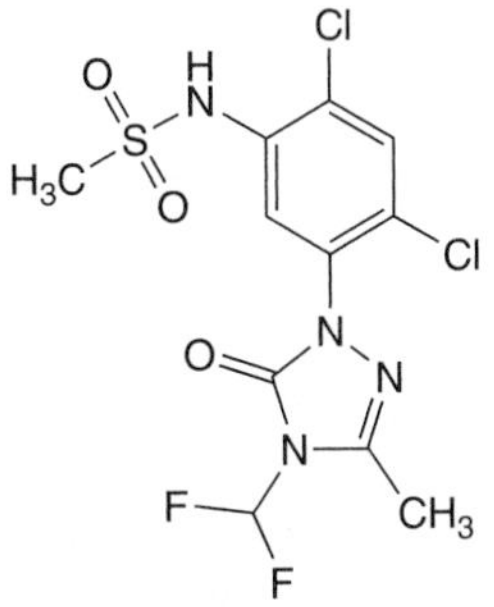

Sulfentrazone

The chemical structure of the triazolinones includes a benzene ring connected to a nitrogen in a five-member triazole ring (in which nitrogen has replaced three of the ring's carbons) with an oxygen atom doubly bonded to one of the ring's carbons. Carfentrazone has a chlorine atom (Cl), a fluorine atom (F), and a chloropropionic acid group ($CH_2CHClCOOH$) as substituents on the benzene ring, and a methyl group (CH_3) and a trifluoromethyl group (CF_3) as substituents on the triazole ring. Sulfentrazone has two chlorine atoms and a methanesulfonamide group (CH_3SO_2NH) as substituents on the benzene ring, and a methyl group (CH_3) and a difluoromethyl group (CHF_2) as substituents on the triazole ring.

Carfentrazone is α,2-dichloro-5-(4-[difluoromethyl]-4,5-dihydro-3-methyl-5-oxo-1H-1,2,4-triazol-1-yl)-4-fluorobenzenepropanoic acid. It is formulated as an emulsifiable concentrate (EC). Sulfentrazone is N-(2,4-dichloro-5-[4-(difluoromethyl)-4,5-dihydro-3-methyl-5-oxo-1H-1,2,4-triazol-1-yl]phenyl)methanesulfonamide. It is formulated as a dry flowable (DF). Carfentrazone (Quicksilver) is primarily used for controlling silvery thread moss in bentgrass greens and tees. Carfentrazone and sulfentrazone are also used alone and in combinations with phenoxy-carboxylic and benzoic acid herbicides for faster burndown of broadleaf weeds in turf. Sulfentrazone also appears to have some activity against sedges in turf.

Oxadiazoles The oxadiazoles include oxadiazon:

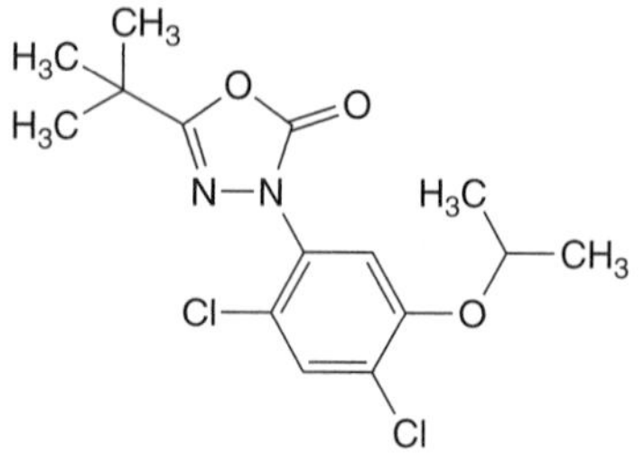

Oxadiazon

The chemical structure of the oxadiazoles includes a five-member heterocyclic ring (in which two nitrogens and one oxygen have replaced three of the ring's carbons to form an oxadiazole ring). The oxadiazole ring in oxadiazon has two substituents: a doubly bonded oxygen atom and a dimethylethyl group ($CH_3[CH_3]_2C$). Also, connected to a nitrogen in the oxadiazole ring is a benzene ring with three substituents: two chlorine atoms (Cl) and a methylethoxy group (CH_3CHCH_3O).

Oxadiazon is 3-(2,4 dichloro-5-[1-methylethoxy]phenyl)-5-(1,1-dimethylethyl)-1, 4,4-oxadiazol-2-(3*H*)-one. It is formulated as a water-soluble packet (WSP) or granule (G). Oxadiazon (Ronstar) is primarily used for preemergence control of annual grasses and some broadleaf weeds in turf and ornamental planting beds.

Phosphinic Acids The phosphinic acids include glufosinate:

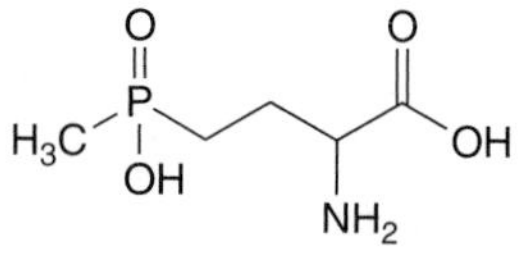

Glufosinate

The chemical structure of phosphinic acid includes a phosphorus atom (P) with one doubly bonded oxygen atom (O), one singly bonded hydroxyl group (OH), and two singly bonded hydrogen atoms (H). Phosphinic acid derivatives have alkyl groups substituting for the hydrogens. The alkyl groups in glufosinate are a methyl group (CH_3) and an aminobutanoic acid group ($CH_2CH_2CHNH_2COOH$). In glufosinate ammonium, the phosphinic acid becomes a phosphinate (anion) with the removal of the hydrogen from the hydroxyl group (OH to O^-) and an ammonium ion (NH_4^+) is used to form an ammonium salt.

Glufosinate is 2-amino-4-(hydroxymethylphosphinly)butanoic acid. It is formulated as an ammonium salt in a liquid water-soluble (SL) form. Glufosinate (Finale) is a nonselective herbicide used for general vegetation control or for controlling winter weeds in dormant bermudagrass turf.

Fatty Acids The fatty acids include pelargonic acid:

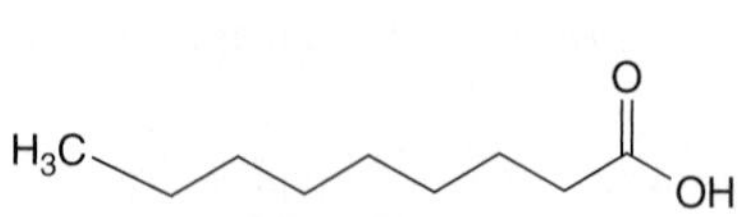

Pelargonic acid

Fatty acids consist of long unbranched hydrocarbons with a carboxylic acid group at one end. Pelargonic acid ($CH_3(CH_2)_7COOH$) is a saturated (i.e., no double bonds between carbons) fatty acid with nine carbons in the chain. The fatty acids are considered biopesticides because they are derived from natural products. The notable

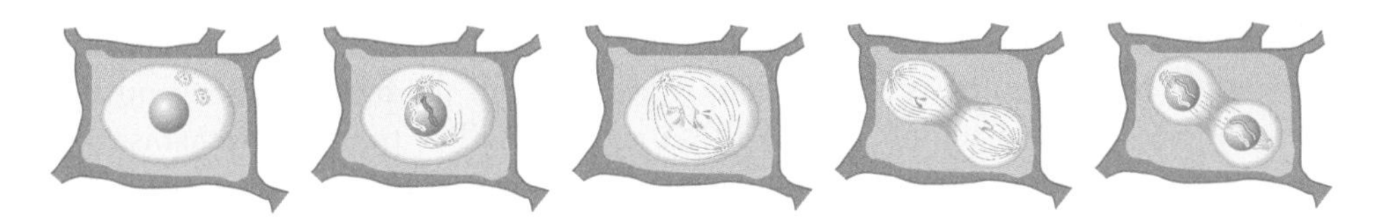

Figure 3.14
Illustration of interphase (left) that precedes mitosis, and the four sequential phases of mitosis: prophase, metaphase, anaphase, and telophase. Mitotic disruptors work by slowing or preventing the microtubule formation in metaphase, halting cell division.

properties of fatty acid herbicides include contact-type efficacy, immediate effect, nonselectivity for weeds, and high biodegradability.

Pelargonic acid (also called nonanoic acid) is a 9-carbon fatty acid derivative. It is very soluble in water and has no soil persistence or activity. It is formulated as an emulsifiable concentrate (EC). Pelargonic acid (Scythe) is a nonselective contact herbicide used for controlling annual and some perennial weeds.

Cell Growth Disruptors and Inhibitors

The cell growth disruptors and inhibitors include these chemical families: dinitroanilines (benefin, oryzalin, pendimethalin, prodiamine, and trifluralin), pyridines (dithiopyr), benzamides (pronamide), phthalic acids (DCPA), carbamates (asulam), chloroacetamides (metolachlor), acetamides (napropamide), phenylureas (siduron), and phosphorodithioates (bensulide). They are usually applied as preemergence herbicides; all inhibit root and/or shoot growth of emerging seedlings in some way, often by inhibiting cell division in embryonic root and shoot meristems. Cell division involves two main processes: mitosis and cytokinesis. Mitosis is the division of the nucleus, and cytokinesis is the division of the cytoplasm to form two cells. Mitosis is the sequence of events by which complete sets of chromosomes are distributed to each daughter nucleus; it consists of four phases: prophase, metaphase, anaphase, and telophase (Figure 3.14).

In prophase, cromatin begins to condense into visible chromosomes (i.e., DNA measuring several cm in length condense into chromosomes 5 to 10 μm long) and the nuclear membranes break down. As each chromosome becomes visible, it appears to be composed of two identical subunits, called chromatids, which are attached to each other at a constricted region of the chromosome called the centromere. In metaphase, the chromosomes line up along the equatorial plane of the cell, and the mitotic spindle—an infrastructure composed of numerous fibers, called microtubules, extending from pole to pole—is visible. In anaphase, each pair of chromatids splits apart at the point of connection, called the centromere, forming two independent chromosomes, which then move along the microtubules toward opposite poles. In telophase, the chromosomes uncoil, elongating into invisible chromatic threads; the spindle fibers disappear; and two separate nuclei are formed. Cytokinesis and telophase occur simultaneously, completing the formation of the two daughter cells. In addition to the mitotic phases, there is the interphase—a period of active growth that precedes mitosis—that is divided into three subphases: G_1, S, and G_2. G_1, the first gap subphase, is the time between the previous cell division and the beginning of DNA replication. S is the subphase during which DNA replication occurs. And G_2, the second gap subphase, is when increased protein synthesis occurs in preparation for the next cell division. Asulam, DCPA, dithiopyr, pronamide, and the dinitroaniline herbicides slow or prevent the formation of the mitotic spindle in metaphase, presumably by binding to tubulin, an enzyme that catalyzes microtubule formation. As a consequence, mitosis is interrupted and cell division does not take place. With growth stoppage at the root and/or shoot meristems, the seedling eventually dies. While microtubule formation occurs in mammalian species as well, the activity of the dinitroaniline herbicides is restricted to plants and protozoa, as these compounds are ineffective against vertebrate or fungal microtubules (Morrissette et al., 2004). This accounts for the low mammalian toxicity of these herbicides. In addition to mitotic inhibition, DCPA also disrupts cell wall formation and asulam inhibits an enzyme involved in the biosynthesis of folic acid, a precursor in the biosynthesis of purine nucleotides, which are components of DNA and RNA. The primary mechanism of action of metolachlor and other chloroacetamides is the inhibition of several long-chain fatty acids required for cell membrane construction.

Dinitroanilines The dinitroanilines include benefin, oryzalin, pendimethalin, prodiamine, and trifluralin:

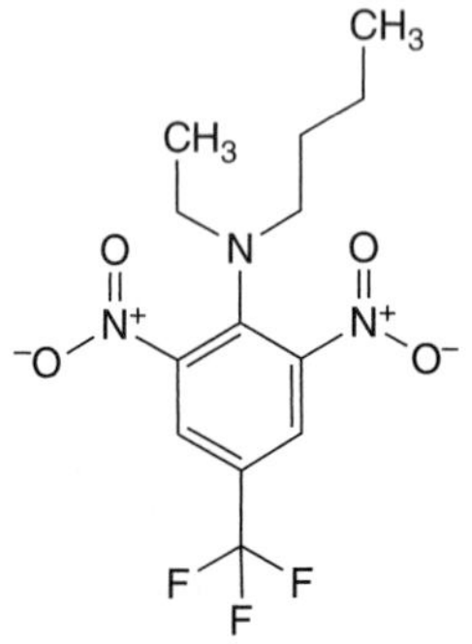

Benefin

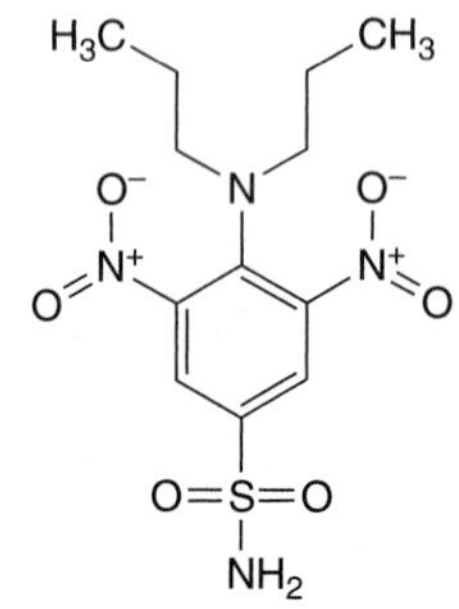

Oryzalin

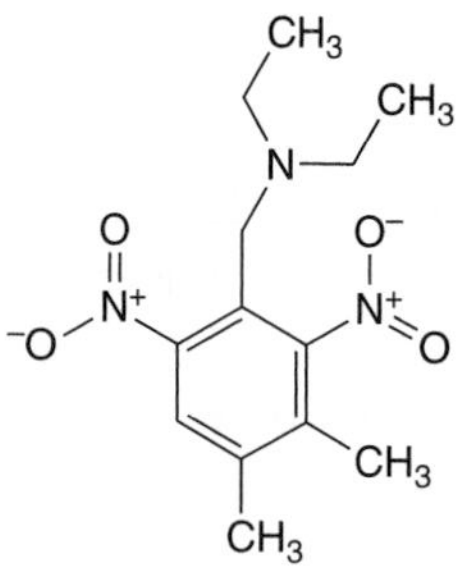

Pendimethalin

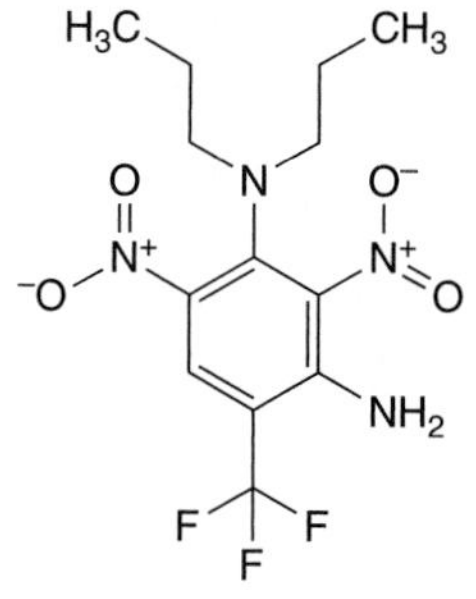

Prodiamine

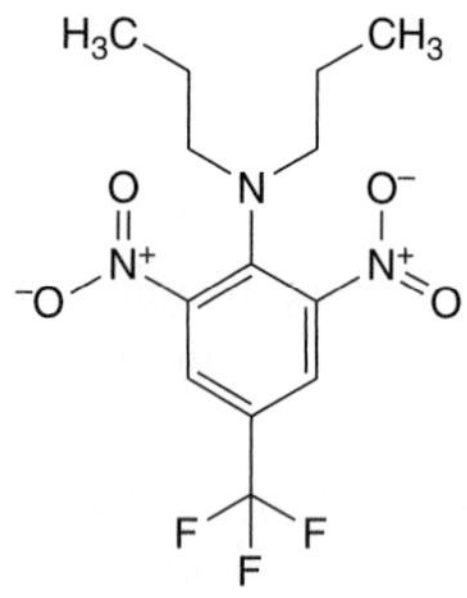

Trifluralin

The common structural features of these herbicides is a benzene ring with a substituted amino (NH_2) group and two nitro (NO_2) groups as ring substituents, making them dinitroanilines. The substituted amino groups are dipropylamino ($[CH_3CH_2CH_2]_2N$) in oryzalin, prodiamine, and trifluralin, diethylamino ($[CH_3CH_2]_2N$) in pendimethalin, and *N*-butyl-*N*-ethyl ($CH_3CH_2CH_2CH_2NCH_2CH_3$) in benefin. There are trifluoromethyl groups (CF_3) as ring substituents in benefin, prodiamine, and trifluralin. Other ring substituents include a sulfonamide group (NH_2SO_2) in oryzalin, two methyl groups (CH_3) in pendimethalin, and an amino group (NH_2) in prodiamine.

Benefin is *N*-butyl-*N*-ethyl-2,6-dinitro-4-(trifluoromethyl)benzeneamine. It is formulated as an emulsifiable concentrate (EC), water-dispersible granule (WG), or granule (G). Benefin (Balan) is used alone or in combination with trifluralin (Team) for preemergence control of crabgrass and other annual weeds in turf. Oryzalin is

4-(dipropylamino)-3,5-dinitrobenzenesulfonamide. It is formulated as a water-dispersible liquid (AS) or granule (G). Oryzalin (Oryzalin) is used alone or in combination with benefin (XL) for preemergence control of crabgrass and other annual weeds in warm-season turfgrass communities and in ornamental planting beds. Pendimethalin is *N*-(1-ethylpropyl)-3,4-dimethyl-2,6-dinitro-benzeneamine. It is formulated as an emulsifiable concentrate (EC), water-dispersible granule (WG), wettable powder (WP), dispersible granule (DG), or granule (G). Pendimethalin (Pendulum) is used for preemergence control of crabgrass and other annual weeds in turf and in ornamental planting beds. Prodiamine is 2,4-dinitro-N^3,N^3-dipropyl-6-(trifluoromethyl)-1,3-benzeneamine. It is formulated as a water-dispersible granule (WG). Prodiamine (Barricade) is used for preemergence control of crabgrass and other annual weeds in turf and in ornamental planting beds. Trifluralin is 2,6-dinitro-*N*,*N*-dipropyl-4-(trifluoromethyl) benzeneamine. It is formulated as an emulsifiable concentrate (EC). Trifluralin in combination with benefin (Team) is used for preemergence control of crabgrass and other annual weeds in turf.

Pyridines The pyridines include dithiopyr:

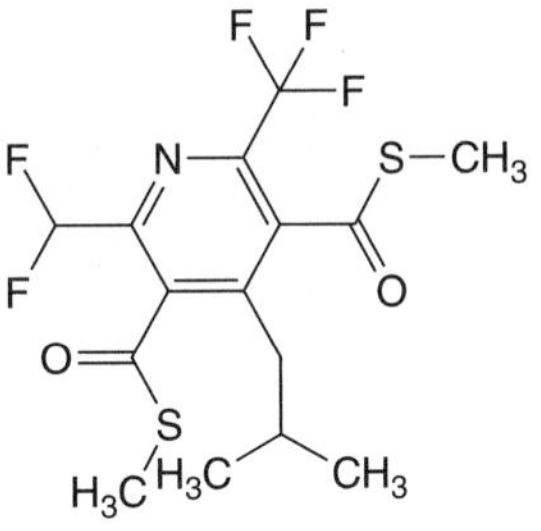

Dithiopyr

The common structural feature of all pyridines is the pyridine ring (in which one of the carbons in the aromatic ring is replaced by nitrogen). Other ring substituents in dithiopyr include difluoromethyl (CHF_2) and trifluoromethyl (CF_3) groups, two methylthiocarboate (CH_3SCO) groups, and a dimethylethyl ($[CH_3]_2CHCH_2$) group.

Dithiopyr is *S*,*S*-dimethyl 2-(difluoromethyl)-4-(2-methylpropyl)-6-(trifluoromethyl)-3,5-pyridinedicarbothioate. It is formulated as an emulsifiable concentrate (EC), a wettable powder (WP), or a granule (G). Dithiopyr (Dimension) is used for preemergence control of crabgrass and other annual weeds in turf and in ornamental planting beds.

Benzamides The benzamides include pronamide:

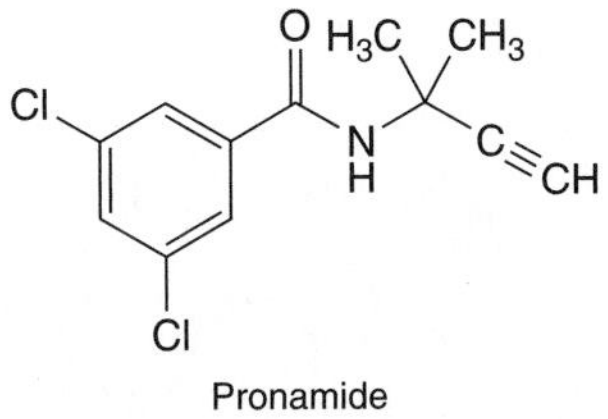

Pronamide

The common feature of these herbicides is the benzamide group composed of a benzene ring, a carbonyl (CO) group, and an amino (NH_2) group. Pronamide also has two chlorine atoms (Cl) as ring substituents and a dimethylpropynyl ($[CH_3]_2CCN$) group substituting for one of the hydrogens in the amino group.

Pronamide is 3,5-dichloro(N-1,1-dimethyl-2-propynyl)benzamide. It is formulated as a wettable powder (WP). Pronamide (Kerb) is used for controlling annual bluegrass in warm-season turfgrass communities and for removing ryegrass to facilitate spring transition of bermudagrass turfs.

Phthalic Acids The phthalic acids include DCPA.

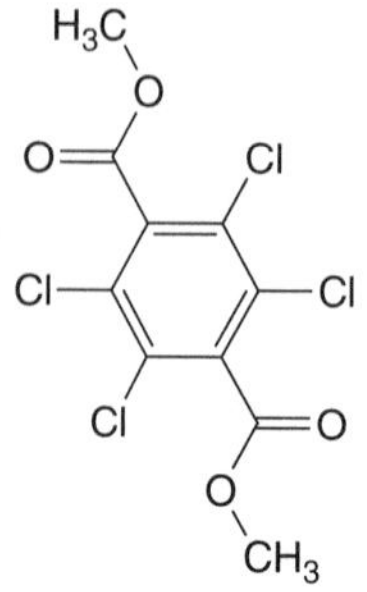

DCPA

The chemical structure of phthalic acids includes a benzene ring with two carboxyl groups (COOH) as ring substituents. DCPA has methylated carboxylate groups ($COOCH_3$) at opposite ends of the ring and chlorine atoms at the four other ring positions.

DCPA is dimethyl 2,3,5,6-tetrachloroterephthalate. It is formulated as a wettable powder (WP) or flowable (F). DCPA (Dacthal) has been used for preemergence control of crabgrass and other annual weeds in turfgrass communities.

Carbamates The carbamates include asulam:

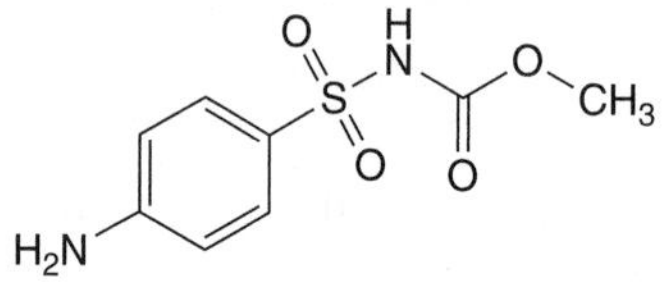

Asulam

The common structural feature of this chemical class is an amino group (NH_2) and a carboxyl group (COOH) linked to form a carbamate group (NH_2COOH). The carboxyl group in asulam is methylated, and the amino group is attached via a sulfonyl group (SO_2) to a benzene ring with an amino group as a ring substituent.

Asulam is methyl[(4-aminophenyl)sulfonyl]carbamate. It is formulated as a water-soluble liquid (SL). Asulam (Asulox) is a professional-use herbicide for controlling crabgrass, goosegrass, and selected other grasses in Tifway bermudagrass and St. Augustinegrass turfs.

Chloroacetamides The chloroacetamides include metolachlor:

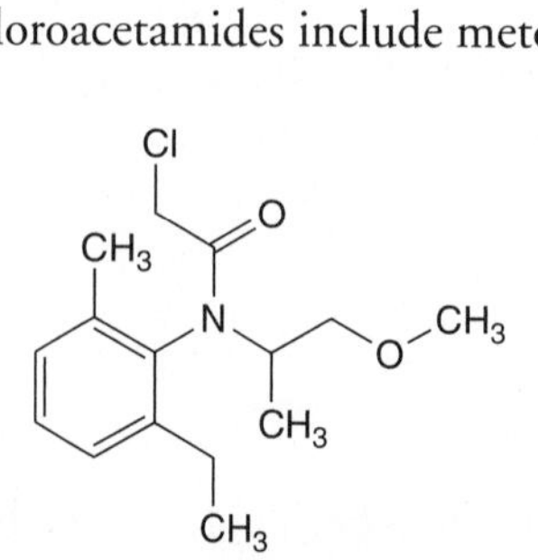

Metolachlor

The common feature of these herbicides is the chloroacetamide group ($ClCH_2CONH_2$). Metolachlor also has a benzene ring with methyl (CH_3) and ethyl

(CH_3CH_2) substituents and methoxypropyl group ($CH_3OCH_2CHCH_3$) attached to the nitrogen in place of two hydrogens.

Metolachlor is 2-chloro-*N*-(2-ethyl-6-methylphenyl)-*N*-(2-methoxy-1-methylethyl acetamide. It is formulated as an emulsifiable concentrate (EC), dry flowable (DF), water-dispersible granule (WG), or granule (G). Metolachlor (Pennant) is used for preemergence control of crabgrass and other annual grasses in warm-season turfgrass communities and in ornamental planting beds.

Acetamides The acetamides include napropamide:

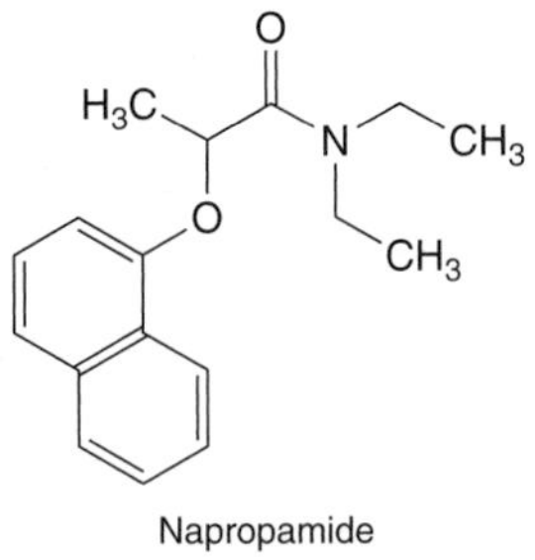

Napropamide

The common feature of these herbicides is the acetamide group (CH_3CONH_2). Napropamide has two methyl groups (CH_3) substituting for hydrogens in its amino group (NH_2), and it has a methyl group attached directly to the acetamide group (making it an propanamide group) and two fused benzene rings attached indirectly via an ether (O) linkage.

Napropamide is *N,N*-diethyl-2-(1-naphthalenyloxy)propanamide. It is formulated as an emulsifiable concentrate (EC), wettable powder (WP), dry flowable (DF), or granule (G). Napropamide (Devrinol) is used for preemergence control of crabgrass and other annual grasses in warm-season turfgrass communities and in ornamental planting beds.

Phenylureas The phenylureas include siduron:

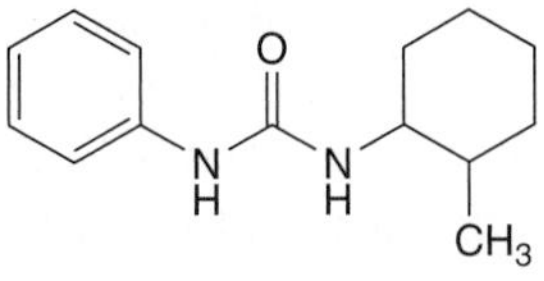

Siduron

The common features of phenylureas are a phenyl group (benzene ring) and a urea group (—NHCONH—) linked via one of the urea nitrogens. Siduron also has a six-member methylcyclohexyl group attached to the other urea nitrogen.

Siduron is *N*-(2-methylcyclohexyl)-*N′*-phenylurea. It is formulated as a wettable powder (WP). Siduron (Tupersan) is used for preemergence control of crabgrass and selected other summer annual grasses in cool-season turfgrasses. Its distinctive feature is that it can be used in conjunction with establishing or renovating cool-season turfgrass communities.

Phosphorodithioates The phosphorodithioates include bensulide:

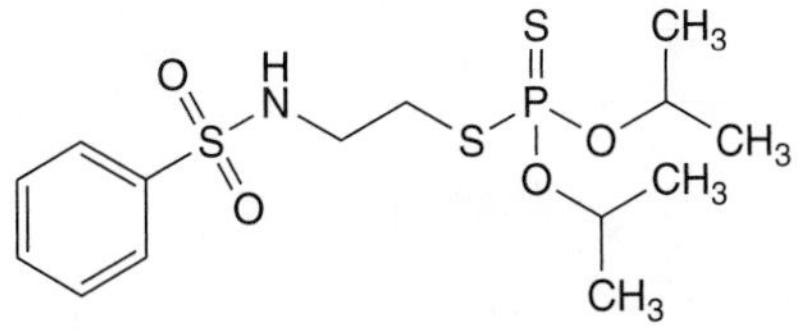

Bensulide

The common feature of phosphorodithioates is a central phosphorus atom (P) bonded to one sulfur (S) and two oxygen (O) atoms. Bensulide also has isopropyl groups ($[CH_3]_2CH$) bonded to each of the two oxygens and a mercaptoethyl group ($—SCH_2CH_2—$) linking phosphorus to a sulfonamide group ($—SO_2NH—$), which links to a benzene ring.

Bensulide is *O,O*-bis(1-methylethyl)S-(2-[phenylsulfonyl]amino)ethylphosphorodithioate It is formulated as an emulsifiable concentrate (EC) and granule (G). Bensulide (Bensumec) is a preemergence herbicide used for controlling annual bluegrass, crabgrass, and other annual grasses in turf.

Cellulose Biosynthesis Inhibitors

The cellulose biosynthesis inhibitors include the benzonitriles (dichlobenil), benzamides (isoxaben), and quinoline-carboxylics (quinclorac; see section on plant-growth regulators). This pathway begins with the splitting of sucrose into fructose and glucose (Figure 3.15).

Since these two simple sugars are interconvertible, some of the fructose is converted to glucose as glucose is transformed to UDP-glucose and incorporated into cellulose. The conversion of sucrose to its component sugars is inhibited by isoxaben, while the incorporation of UDP-glucose is inhibited by dichlobenil and, presumably, quinclorac. This, in turn, results in the diversion of UDP-glucose to the formation of alternate compounds. With the inhibition of cellulose biosynthesis, there is no cell wall development. The resultant lack of cellular integrity leads to arrested or abnormal growth and, eventually, death of the plant. Cell walls enclose individual cells and constitute the *exoskeleton* of plants, controlling cell shape and allowing high turgor pressures to develop. Cell walls are composed primarily of cellulose, a structural carbohydrate consisting of thousands of glucose molecules linked end to end. A thousand or more of these chains are twisted together to form microfibrils, which are like tiny cables of very high tensile strength. Several dozen microfibrils are intertwined to make fibrils. Layers of fibrils are cemented together into strong three-dimensional grids by other structural carbohydrates called pectins and hemicelluloses. The cellulose biosynthesis inhibitors each act at a different stage in the cellulose biosynthesis pathway. Dichlobenil inhibits cell wall formation in telophase, presumably by interfering with the shunting of UDP-glucose in the construction of cellulose. Isoxaben acts earlier in the cellulose biosynthesis pathway by inhibiting the conversion of sucrose to UDP-glucose. Quinclorac inhibits the incorporation of glucose into cellulose and hemicellulose.

Benzonitriles The benzonitriles include dichlobenil:

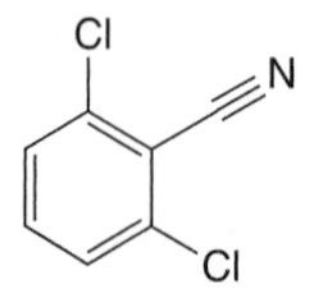

Dichlobenil

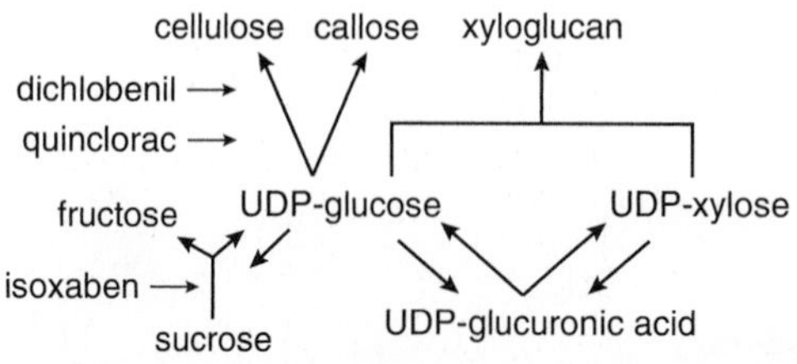

Figure 3.15
Illustration of the cellulose biosynthesis pathway involving the conversion of sucrose into fructose and glucose, which is then used to form cellulose.

The molecular structure of the benzonitriles includes a benzene ring with a nitrile group (CN) as a ring substituent. Dichlobenil also has two chlorine atoms (Cl) as ring substituents.

Dichlobenil is 2,6-dichlorobenzonitrile. It is formulated as a granule (G). Dichlobenil (Casoron) is used for weed control in woody ornamental planting beds.

Benzamides The benzamides include isoxaben:

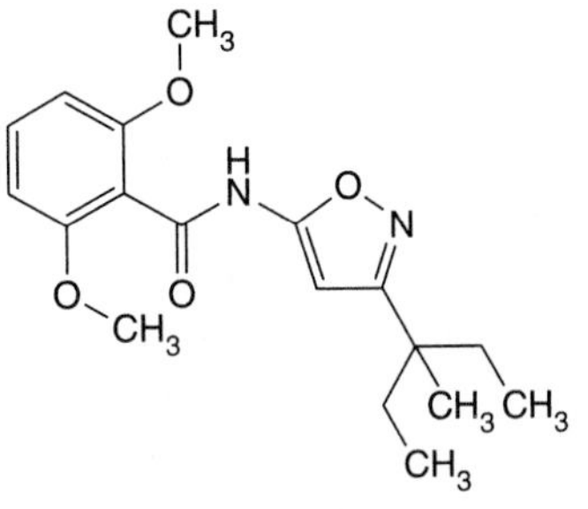

Isoxaben

The common feature of these herbicides is the benzamide group composed of a benzene ring, a carbonyl (CO) group, and an amino (NH_2) group. Isoxaben also has two methoxy groups (CH_3O) as ring substituents and an isobutyl group ($[CH_3]_3C$) connected to an oxazole group (five-member heterocyclic ring in which one nitrogen and one oxygen substitute for ring carbons), which substitutes for one of the hydrogens in the amino group.

Isoxaben is *N*-[3-(1-ethyl-1-methylpropyl)-5-isoxazoly]-2,6-dimethoxy-benzamide. It is formulated as a dry flowable (DF). Isoxaben (Gallery) is a preemergence herbicide used for broadleaf weed control in turf.

Soil Fumigants

The soil fumigants include the dithiocarbamates (metham) and thiadiazines (dazomet). These are preplant pesticides used to kill weed seeds and other soil organisms. They are highly volatile compounds that rapidly decompose to methyisothiocyanate, or MITC ($CH_3N{=}C{=}S$), which diffuses through the soil at rates that vary with the soil texture and structure.

Dithiocarbamates The dithiocarbamates include metham:

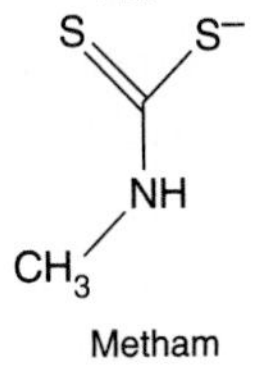

Metham

The common structural feature of this chemical class is an amino group (NH_2) and a thioate group (CSS^-) that are linked to form a dithiocarbamate group ($NH_2CS_2^-$). Metham also has a methyl group (CH_3) substituting for one of the hydrogens in the amino group.

Metham (Vapam) is sodium methyl dithiocarbamate. It is formulated as a solution concentrate (SC). In addition to methyisothiocyanate ($CH_3N{=}C{=}S$), metham-sodium breaks down to carbon disulfide (CS_2), dimethylthiourea ($[CH_3]_2$ $NHSONH_2$), hydrogen sulfide (H_2S), and elemental sulfur (S) in the soil soon after it is activated by water.

Thiadiazines The thiadiazines include dazomet:

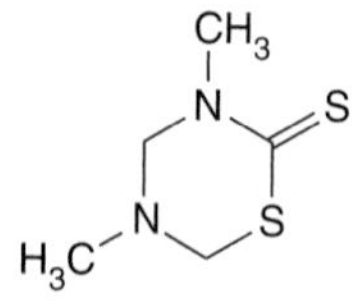

CH_3 N S N S H_3C

Dazomet

The common feature of members of this chemical family is the thiadiazine ring—a six-member heterocyclic compound in which two nitrogens and one sulfur replaced ring carbons. Dazomet also has two methyl groups (CH_3) and one doubly bonded sulfur as ring substituents.

Dazomet (Basamid) is tetrahydro-3,5-dimethyl-H-1,3,5-thiadiazine-2-thione. It is formulated as a granule (G). In addition to methyisothiocyanate ($CH_3C{=}O{=}N$), dazomet breaks down to formaldehyde ($H_2C{=}O$), hydrogen sulfide (H_2S), and methylamine (CH_3NH_2) in the soil as soon as it is activated by water.

HERBICIDE RESISTANCE

Herbicide *resistance* is usually characterized as "the inherited ability of a plant to survive and reproduce following exposure to a dose of herbicide normally lethal to the wild type" (Banks et al., 2004). Since not all herbicide-resistant *plants* are herbicide-resistant *weeds*, Heap (2004) proposed a definition specific to weeds: "the evolved capacity of a previously herbicide-susceptible weed population to withstand an herbicide and complete its life cycle when the herbicide is used at its normal rate in an agricultural situation." *Tolerance,* meanwhile, is a plant's ability to remain uninjured by herbicide doses normally lethal to other plant species.

Production practices employed with some monocultural crops, including turfgrasses and ornamental plants, favor the development of herbicide resistance because of the lack of crop rotation, which would allow substituting crops tolerant of herbicides that attack different sites of action in weed species. Resistance pressures are even greater where the soil between ornamental plants is untilled, requiring complete reliance on herbicides for weed control. Where herbicides that attack weeds at the same site of action are used, there is no competition between tolerant and resistant weeds, allowing the resistant ones to quickly dominate. Furthermore, herbicide resistance appears more frequently in plants with an annual life cycle. Since seeds are the major means by which annual plants reproduce, the potential to transfer resistance to the next generation is relatively high, compared with that of perennial plants, which can persist for years without producing viable seed.

History and Severity

Arguably, Switzer (1957) was the first to report the incidence of herbicide resistance in weed species: the existence of wild carrot strains that were resistant to 2,4-D, an auxinic plant-growth regulator. In 1968, common groundsel was reported to be triazine resistant (LeBaron and Gressel, 1982). Subsequently, resistance has been reported for herbicides representing nearly all chemical families and sites of action (Prather et al., 2000); however, most herbicide-resistant weeds (182 out of 303) represent just two sites of action: ALS and PSII inhibition (Table 3.2).

Weed resistance to a particular herbicide may be induced by selection pressure, genetic engineering, or other means. Through herbicide-induced selection pressure, susceptible plants within a weed population are killed while resistant plants survive to reproduce without competition from susceptible plants. With

Table 3.2
SUMMARY OF HERBICIDE-RESISTANT WEEDS BY SITE OF ACTION[1]

Mechanism of Action	Site of Action	Weed Number
Plant Growth Regulators (PGR)	IAA-like	24
Amino Acid Biosynthesis	EPSP Enzyme	8
Inhibitors	ALS Enzyme	95
Pigment Inhibitors	Carotenoid Biosynthesis	6
Lipid Biosynthesis Inhibitors	ACCase Enzyme	35
	Not ACCase Enzyme	8
Cell Membrane Disruptors	Photosystem II (PSII)	87
	Photosystem I (PSI)	23
	Protox (PPO) Enzyme	3
Cell Growth Disruptors and	Mitotic Disruptors	11
Inhibitors	Shoot and/or Root Inhibitors	2
Cell Wall Formation Inhibitors	Cellulose Biosynthesis	1

[1]http://www.weedscience.org/summary/MOASummary.asp; June 20, 2006.

continued use of the herbicide, resistant plants eventually become dominant within the population.

The potential for resistance development varies, depending upon several herbicide-related factors (Murphy, 2004; Prather et al., 2000). For example, preemergence herbicides with prolonged soil residual activity will be more effective than less persistent herbicides in selecting for a resistance trait, as more of the susceptible plants germinating during the growing season will be killed. Frequent use of nonresidual postemergence herbicides can have a similar effect. Herbicides that work by attacking a single target site are more likely to encounter resistant plants than are herbicides that attack multiple target sites. Highly effective herbicides can rapidly kill susceptible weeds within a population, leaving the few remaining plants that are resistant to the herbicide to repopulate the site and, through their seeds or vegetative propagules, populate other sites. Plant-related factors that influence the evolution of herbicide resistance include the number of alleles (versions of genes) involved in the expression of functional resistance, the frequency of resistance alleles in natural populations of weed species, the mode of inheritance of the resistance allele(s), the reproductive and breeding characteristics of the weed species, the intensity of selection differentiating resistant genotypes from susceptible ones, and the biological fitness of resistant and susceptible genotypes (Mortimer, 2005).

Some turfgrass weeds have developed herbicide resistance. Among them are goosegrass (*Elusine indica* [L.] Gaertn.) resistance to dithiopyr, diclofop-methyl, and dinitroanilines; crabgrass (*Digitaria ischaemum* [Schred.] Muhl.) resistance to fenoxyprop-ethyl; and annual bluegrass (*Poa annua* [L.]) resistance to triazines, ethofumesate, dithiopyr, and dinitroanilines (Murphy, 2004; Yelverton, 1998), as well as to glyphosate (Gaussoin, 2006).

Resistance becomes evident only after several years (typically, 7 to 10) of repeated applications of herbicides with the same or similar sites of action when resistant plants have survived a dose several to many times (20 to 1,000 times) greater than that normally needed for control (Figure 3.16).

The most common mechanism of herbicide resistance in weeds is an alteration of the site of action, resulting in reduced herbicide binding on an enzyme or protein, thus reducing the sensitivity of the inhibited enzyme or protein. This may reflect

Figure 3.16
Diclofop-methyl susceptible (top) ryegrass (Lolium spp.) at 0, 0.42, 0.63, 0.84, and 1.68 kg ai/ha (left to right), compared with a resistant biotype (bottom) at 0, 1.22, 2.45, 3.67, and 4.89 kg ai/ha.
(Courtesy Lane Crooks.)

variations in—or mutations of—the gene that produces the enzyme or protein. An alternative mechanism is an enhanced ability of the plant to metabolize and detoxify the herbicide or to possess an uptake or translocation mechanism different from that of susceptible plants (Holt et al., 1993). As with target site changes, selection for enhanced metabolism can occur with repeated application of the same herbicide or from the use of herbicides from different chemical families that attack the same site of action. It is more likely to develop where an herbicide is used continuously at lower-than-recommended rates of application. Weed plants with enhanced metabolism tend to have a substantially lower level of resistance than that encountered in plants expressing resistance through altered sites of action.

Time Needed for Resistant Plants to Appear

Herbicide applications do not induce gene mutations resulting in herbicide resistance; rather, mutations occur naturally within a weed population and, through the repeated use of a particular herbicide, susceptible plants are removed, leaving the herbicide-resistant ones to dominate. Normally, herbicide-resistant plants are not fit enough to outcompete herbicide-susceptible plants; thus, they remain a minor component (often less than 1 percent) of the weed population until repeated use of a particular herbicide releases it to quickly spread (Figure 3.17).

The time needed for a proportion of resistant weeds to dominate a population varies between plants and herbicide(s) used. For example, the appearances of the first resistant populations to herbicides with various sites of action are listed in Table 3.3. With continuous herbicide use over a sufficiently long period, a single additional application may result in a major control failure. For example, it has been predicted that after 6 years of continuous use, 8 percent of the plant population will express resistance; however, this will increase to 55 percent in year 7 and 93 percent in year 8 (Maxwell and Mortimer, 1994).

Managing Resistant Weeds

Methods for reducing the potential for developing herbicide resistance include rotating or tank mixing herbicides that attack different sites of action and using both

Figure 3.17
*Fitness (vigor) of triazine-resistant annual bluegrass (*Poa annua*), left, compared with a triazine-susceptible biotype, right.* (Courtesy Fred Yelverton.)

preemergence and postemergence herbicides against the same weed populations. Another, perhaps more effective, approach would be to implement cultural practices that promote healthy, vigorous turfgrass and ornamental-plant growth so that weed pressure and the use of herbicides are minimized.

An example of estimating the effects of different herbicide rotations on occurrence of resistant weeds involves the triazine herbicides. It's been estimated that 9.5 years of continued use of triazine herbicides is required for the buildup of 90-percent-resistant plant populations. If the population were treated every third year with an herbicide having a different site of action, 45 years would be required for a similar level of triazine-resistance buildup. If the weed population were rotated every other year with herbicides possessing different sites of action, 60 years would be needed for triazine-resistance buildup. If a sequence of treating the weed population with triazine herbicides one year and then rotating with herbicides that attack different sites of action for the next two years were implemented, 90 years would be required for 90 percent triazine-herbicide resistant weed populations to occur (Gressel and Segel, 1982).

Table 3.3
TYPICAL APPEARANCE OF FIRST RESISTANT WEED POPULATIONS TO VARIOUS CLASSES OF HERBICIDES (ROSS AND LEMBI, 1999)

Herbicide Site of Action (and Family)	Years of Continuous Use before Resistant Populations Appear
ALS Inhibitors (Sulfonylureas) ACCase Inhibitors (Aryloxyphenoxypropionates)	3 to 5
Photosystem II Inhibitors (Triazines)	10+
Photosystem I Inhibitors (Bipyridiliums) Mitotic Disruptors (Dinitroanilines)	Only occasionally

4

Herbicides in the Environment

The environmental fate of herbicides used for agriculture—including turfgrass and ornamental plant culture—is a serious concern, not only for the landscape manager, but for society as a whole. An environmentally responsible landscape manager understands that herbicides and other agrichemicals should be handled in ways that ensure no harm is done to people, animals, or the environment, including our air, water, and soil resources. In this respect, the landscape manager is bound by the same Hippocratic oath as the Greek physician: *primum non nocere*, "above all, not knowingly to do harm." But to avoid harm, one must understand the circumstances under which it can result and the means by which to avoid it. Specifically, one must know how much of the herbicide to apply and how to apply it, how far it will move from the site of application, and how long it will persist.

Herbicides applied to turfgrass and ornamental plant communities are subject to many forces that result in their dissipation over time. Drift, volatilization, photolysis, and runoff are processes that can occur above the soil surface during or shortly after herbicide application. Adsorption onto soil surfaces, leaching and dilution through the soil profile, absorption by plants, chemical reaction, and microbial degradation are processes that can occur below the soil surface some time following herbicide application (Figure 4.1).

DRIFT

Drift is the wind-induced movement of airborne herbicide particles from the target site to a nontarget site. Herbicide drift is usually associated with spray particles; granules (dry particles) may move from a target site in high winds, but are not generally considered important sources of herbicide drift. Depending on the specific herbicide and the level of susceptibility of nontarget plants, drift can result in significant damage to nearby crops and ornamental plants. It can also result in the contamination of adjacent land and surface water resources.

Spray particle size is measured as the diameter of the droplet, in microns (μm). The drift potential is higher with small droplets and lower—but still possible—with large droplets. While spray nozzles with large orifices produce larger droplets than do nozzles with small orifices, the droplets produced by each can vary substantially in size and include some that are susceptible to drift. A common term used to characterize droplet size is the *volume median diameter* (VMD). This is the droplet size at the median of the droplet-size spectrum, so that half of the total spray volume released by the nozzle is in droplets larger than the VMD and half is in smaller droplets.

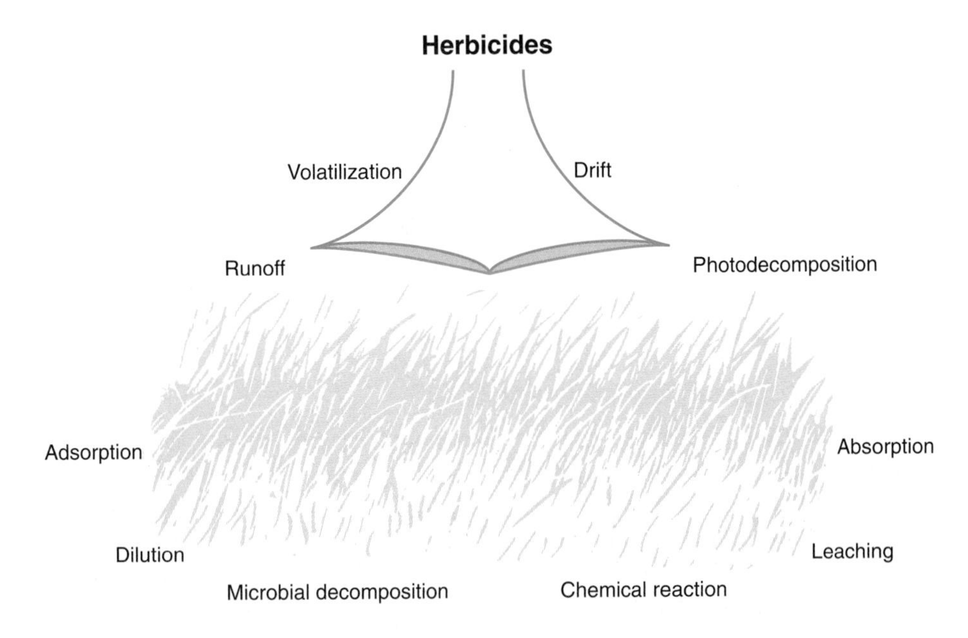

Figure 4.1
Illustration of the environmental fate of herbicides, including drift, volatilization, photolysis, runoff, adsorption, leaching, dilution, absorption, chemical reaction, and microbial degradation.

Nozzle manufacturers use a standardized droplet-size classification system (ASAE Standard S-572) to indicate the droplet sizes of their nozzles for different size and pressure combinations. This classification system has six categories of droplet sizes: very fine (VF, <150 μm), fine (F, 150–250 μm), medium (M, 250–350 μm), coarse (C, 350–450 μm), very coarse (VC, 450–550 μm), and extremely coarse (XC, >550 μm).

While the larger spray droplets may fall rapidly to the ground, some of the smaller ones may move laterally, especially during windy periods (Table 4.1). Very small droplets can remain in the air for relatively long periods and travel great distances during periods of little or no wind (usually less than four to five miles [6.4 to 8.0 kilometers] per hour, depending on particle size). Excessive spray-tank pressures should be avoided, and spray volumes (i.e., amounts of water applied per unit area) should be sufficient to provide spray particles that are large enough (>200 μm) to resist drift.

In addition to spray droplet size, factors that influence the potential for drift include atmospheric conditions and characteristics of the spray equipment.

Table 4.1
RELATIONSHIP BETWEEN SPRAY DROPLET SIZE AND DRIFT POTENTIAL[1]

Droplet Diameter (μm)	Fall Rate (m/s)	Time to Fall 3 m in Still Air (s)	Drift Distance in 3-m Fall with 5-km/hr Wind (m)	Droplet Lifetime (s)
5	7.6×10^{-4}	3,960	4,816	0.04
10	3.0×10^{-3}	1,020	1,372	0.16
20	1.2×10^{-3}	230	338	0.64
50	7.5×10^{-2}	40	54	3.5
10	2.3×10^{-1}	11	15	1.4
150	4.6×10^{-1}	8.5	8	36
200	7.2×10^{-1}	5.4	5	56
500	2.1	1.6	2	400
1,000	4.0	1.1	1	1,620

[1]Bode, 1987.

Atmospheric Conditions

Wind velocity will determine the distance the herbicide droplets will move downwind; however, unstable air with substantial vertical movement can carry herbicide droplets to great heights, potentially resulting in much longer drift distances. Low relative humidity and high temperatures cause more rapid evaporation of spray droplets than that which occurs with high relative humidity and low temperatures; evaporation of spray droplets reduces droplet size, increasing the drift potential. For example, if a 200-μm droplet evaporates down to 150 μm, its drift distance in a 5-km (3.1-mile) per hour wind and 3-m (9.8-ft) fall increases from 5 to 8 m (15.4 to 26.3 ft) (Table 4.1).

Spray Equipment Characteristics

The longer the distance between the nozzle and the ground, the longer it takes a droplet to reach the ground and the greater is the potential for spray drift. Also, since wind velocities are often lower close to the ground than at higher elevations, the drift potential varies accordingly. For a standard 80° nozzle with 20-in (51-cm) spacing, boom height should be between 17 and 19 in (43.2 and 48.3 cm) above the target. Excess boom height reduces the likelihood that small spray droplets will reach the target before decelerating or evaporating.

Spray pressure influences the size of droplets. The spray emerges from the nozzle as a sheet, with droplets forming at the edge of the sheet. Higher spray pressures result in thinner sheets, which break up into smaller droplets. Thinner sheets also form with smaller nozzle orifice sizes. Therefore, employing relatively low spray pressures and relatively large nozzle orifice sizes can minimize drift potential.

Shields constructed around spray booms can partially protect spray droplets from wind, reducing spray drift by as much as 85 percent. As shields are most effective where wind velocities are low and spray droplets are relatively large, shields should be used as a complement to other methods for reducing spray drift.

RUNOFF

Runoff is the surface movement of an herbicide across the ground surface. When runoff extends to a nontarget site, serious consequences can result with some pesticides. Runoff usually occurs as a result of surface water movement during irrigation or natural rainfall. The susceptibility of an herbicide to being carried in runoff increases in proportion to its solubility in water. However, even insoluble herbicide particles can move downslope where surface flow of water is sufficient. On bare soil, severe runoff can occur where adsorbed herbicides are carried off on eroding soil particles. *Washoff* is the term used to describe such losses.

VOLATILIZATION

Volatilization is the process by which an herbicide changes from a liquid (or a solid) to a gas and is lost to the atmosphere. Volatile herbicides can travel long distances in the gaseous state, causing damage to sensitive plants located downwind. Volatilization can occur during or after application and is therefore a more serious concern than drift. The volatilization potential is reflected in the vapor pressure of the herbicide (Table 4.2). Herbicides with low vapor pressures (i.e., 10^{-7} to 10^{-8} mm Hg at 25°C) have low volatility, while those with high vapor pressures (i.e., 10^{-2} to 10^{-4} mm Hg at 25°C) are very volatile. Volatility increases substantially with increasing temperature; thus, applications of volatile herbicides made in the late morning or early afternoon when temperatures are increasing pose a greater hazard to nontarget species than those made in the late afternoon or evening hours when temperatures are decreasing.

Table 4.2
PHYSICAL PROPERTIES OF HERBICIDES INFLUENCING THEIR ENVIRONMENTAL FATE

Herbicide	Vapor Pressure[1] (mm Hg @ 25°C)	Water Solubility (ppm)	Sorption Index[2] (K_{oc})	Soil Half-Life (Days, $T_{1/2}$)
2,4-D (acid)	1.4×10^{-7}	890	20	10
Asulam (Na salt)	1.0×10^{-7}	5,34,000	60–120	7
Atrazine	2.9×10^{-7}	33	100	60
Benefin	7.8×10^{-5}	0.1	9,000	40
Bensulide	8.0×10^{-7}	25	1,000	120
Bentazon (Na salt)	7.5×10^{-9}	500	34	20
Bispyribac (Na salt)	3.9×10^{-5}	73.3	852 to 1793	<10
Bromoxynil (acid)	4.8×10^{-6}	130	32	7
Carfentrazone	1.2×10^{-7}	22	750	0.1
Chlorsulfuron	2.3×10^{-11}	7,800	40	40
Clethodim	$<1.0 \times 10^{-7}$	6,634	116	3
Clopryalid (salt)	1.0×10^{-5}	3,00,000	6	40
Dazomet	2.8×10^{-6}	2,000	10	7
DCPA	2.5×10^{-6}	0.5	5,000	60–100
Dicamba (salt)	9.2×10^{-6}	4,00,000	2	<14
Dichlobenil	5.5×10^{-4}	21	400	60
Dichloroprop (ester)	3.0×10^{-6}	50	1,000	10
Diclofop-methyl	3.5×10^{-6}	0.8	16,000	30
Diquat (salt)	$<1.0 \times 10^{-8}$	7,18,000	1,000,000	1000
Dithiopyr	4.0×10^{-6}	1.38	1,638	17
DSMA (salt)	1.0×10^{-7}	2,69,000	7,000	180
Ethofumesate	6.5×10^{-7}	110	340	30
Fenarimol	2.2×10^{-8}	14	600	360
Fenoxyprop-ethyl	1.4×10^{-7}	0.8	9,490	9
Fluazifop-butyl	2.5×10^{-7}	1.1	5,700	15
Fluroxypyr	3.8×10^{-7}	4,000	39–71	25
Flurprimidol	3.6×10^{-7}	130	314	5 to 21
Foramsulfuron	9.8×10^{-13}	3,293	38 to 151	15
Glufosinate (NH_4 salt)	7.5×10^{-7}	1,370,000	0	7
Glyphosate (amine salt)	1.8×10^{-7}	9,00,000	24,000	47
Halosulfuron	2.8×10^{-12}	15	93.5	4 to 34
Imazaquin (acid)	2.0×10^{-8}	60	20	60
Isoxaben	3.9×10^{-7}	1	190-570	50–120
MCPA (salt)	1.5×10^{-6}	8,66,000	20	5 to 6
Mecoprop (salt)	2.3×10^{-6}	6,60,000	20	21
Mesotrione	4.4×10^{-8}	15	14 to 390	5 to 15
Metham Na (salt)	24	7,22,000	10	7
Metolachlor	1.3×10^{-5}	488	200	90
Metribuzin	1.2×10^{-7}	1,100	60	14 to 28
Metsulfuron-methyl	2.5×10^{-12}	2,790	35	30
MSMA (Na salt)	7.5×10^{-8}	1,040,000	7,000	180
Napropamide	4.0×10^{-6}	73	700	70
Norflurazon	2.9×10^{-8}	28	700	45 to 180

(continued)

Table 4.2
PHYSICAL PROPERTIES OF HERBICIDES INFLUENCING THEIR ENVIRONMENTAL FATE *(CONTINUED)*

Herbicide	Vapor Pressure[1] (mm Hg @ 25°C)	Water Solubility (ppm)	Sorption Index[2] (Koc)	Soil Half-Life (Days, $T_{1/2}$)
Oryzalin	1.0×10^{-8}	2.6	600	20
Oxidiazon	7.8×10^{-7}	0.7	3,200	60
Oxyfluorfen	2.0×10^{-6}	0.1	1,00,000	35
Paclobutrazol	7.5×10^{-9}	26	400	200
Paraquat (salt)	1.0×10^{-7}	6,20,000	1,000,000	1,000
Pendimethalin	9.4×10^{-6}	0.275	17,200	45
Prodiamine	2.5×10^{-8}	0.013	13,000	120
Pronamide	8.5×10^{-5}	15	800	60
Quinclorac	1.0×10^{-7}	62	36	168 to 391
Rimsulfuron	1.1×10^{-5}	7,300	51	11
Sethoxydim	1.6×10^{-7}	4,390	100	5
Siduron	4.0×10^{-9}	18	420	90
Simazine	2.2×10^{-8}	6.2	130	60
Sulfentrazone	8.0×10^{-10}	400		121 to 302
Sulfometuron-methyl	5.5×10^{-16}	300	78	20 to 28
Sulfosulfuron	6.6×10^{-10}	1,627	5 to 89	14 to 75
Triclopyr (amine salt)	1.3×10^{-6}	2,100,000	20	10 to 46
Triclopyr (ester)	1.3×10^{-6}	23	780	10 to 46
Trifloxysulfuron-sodium	7.5×10^{-8}	5,016	29 to 574	5 to 15
Trifluralin	1.1×10^{-4}	0.3	7000	45
Trinexapac-ethyl	1.6×10^{-5}	1,100	−0.38	1.1

[1]Herbicides with low (10^{-7} to 10^{-8}), medium (10^{-5} to 10^{-6}), and high (10^{-2} to 10^{-4}) vapor pressures have low, medium, and high volatility, respectively. Can be converted to Pascals (Pa) by multiplying by 133.
[2]Adsorptive strength of herbicides varies from very strong ($K_{OC} > 5000$), to strong (K_{OC} 600 to 4999), moderate (K_{OC} 100 to 599), and weak (K_{OC} 0.5 to 99).

Short-chain esters of 2,4-D are highly volatile; they are not generally used on turf because of their great potential for injuring nontarget plants. Low-volatile ester formulations are available, however, and should be used when an ester formulation of a phenoxy carboxylic herbicide, such as dichlorprop, is desired in order to penetrate the cuticle on difficult-to-control broadleaf species. Losses from volatilization of soil-applied herbicides can be reduced by watering them into the turf or soil following application.

PHOTODECOMPOSITION

Photodecomposition is the breakdown of an herbicide under the influence of light. This is initiated when the herbicide molecule absorbs light energy, which causes excitation of electrons, resulting in the formation or breakage of chemical bonds within the molecule. Photodecomposition-induced transformations may produce metabolites similar to those occurring from microbial or chemical decomposition. Photodecomposition has been more of a concern for insecticides than for herbicides. Foliar-applied insecticides that adhere to turfgrass leaves and are consumed by nocturnal insects should be applied late in the day to minimize losses in efficacy due to photolysis. Foliar-applied herbicides are primarily intended for rapid absorption by the leaves; thus, as long as absorption occurs fairly rapidly, photodecomposition is less of a concern. One group of herbicides that is highly susceptible to photodecomposition is the cyclohexanediones, including clethodim and sethoxydim. If substantial degradation occurs before one of these substances enters the leaf, activity can be lost. Formulations currently used for these herbicides usually contain adjuvants (see Chapter 5) that hasten their uptake, thus reducing loss from photodecomposition. As a general principle, soil-applied pesticides, including herbicides, should be washed off of the foliage and into the turf immediately after application in order to reduce the amount of chemical subject to photodecomposition.

ADSORPTION

Adsorption is the physical binding of an herbicide onto soil surfaces. It is the process of accumulation *at* a surface and is contrasted with absorption, which is passage *through* a surface. The reverse of adsorption is desorption, the movement of an herbicide away from the soil particle surfaces and into the soil solution. The terms *adsorption* and *desorption* are collectively referred to as *sorption.* Between the amount of herbicide adsorbed and the amount in solution, an equilibrium exists that reflects the specific properties of the herbicide, including its solubility in water and potential for leaching. With many herbicides, adsorption increases in proportion to the amount of clay and organic colloids in the soil. Adsorption onto soil particles can be quantified with the *soil distribution coefficient* (K_D), which is defined as the herbicide concentration adsorbed onto soil particle surfaces, divided by the herbicide concentration dissolved in the water surrounding the soil particles:

$$K_D = (\text{mg herbicide adsorbed/kg soil}) \div (\text{mg herbicide in solution/L solution})$$

Therefore, the unit of measure for K_D is L/kg. The K_D for a particular herbicide may vary widely, depending on the nature of the soil in which the herbicide is distributed. This variation is primarily due to the amount of organic matter (i.e., organic carbon) in the soil. Therefore, a more useful value is the *organic carbon distribution coefficient* (K_{OC}). This term is usually shortened to the *soil sorption index:*

$$K_{OC} = (\text{mg herbicide adsorbed/kg organic carbon}) \div (\text{mg herbicide in solution/L solution})$$

The K_{OC} can also be calculated by dividing the K_D by the fraction of organic carbon (g/g) in the soil. K_{OC} values for herbicides used in turf and ornamentals are provided in Table 4.2.

The bipyridilium herbicides—paraquat and diquat—are true cations that are strongly attracted to the negatively charged surfaces of clay colloids (Figure 4.2). Other herbicides, including the triazines, triazinones, and imidazolinones, are weak bases that can accept protons at one or more of their amino ($-NH_2$) groups, resulting in positive charges ($-NH_3^+$) at those locations. As a consequence, they can also be attracted to clay colloids. As protonation increases with decreasing pH (i.e., higher soil H^+

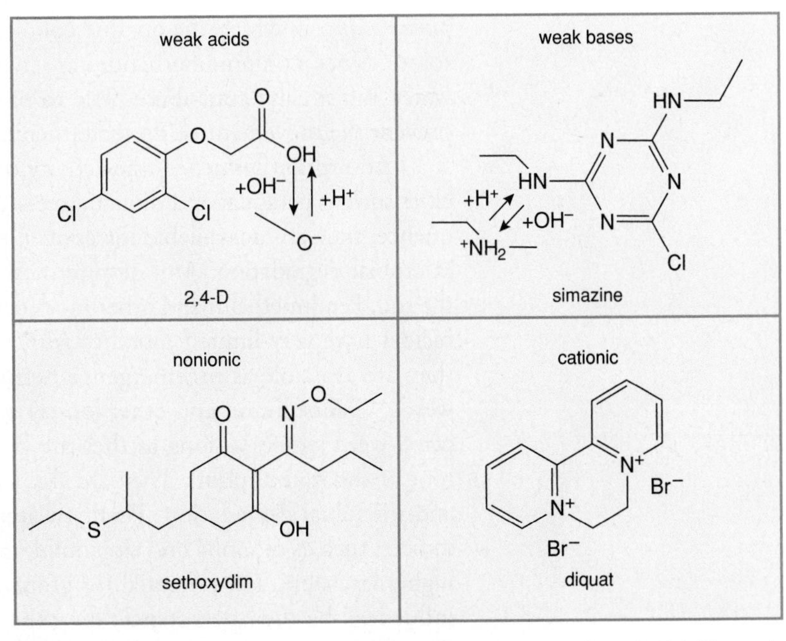

Figure 4.2. *Illustration of specific herbicides that are weak acids (2,4-D), weak bases (simazine), nonionic (sethoxydim), and cationic (diquat).*

concentration), the attraction of these herbicides to clay colloids increases proportionately. This accounts for the fact that atrazine, for example, is more tightly bound to the soil at lower pHs than at higher pHs; however, this attraction is never as strong as that for the bipyridilium herbicides.

Some herbicides, including the phenoxy-, pyridine-, and quinoline-carboxylics, are weak acids that will release protons (H^+) from carboxyl (—COOH) and hydroxyl (—OH) groups (i.e., —COO^- in place of —COOH; and —O^- in place of —OH), resulting in negative charges at these locations. As a consequence, they can be repelled by the negatively charged clay colloids. As the release of protons increases with increasing pH (i.e., lower soil H^+ concentration), this repulsion can increase proportionately, resulting in a higher potential for leaching of these herbicides. However, not all negatively charged herbicide ions are susceptible to leaching. Glyphosate is anionic, in part because of its phosphonate (—$PO_3H_2^-$) group. As with phosphate (—$PO_4H_2^-$) fertilizers, it can bind to cationic metals such as calcium (Ca^{2+}), iron (Fe^{3+}, Fe^{2+}), and aluminum (Al^{3+}) that are associated with negatively charged colloids. This interaction results in the formation of colloid-metal-phosphate (or –phosphonate, in the case of glyphosate) complexes, effectively binding the phosphate or phosphonate to the colloid. Additionally, insoluble calcium phosphates/phosphonates can form at high soil pHs, while insoluble iron and aluminum phosphates/phosphonates can form at low pHs. The potential for glyphosate adsorption and inactivation increases with increasing concentrations of these cations in the soil. This is largely true for the organic arsenical herbicides as well.

Ion adsorption does not account for all herbicide binding, especially where organic matter is involved. Highly decomposed organic residues, called humus, coat some or all of the mineral components of many soils. Humus molecules are very large and complex, consisting of numerous aromatic rings and aliphatic chains with attached carboxyl, hydroxyl, and other ionizable groups. Within the normal range of soil pHs, many of these groups are ionized, resulting in an overall negative charge that can substantially add to the cation exchange capacity of the soil. The presence of so many aromatic rings in these molecules also imparts a nonpolar property to humus, enabling it to interact with nonpolar chemicals. Many herbicides are nonpolar (nonionic) and, as a consequence, are unaffected by soil pH or surface charges. These herbicides can bind to soil organic colloids through a process called *partitioning*, in which the organic colloids act as the solvent and the herbicide as the solute. Since nonionic herbicides are hydrophobic and thus are actually repelled by water, moist clays contribute little to partitioning, as the adsorbed water molecules prevent the movement of these herbicides into or around the clay particles.

Adsorption influences the activity of herbicides in soil. Strongly adsorbed herbicides such as paraquat and diquat are essentially inactivated by adsorption. As a consequence, they are unavailable for controlling weeds and inaccessible for chemical and microbial degradation. And despite their high water solubility, they are immobile in the soil. Pendimethalin and other moderately adsorbed, water-insoluble, nonionic herbicides have very limited mobility within the soil. Because they are degraded slowly, they are available as preemergence herbicides for extended control of germinating weeds. Chlorsulfuron and other loosely adsorbed herbicides can be active against targeted weed species as long as they are positioned properly within the soil profile for root or shoot absorption. They are also susceptible to some leaching and to chemical and microbial degradation. Finally, water-soluble herbicides with very low sorption indices, such as dicamba and clopyralid, can be both shoot- and root-absorbed and are highly leachable. The potential for groundwater contamination by these herbicides is influenced by their persistence; dicamba has a relatively short half-life (indicated as $T_{1/2}$) in the soil ($<$14 days), while clopyralid persists longer ($T_{1/2}$ = 40 days) and thus poses a slightly greater hazard to groundwater.

DILUTION AND LEACHING

Dilution and leaching are processes by which an herbicide is carried downward in the soil profile with water. As explained earlier, a relatively high concentration of herbicide in the soil solution surrounding soil particles, compared with the concentration adsorbed onto soil particles, influences the potential for downward movement. In the case of preemergence herbicides, which work by controlling germinating weed seeds, dilution of the herbicide concentration within a larger volume of soil may reduce the surface herbicide concentration to below that required for effective control. Further leaching may carry the herbicide down to the water table, where it can subsequently move in drain tiles to ponds, streams, storm sewers, or other receptacles.

The K_{OC} values for herbicides, along with their water solubility (ppm) and soil half-life ($T_{1/2}$), as shown in Table 4.2, are useful for predicting the potential for leaching and dilution within the soil profile. A highly leachable herbicide would have a low K_{OC} value, high water solubility, and a relatively long soil half-life. The relative mobility of selected herbicides can be expressed as Rf values, which are then divided among five mobility classes; herbicides with the lowest Rf values are tightly adsorbed to soil colloids and show little or no movement (Class 1), while herbicides with the highest Rf values are highly mobile (Class 5), as shown in Table 4.3. Soil mobility is generally consistent among herbicides within the same chemical family. The aryloxyphenoxypropionates, bipyridiliums, dinitroanalines, and organic arsenicals are Class-1 herbicides and thus are not mobile in soil. While they vary widely in water solubility and soil half-life, all have high K_{OC} values.

The diphenylethers and imidazolinones have low to moderate mobility (Class 2), and the triazines have moderate mobility (Class 3). The sulfonylureas are distributed

Table 4.3
RELATIVE MOBILITY OF HERBICIDES THROUGH A SOIL PROFILE BASED ON HELLING'S CLASSIFICATION SYSTEM[1]

1	2	3	4	5
Benefin	Bensulide	Atrazine	2,4-D	Bentazon
Bromoxynil	Dichlobenil	Chlorsulfuron	MCPA	Clopyralid
CAMA	Imazaquin	Ethofumesate	Metribuzin	Dicamba
DCPA	Napropamide	Glufosinate	Triclopyr	Mecoprop
Diclofop	Norflurazon	Isoxaben		Sethoxydim
Diquat	Oxadiazon	Metolachlor		
DSMA	Oxyfluorfen	Metsulfuron		
Fluazifop	Siduron	Simazine		
Glyphosate	Sulfometuron			
MSMA				
Oryzalin				
Paraquat				
Pendimethalin				
Prodiamine				
Trifluralin				

Class 1 = essentially immobile (Rf @ 0.0–0.9); Class 2 = low to moderate movement (Rf @ 0.1–0.34); Class 3 = moderate movement (Rf @ 0.35–0.64); Class 4 = moderate to high movement (Rf @ 0.65–0.89); and Class 5 = high movement (Rf @ 0.9–1.00).

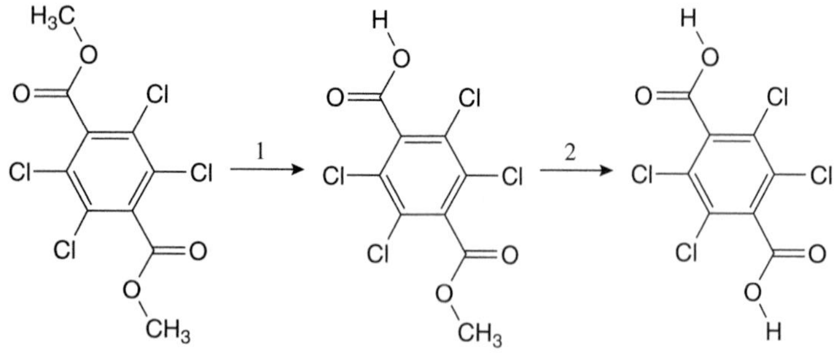

between the low-to-moderate (Class 2) and moderate (Class 3) movement classes, and the auxinic growth regulators (phenoxy- and pyridine-carboxylic acids and benzoic acids) are distributed between the moderate-to-high (Class 4) and high (Class 5) movement classes.

While DCPA is one of the preemergence herbicides in Class 1, its demethylated metabolites are easily ionized, making them water soluble and highly susceptible to leaching. Because of the stabilizing effect of the chlorines and carboxyl groups, they can accumulate and persist for very long periods in groundwater (Figure 4.3).

Figure 4.3.
Illustration of DCPA (left), then the metabolite resulting from the first (1) demethylation (center), and finally the metabolite resulting from the second (2) demethylation (right). The demethylated metabolites are easily ionized, making them highly susceptible to leaching into the groundwater.

ABSORPTION

Absorption is the uptake of an herbicide by living organisms. Once absorbed, contact herbicides are usually immobile, while systemic herbicides are translocated within receptor plants. Absorption of a sufficient amount of an herbicide by germination or by mature weeds is essential for effective control. While an absorbed herbicide may be stored or discharged in its original form, many herbicides are altered in plants by metabolism, and the breakdown products, called *metabolites*, are either used by the plant or discharged. The discharging of herbicides or their metabolites usually occurs from the roots to the soil solution. Refer to Chapter 3 for a detailed discussion of herbicide absorption.

CHEMICAL REACTION

Chemical reaction results in the nonenzymatic transformation of herbicide molecules to usually less active or inactive forms. Hydrolysis, oxidation, reduction, and the formation of water-insoluble salts or chemical complexes are types of chemical reactions that can occur with some herbicides in some soils. *Hydrolysis* occurs when herbicide molecules react with water, resulting in the breaking of chemical bonds within the herbicide molecule and the formation of a bond between one of the ions of water, usually hydroxide (OH^-). For example, triazines are commonly degraded by hydroxylation and removal of chlorine, converting them to nonphytotoxic hydroxy derivatives through a purely chemical, nonenzymatic process. Triazines can also be degraded to their hydroxy forms by soil microorganisms. These two types of degradation—chemical and microbial—may work in concert on the same herbicide molecule. The sulfonylureas can also be degraded in soil by hydrolysis if the pH is acidic; however, as the pH becomes more alkaline, approaching or exceeding neutrality, microbial degradation tends to dominate.

Oxidation and *reduction* reactions result in the transfer of electrons between reactants, resulting in the formation of ionized compounds. In some soils, ionized herbicides may react with calcium ions to form insoluble calcium salts. Herbicides can also form stable complexes with cobalt, iron, magnesium, and nickel. These chemical reactions with soil constituents usually result in the deactivation of the herbicide.

MICROBIAL DEGRADATION

Microbial degradation is the breakdown of herbicide molecules by microorganisms in the soil. Algae, fungi, actinomycetes, and bacteria make up the highly diverse community of soil microorganisms; for herbicide degradation, however, bacteria and fungi are the most important members of this community. Many herbicides serve as a food source from which microorganisms derive energy and carbon units for growth and respiration. Optimum temperatures (75° to 90°F, 24° to 32°C) and pH (moderately acid for fungi, near neutral for bacteria), along with an adequate supply of soil moisture, oxygen, and mineral nutrients, are conditions that favor microbial degradation of herbicides. Under similar temperatures and moisture levels, herbicide degradation occurs more rapidly in soils that are rich in organic matter. In soils with high adsorptive capacity (i.e., with high K_{OC} values), herbicides tend to persist longer and are less available for microbial degradation. Various herbicide constituents can affect the capacity of microorganisms to degrade an herbicide. For example, the halogens (Cl, Br, F, I) on a benzene ring can stabilize the ring, making it more resistant to microbial degradation. The persistence of herbicides used for weed control in turf and ornamental plant beds is provided in Table 4.2 as estimations of soil half-life ($T_{1/2}$) in days.

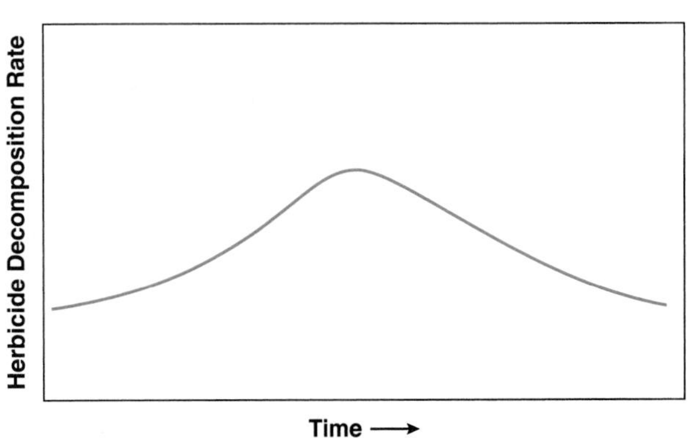

Figure 4.4.
Illustration of the microbial population over time, reflecting the introduction of an herbicide in the soil. After an initial lag phase, the microbial population increases rapidly until the herbicide food source becomes limiting; then the population declines to its original level.

Once an herbicide has been introduced, microbial populations with the capacity to utilize the herbicide as a food source tend to increase until much of the herbicide residue has been degraded; then these populations eventually return to normal levels (Figure 4.4). Microbial degradation proceeds by many pathways, including the following: *dehalogenation* (removal of a halogen atom, including chlorine [Cl], fluorine [F], and iodine [I]); *dealkylation* (removal of alkyl group, including methyl [—CH_3], methylene [—CH_2–], ethyl [—CH_2CH_3], and ethylene [—CH_2CH_2–] groups); *decarboxylation* (removal of a carboxyl [COOH] group); *oxidation* (loss of electrons, often accompanied by the addition of oxygen [O]); *reduction* (gain of electrons, often accompanied by the loss of oxygen); *hydrolysis* (reaction with water [H_2O], usually resulting in the addition of hydrogen [—H] or a hydroxide [—OH] group); *hydroxylation* (addition of a hydroxide [—OH] group); *ether cleavage* (breaking an R—O—R link); *conjugation* (formation of a bond with another compound, usually a sugar or an amino acid); and *ring cleavage* (breaking an aromatic [i.e., benzene, pyridine, etc.] or aliphatic [i.e., cyclohexane] ring). In the vast majority of cases, microbial decomposition leads to detoxification—that is, the loss of herbicide activity. Many herbicides are degraded to carbon dioxide or to compounds that naturally occur in biological systems and therefore pose no long-term threat to the environment or to the organisms residing in or above the soil.

5

Herbicide Formulations and Application

The safe and effective use of an herbicide first requires its preparation in a formulation that facilitates its use; then the formulated product must be applied uniformly and at the proper rate to the target site. This chapter covers the array of herbicide formulations currently in use and the methods and equipment used for application.

HERBICIDE FORMULATIONS

To *formulate* means to prepare in accordance with a specified formula for a particular purpose. Herbicides are formulated with various constituents to ensure that they can be handled, transported, and applied safely and effectively. The term *formulation* refers to both a product and a process. An herbicide formulation is a preparation that includes the active ingredient (a.i., the actual herbicide) along with other materials, including solvents, diluents, and adjuvants, that facilitate its use. Herbicide formulation is also the process the manufacturer carries out in order to prepare an herbicide for practical use. Some herbicides are formulated as different products that vary in their selectivity, efficacy, concentration, handling characteristics, and cost, as well as the hazards associated with their use and methods of application. Herbicide formulations are grouped into two broad types: those intended for application as liquids (sprayable formulations) and those intended for direct, dry application.

Liquid Formulations

These include water-soluble, emulsifiable, and suspendable formulations. All can be applied in a liquid carrier (typically water) by the use of a sprayer.

Water-soluble formulations mix with water, forming clear—often colored—solutions. *Soluble liquids* (designated as SL) include active ingredients dissolved in water. Carboxylic, benzoic, and other acids are often converted to salts (by reacting them with bases) to maximize their solubility in water. SL formulations require little agitation to dissolve and need no additional agitation to maintain a uniform distribution of the herbicide in solution. Wetting agents are usually required to maximize foliar coverage and absorption of the herbicide; these may be included in the formulation or added to the spray tank. Most SL formulations contain from 2 to 4 lb of active ingredient per gallon (240 to 480 g/L). *Soluble powders* (SP) and *soluble granules* (SG) are dry solids, usually salt concentrates. The powders are very fine textured, while the granules are much coarser in texture. SP may also refer to *soluble packets,* where the powder is available in water-soluble packets that can be added directly to the spray tank. The SP and SG formulations may require considerable agitation to dissolve them in water; however, once fully dissolved, they form stable solutions and require no further agitation. As with SL formulations, wetting agents are usually required to maximize foliar coverage and absorption of the herbicide. SP and SG formulations typically contain from 40 to 95 percent active ingredient.

Emulsifiable formulations contain oil-soluble herbicides in oil. Emulsifiers are added to this mixture, forming *emulsifiable concentrates* (E or EC). When added to the spray tank, the oil disperses in the water, forming an emulsion. Normally, oil and water do not mix, as they are immiscible liquids; however, the addition of an emulsifier—a type of surfactant—enables the formation of an emulsion, which is tiny oil droplets (called the dispersed phase) suspended throughout the water (continuous phase). The emulsifier molecule has both hydrophilic (polar) and lipophilic (nonpolar) ends; the polar end dissolves in water while the nonpolar end dissolves in oil, stabilizing the oil-water mixture. Emulsions are milky in appearance; in fact, milk is a naturally occurring emulsion. This formulation of herbicide requires some agitation in the spray tank to prevent the oil and water from separating. Since a surfactant—the emulsifier—is an essential component of this formulation, additional surfactant, as wetting agent, is usually not required for foliar applications. Most E or EC formulations contain from 2 to 6 lb of active ingredient per gallon (240 to 720 g/L). *Gels* (GL) are thickened emulsifiable concentrates available in water-soluble packets that can be added directly to the spray tank.

Suspendable formulations include a variety of liquids and dry solids that can be suspended in water. The liquid forms include finely ground water-insoluble powders suspended in a liquid system, usually water, and are called *liquids* (L), *water-dispersible liquids* (WDL), or *flowables* (F); however, other names are sometimes used, including *suspension concentrates* (SC) and *aqueous suspensions* (AS). When added to the spray tank, they form suspensions in the water. Agitation is required to maintain uniform suspensions; otherwise, the particles will settle out. *Capsule suspensions* (CS) or *microencapsilated* (ME) formulations are small particles of an herbicide that are surrounded by a barrier layer, usually a polymer, and concentrated in a liquid carrier. Because microencapsulation substantially reduces the amount of liquid carrier required, this formulation tends to be more concentrated and lower in cost per unit of active ingredient than some other formulations, including emulsifiable concentrates. It is also safer and more efficient, as the polymer barrier protects the user from direct contact with the herbicide and reduces the potential for photodecomposition, volatilization, and leaching following application.

Dry Formulations

These include dry solids that are suspended in water for spraying and formulations for dry application. *Wettable powders* (WP) are finely ground solids that can be suspended in water, but are often coarser than the particles in flowable (F) formulations. It is usually recommended that they be mixed with a small amount of water to form a slurry before being added to the spray tank. Moderate to vigorous agitation is required to maintain a uniform suspension. Most WP formulations contain from 50 to 80 percent active ingredient. *Water-dispersible granules* (WDG), also known as *dry flowables* (DF) or *dry granules* (DG), are dry solids combined into larger granule-size particles. The granules usually contain suspending agents; they can be added to the spray tank directly, where they disperse in the water without clumping. As with flowable formulations, some agitation is required to prevent the suspended particles from settling out.

The primary dry formulation used for direct application to a turf or ornamental bed is called *granules* (G). Particle-size diameter is usually 10 mm^3 or smaller. The particles are typically composed of clay minerals, dry fertilizers, or ground plant residues. The herbicide is coated on the outside of the particles, with herbicide concentrations ranging from 2 to 20 percent. *Matrix granules* (MG) composed of starch have been used to entrap the herbicide within the starch matrix to slow the release of the herbicide into the soil; however, these have not been widely adopted. *Pellets* (P) are dry formulations in which the particle size measures 100 mm^3 or larger. In contrast to granules, which are usually applied mechanically by a spreader, pellets are usually applied by hand as spot treatments. Among the advantages of dry application over sprayable formulations are that water is not needed for application, application equipment is less expensive to purchase and maintain, and granules can pass through a dry turfgrass canopy—or bypass dry ornamental plants in a bed—for direct soil application. The disadvantages include higher weight and cost, and the difficulties encountered in calibrating application equipment and in uniformly applying the herbicide.

Changing Formulation Technologies

Early formulations of herbicides were developed primarily to facilitate their application; water-soluble and oil-soluble chemicals were concentrated in water and oil, respectively, and insoluble chemicals were finely ground for suspension by spray-tank agitation. The problems associated with these formulations included splashes and spills from liquid formulations, toxic vapors from oil solvents, and dusts from wettable powders. Recent efforts in formulation technology have been directed at addressing these and other problems in order to improve the handling characteristics, safety, and effectiveness of herbicides. For example, emulsifiable concentrates are being replaced by microencapsulated formulations to minimize the vapor problems associated with oil carriers. Additional gains from ME formulations include increased efficiency by the minimizing of volatilization, photodecomposition, and leaching; and improved safety by isolation of the user from direct contact with the herbicide. Wettable powders and other suspendable formulations are being replaced by water-dispersible granules to eliminate the hazards associated with the dusts and spills from these formulations. In addition, dry WDG formulations are less likely to be damaged by freezing temperatures during storage than their liquid versions, and they are easier to clean up if spilled.

Spray Adjuvants

An *adjuvant* is a material that is added to a spray tank to enhance herbicidal activity or to modify the physical characteristics of the chemical mixture. Most herbicides require adjuvants, either in the commercial formulation—called *formulation adjuvants*—or in the spray tank—called *application adjuvants*—or both. Adjuvants include a wide array of materials designed for specific purposes. For example, a *surfactant* is an adjuvant for improving the surface properties of liquids. The term *surfactant* is an acronym for *surface-active-agent.* Surfactant molecules have a hydrophilic (water-soluble) end and a lipophilic (oil-soluble) end. Surfactants include wetting and dispersing agents, as well as stickers and emulsifiers.

Wetting agents reduce the interfacial tension between the surfaces of a water droplet and a waxy leaf cuticle. The lipophilic end of the wetting agent molecule partitions into the waxy cuticle, while the hydrophilic end partitions into the water droplet, reducing the water tension and enabling the droplet to spread and make better contact with the leaf surface. Wetting agents are usually added to the spray tank to constitute 0.25 to 0.50 percent of the spray volume. The proper concentration of a wetting agent in the spray tank is essential to its performance; adding too little or too much could substantially reduce the herbicide's efficacy.

Wetting agents can be divided into four groups based on the ionization characteristics of the hydrophilic end of the molecule; they are anionic, cationic, nonionic, and amphoteric. Anionic and cationic wetting agents have negative and positive charges, respectively, in water. Amphoteric wetting agents have varying amounts of charge, depending on the pH of the water. And nonionic wetting agents, as the name implies, do not have an electrical charge in water. Because anionic, cationic, and amphoteric wetting agents can react with other ions in solution, including ionized herbicides, they are not as popular as the nonionic wetting agents for use with herbicides. In contrast, nonionic wetting agents are compatible with most herbicides and do not precipitate out (as calcium and magnesium salts) in hard water.

Dispersing agents are surfactants that enhance the dispersal of a powder in a solid-liquid suspension. As with wetting agents, the lipophilic end of the dispersing-agent molecule partitions into the suspended particle while the hydrophilic end partitions into the surrounding water medium, reducing the water tension and enabling the particle to repel other particles and remain in suspension.

Stickers are used to reduce losses of spray droplets from leaf surfaces by simultaneously increasing the viscosity of the spray droplets and reducing the interfacial tension between the spray droplet and the leaf surface. As a consequence, the droplets resist being washed off by rain or knocked off by physical contact and, at the same time, spread out on the leaf surface. Sometimes the term *spreader-sticker* is used for a surfactant that functions as both a wetting agent and a sticker.

Emulsifiers are used to disperse and stabilize tiny oil droplets (in which the oil-soluble herbicide is dissolved) in an aqueous medium. The lipophilic (oil-soluble) end of the emulsifier molecule dissolves in oil while the hydrophilic (water-soluble) end dissolves in water, stabilizing the oil-water mixture.

Other spray-tank adjuvants include *acidifiers* for reducing the pH of spray tank solutions, *buffers* that enable spray-tank solutions to resist pH changes, *compatibility agents* for allowing the mixing of two or more herbicide formulations or herbicide-fertilizer combinations in tank mixes, *crop oil concentrates* for enhancing herbicide penetration through leaf cuticles, *defoamers* for suppressing spray-tank foaming problems, *drift control* agents for reducing spray drift, *humectants* for slowing the drying of spray droplets on leaf surfaces, *penetrants* for enhancing the penetration of leaf surfaces by herbicides, and *safeners* for reducing herbicide toxicity to turfgrasses by a physiological mechanism.

HERBICIDE APPLICATION

The safe and effective use of herbicides requires their uniform application at the proper rate. This requires the proper selection, calibration, and operation of application equipment. Common causes of improper application with sprayers include inaccurate measurement of the amount of water in the spray tank; nozzle-related problems (e.g., worn, damaged, or plugged nozzles; different nozzles on the same spray boom; incorrect nozzle height above the surface; and nozzles without screens), and broken or inaccurate pressure gauges. Improper application with spreaders is often due to poor maintenance; this is reflected in worn hopper agitators, bent or plugged delivery tubes, and plugged or rusted hopper outlets. Inaccurate application of herbicides can result in injured turf, ineffective weed control, and wasted money. It can also invite criticism from environmental organizations and more regulation from government agencies.

The assortment of equipment available for applying herbicides to turf and ornamentals ranges from simple hand-held pump sprayers and push spreaders to highly sophisticated automated systems for liquid and granular application.

Sprayers

Spraying is the most common method for applying herbicides. Liquid sprays can be applied uniformly and accurately, as long as the sprayer has been properly designed, calibrated, and operated, and the herbicide formulation in the spray tank has been mixed correctly.

A typical sprayer consists of the following components:

- a *tank* to hold the spray mixture
- a *pump* to deliver the mixture through the sprayer
- a *pressure regulator* to maintain constant pressure
- a *bypass line* to return excess spray to the tank (so that the pressure regulator can function)
- a *pressure gauge* to provide pressure readings
- a *boom with nozzles* or a *handgun* to deliver the spray to the application site
- a *shutoff valve* located upstream from the boom (or in the handgun) to control the flow of the spray mixture
- a mechanical or hydraulic *agitator* to keep suspended particles or oil droplets uniformly dispersed within the spray tank
- *strainers* and/or *screens* to prevent large particles from clogging the sprayer
- a *suction line* to convey the spray mixture from the tank to the pump
- various *hoses, fittings, clamps,* and other devices—perhaps including governors and controllers—that enable or enhance the overall functionality of the sprayer

Sprayer Tanks, Pumps, and Pressure Regulators and Gauges Sprayer *tanks* can be constructed from a variety of materials, including stainless steel, aluminum, aluminized steel, galvanized steel, fibergrass, and plastic. Stainless steel is the most adaptable, but also the most expensive. Galvanized steel parts should be avoided, as they tend to react with some herbicides and are susceptible to rust, generating rust flakes that can clog sprayers. Some herbicides can also react with aluminum and aluminized steel. Fiberglass and plastic resist chemical reactions with herbicides and do not rust; they are lighter in weight than metal tanks are, but are more susceptible to breakage and possible degradation from long-term exposure to sunlight. Tanks mounted on turf utility vehicles typically range in size from 100 to 300 gal (379 to 1136 L). The largest tanks usually have baffles to reduce liquid shifting during operation.

The most common *pumps* used for spraying herbicides are the centrifugal, roller, piston, and diaphragm types (Figure 5.1). The power supply for operating pumps include tractor power takeoff (PTO), direct-drive gasoline engines or electric motors, and ground-wheel tractor drive. Centrifugal pumps, which develop pressure as a result of centrifugal force, are the most commonly used pumps for low-pressure sprayers. The spray mixture enters the pump through the center of the impeller, forcing the liquid outward to the wall of the pump housing and releasing it through an outlet. Centrifugal pumps wear well, even with coarse wettable powders, and deliver 70 to 130 gal/min (265 to 492 L/min) at 30 to 40 lb/in^2 (207 to 276 kPa). Because they operate at speeds of 3,000 to 4,500 rev/min, a step-up from the PTO is required.

Roller pumps have loose rollers fitted into slots in an impeller that is mounted off-center within the pump housing. The rollers are forced to follow the pump housing by the rotating impeller, propelling the spray mixture ahead of the rollers and through the outlet. Roller pumps operate at PTO rev/min. Because wettable powders and other abrasive formulations can cause the rollers to wear quickly, roller pumps, while inexpensive, may be relatively short-lived. However, they are easily repaired. As the name implies, a piston pump employs a piston that pulls the spray mixture into the chamber through a valve that opens under the negative pressure caused by the receding piston, and then forces the mixture out through another valve that opens as

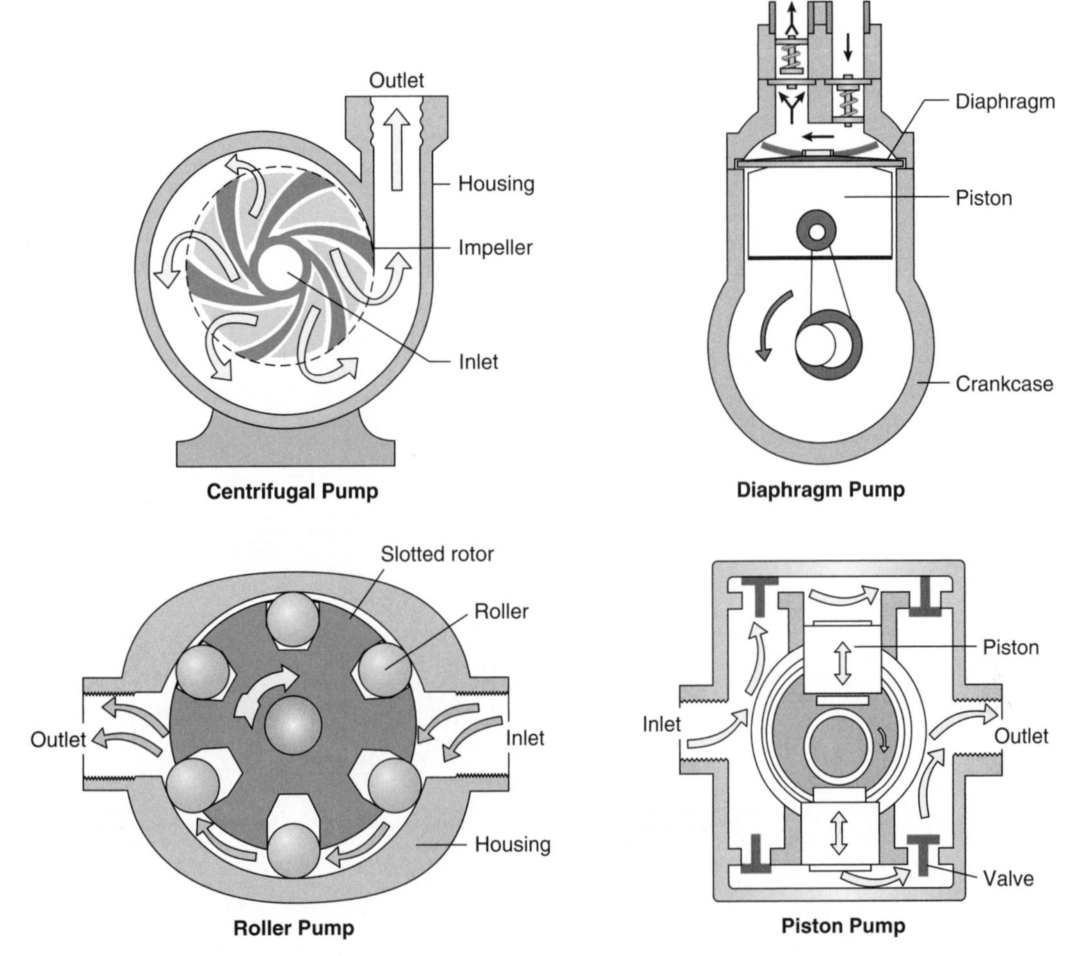

Figure 5.1 *Illustrations of centrifugal, diaphragm, roller, and piston pumps.* (Courtesy American Society of Agronomy and Crop Science Society of America.)

the piston advances. Piston pumps deliver an exact volume of spray each time the piston makes a cycle. Diaphragm pumps employ a flexible diaphragm that moves up and down within a chamber, alternatively creating pressure and suction. Operating separately from the diaphragm are intake and discharge valves that open and close, forcing the spray mixture through the pump. Diaphragm pumps require less power to operate than other pumps. They develop moderate pressures and deliver 15 to 50 gal/min. Centrifugal and roller pumps are often employed for spraying herbicides because they are designed to operate at relatively low pressures (30 to 60 lb/in^2, 207 to 414 kPa) and thus are relatively inexpensive. Sprayers used for applying insecticides and fungicides at higher pressures (100 to 500 lb/in^2, 690 to 3450 kPa) are usually equipped with the more expensive piston and diaphragm pumps. A pump should fill with spray mixture as soon as it starts, in order to minimize wear. Roller, piston, and diaphragm pumps are self-priming positive-displacement pumps that move a constant volume of liquid during each pump cycle. Centrifugal pumps are nonpositive-displacement pumps and, therefore, must be primed (filled with spray mixture) in order to function; this can be accommodated by placing them below the spray tank, so that valves at the top of these pumps allow trapped air to escape as the spray mixture is pulled in.

A *pressure regulator* is needed in systems with positive displacement pumps (e.g., roller, piston, and diaphragm pumps) to control operating pressure (Figures 5.2 and 5.3). The valve opens when pressure increases to some adjustable level, relieving the pressure

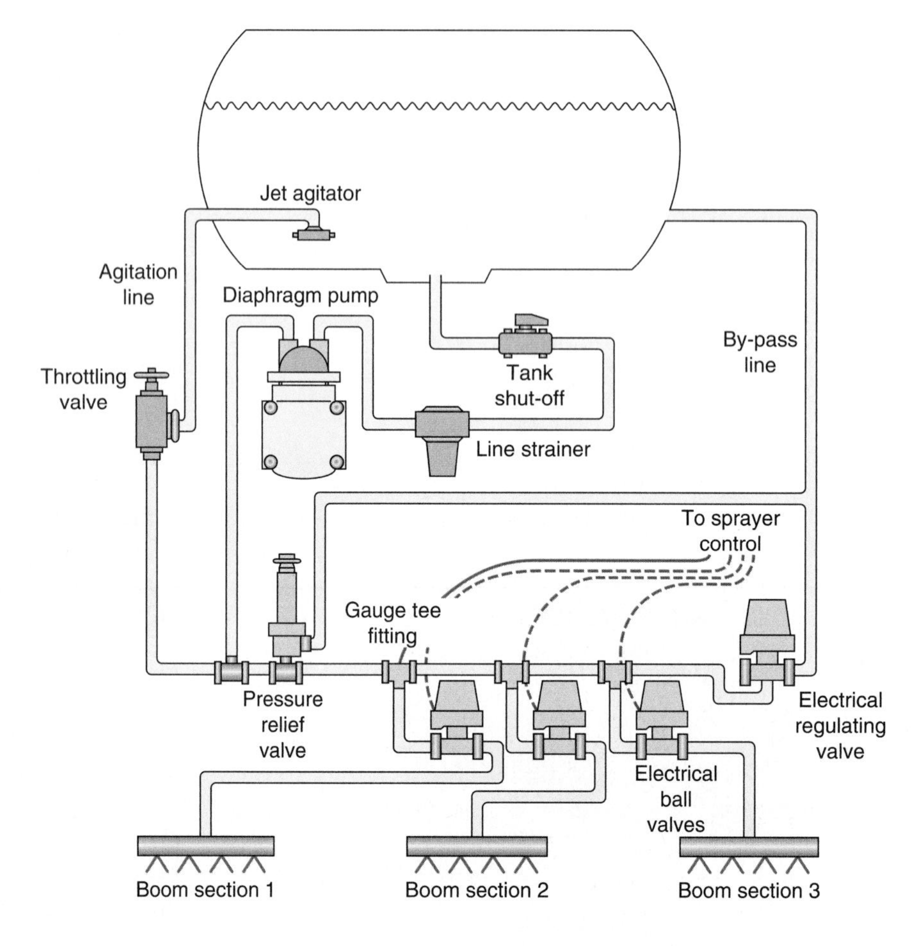

Figure 5.2
Illustration of plumbing system for a positive displacement pump. (Courtesy American Society of Agronomy and Crop Science Society of America.)

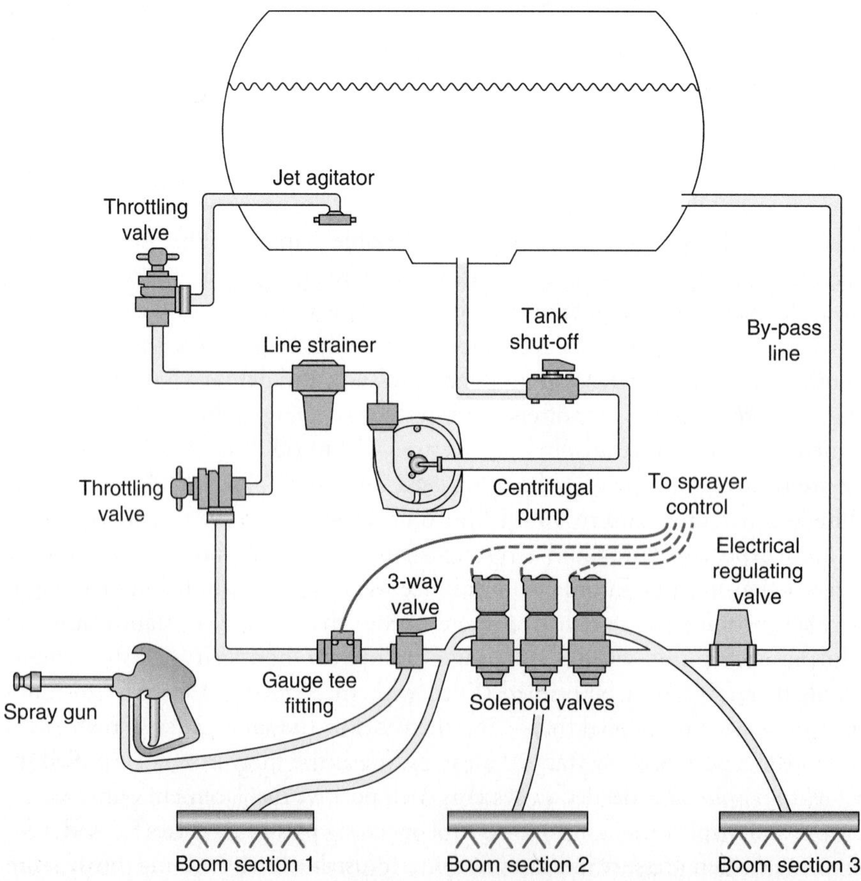

Figure 5.3
Illustration of plumbing system for a nonpositive displacement pump. (Courtesy American Society of Agronomy and Crop Science Society of America.)

and conducting some of the flow back to the spray tank. In some systems, the same *bypass line* is used for hydraulic agitation of the tank spray mixture. The disadvantage of linking pressure relief and hydraulic agitation, however, is that the flow through the bypass line can fluctuate due to changes in the system's pressure requirements.

A *pressure gauge* is used to measure liquid pressure. The range should be about twice the maximum anticipated pressure encountered in the operation of the sprayer. Liquid-filled pressure gauges are preferable to dry gauges, as they dampen vibrations caused by pressure spikes in the system. It should be installed in a position that is easily observed by the sprayer operator.

Booms Spray booms include those mounted onto the sprayer and walking booms that are attached to the sprayer by a long flexible hose. Booms can also be wet or dry; wet booms actually carry the spray mixture through a pipe—usually stainless steel—to the nozzles, while dry booms carry the spray mixture through flexible tubing to the nozzles, which are attached to a rigid frame. Mounted booms are usually 15 to 20 ft (4.6 to 6.1 m) in length and often come in three sections, with each section operating independently. The lateral sections can be folded up for transport. Walking booms measure up to 80 in (2 m), sometimes with fold-up lateral sections. These are used where compaction and wear from vehicles precludes the use of mounted booms, such as on golf greens or newly established areas.

Nozzles The *nozzles* mounted on a spray boom convert the spray mixture into spray droplets for delivery to the plant or soil surface. Droplets of varying sizes form from the tearing action of air on the liquid emerging from the nozzles. Depending upon their size and condition, the nozzles, in conjunction with pressure and other factors, determine the droplet size spectrum, the amount of spray delivered (spray volume), the spray pattern, and the distribution of the herbicide. As droplet size is inversely related to pressure and spray angle, increasing pressure reduces the median size of the spray droplets while increasing the spray volume. The flow rate from a nozzle conforms to the following formula:

$$\text{gal/min}_1 \div (\text{lb/in}^2{}_1)^{1/2} = \text{gal/min}_2 \div (\text{lb/in}^2{}_2)^{1/2}$$

or

$$\text{L/min}_1 \div (\text{kPa}_1)^{1/2} = \text{L/min}_2 \div (\text{kPa}_2)^{1/2}$$

If a particular nozzle delivers 20 gal/min at 30 lb/in^2, the pressure required to increase the flow rate to 40 gal/min would be calculated as follows:

$$20 \text{ gal/min} \div (30 \text{ lb/in}^2)^{1/2} = 40 \text{ gal/min} \div (x \text{ lb/in}^2)^{1/2}$$
$$x = [40 * (30)^{1/2} \div 20]^2 = 120$$

Therefore, doubling the flow rate from 20 to 40 gal/min (76 to 151 L/min) would require increasing the pressure from 30 lb/in^2 (138 kPa) to 120 lb/in^2 (552 kPa), or fourfold. This would substantially affect the atomization of the spray, reducing droplet size and dramatically increasing the drift potential. Thus, the proper way to substantially increase flow rate, or spray volume, would be to change to a larger nozzle size and not to increase pressure.

A typical nozzle has four parts: body, strainer, tip, and cap. The cap screws onto the body and holds the tip and strainer in place. The strainer is placed immediately behind the nozzle tip to filter the liquid and prevent nozzle clogging. Nozzle bodies are constructed of stainless steel, brass, aluminum, nylon, and plastic. Nozzle tips are made from these materials, as well as from hardened stainless steel and ceramic material. Stainless steel, hardened stainless steel, and ceramic nozzles are the most expensive, but

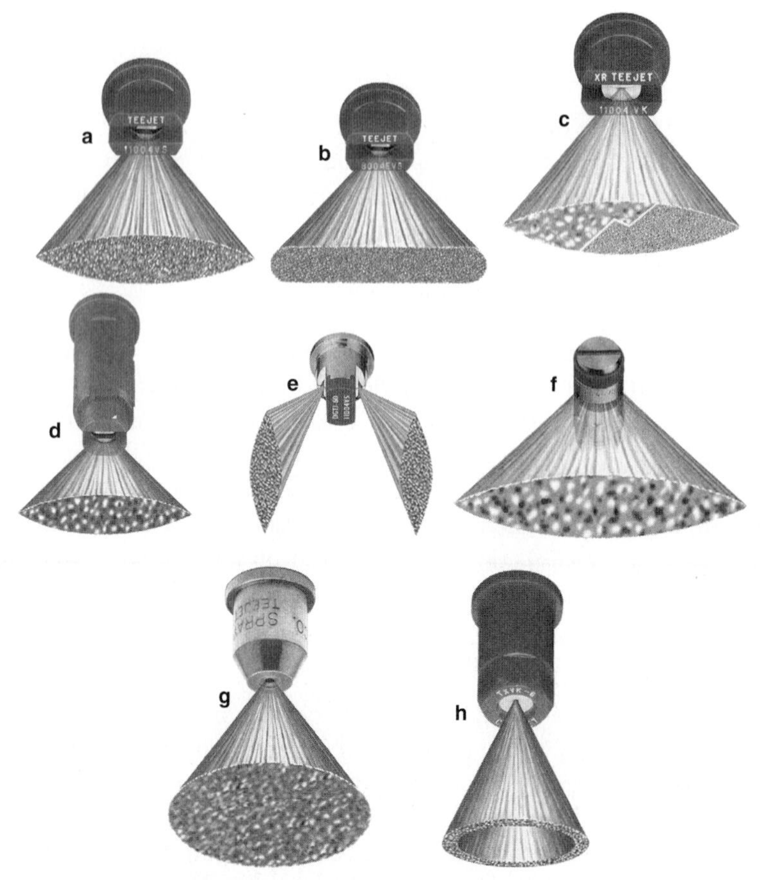

Figure 5.4 *Illustrations of nozzle types: a. tapered flat-fan; b. even flat-fan; c. extended-range flat-fan; d. air-induction; e. twin-orifice flat-fan; f. flooding; g. solid-cone; and h. hollow-cone.* (Courtesy TeeJet® Image Library, www.teejet.com.)

are the most resistant to wear and corrosion. Thermoplastic nozzle tips are also wear resistant, but may be easily damaged during cleaning. Brass tips, while relatively inexpensive, are the least wear resistant. Nozzle output should be checked regularly and worn nozzles replaced as needed to ensure accurate and uniform herbicide application.

There are three basic types of nozzle spray patterns: flat, solid cone, and hollow cone. Each pattern has characteristics that are favorable for specific applications. The spray droplets from a flat nozzle tip form a fan-shaped pattern as they emerge from the elliptical orifice of the nozzle. Some fan-shaped nozzles produce a tapered pattern (Figure 5.4a), while others produce an even pattern (Figure 5.4b). The tapered pattern is well suited for broadcast application of herbicides, with adjacent nozzles set to provide overlapping patterns of application. As the spray volume tapers off at the edges of the fan, uniform delivery of spray mixtures across the entire boom requires overlapping the spray patterns of adjacent nozzles by at least 30 percent. Several spray angles are available, including 65°, 73°, 80°, and 105°. Selection is determined by the spacing of nozzles along the boom and by the nozzle tip height above the ground. Nozzle tips with wider angles produce smaller droplets and are more susceptible to clogging, but they can be spaced farther apart along the spray boom and operated closer to the ground. Nozzle tips with narrower angles produce larger droplets and a more penetrating spray, but they must be spaced closer together along the spray boom and operated farther from the ground. Operating pressure should be between 30 and 60 lb/in^2 (psi) (207 to 414 kPa), as uniformity of coverage along the boom may be reduced below this range, while spray drift potential is increased above this range.

Flat-fan nozzles that produce an even pattern of distribution at operating pressures of between 15 and 20 lb/in^2 (104 to 207 kPa) are used for single-nozzle sprayers for spot treating weeds in lawns and ornamental beds.

In addition to conventional flat-fan nozzles, there are several modified versions that may be chosen for specific applications. The extended-range flat spray tip (Figure 5.4c) was designed to provide better spray droplet distribution over a broader range (15 to 60 lb/in^2, 104 to 414 kPa) of operating pressures. At the low end of this range, systemic herbicides can be uniformly applied in a smaller number of larger droplets with substantially reduced drift potential, and higher pressures can be used for applying a larger number of smaller droplets containing contact herbicides, which require more complete foliar coverage for maximizing efficacy. The extended-range flat spray tip is available for both tapered and even flat-fan nozzles. Another version of the flat-fan nozzle, called the air-induction nozzle (Figure 5.4d), employs a preorifice to reduce spray velocity and pressure and to produce larger droplets. This can be used for applying systemic herbicides at pressures of 30 to 40 lb/in^2 (207 to 276 kPa) and contact herbicides at up to 80 lb/in^2 (552 kPa), to provide adequate foliar coverage while minimizing drift potential. Conversely, the twin-orifice flat-fan nozzle (Figure 5.4e) is designed for applications in which thorough foliar coverage is required. This nozzle tip has two orifices: one that directs a flat-fan pattern 30° forward and another that directs another flat-fan pattern 30° to the rear. The smaller droplets emerging from the two orifices provide thorough coverage of contact herbicides at pressures of 30 to 60 lb/in^2 (207 to 414 kPa), but with substantially increased drift potential.

Flooding nozzle tips (Figure 5.4f) also produce a fan-shaped pattern after the spray leaves the surface of the deflector; however, the spray pattern typically has higher spray volumes along the edges than at the center, producing the characteristic "spray horns." As a consequence, 100 percent overlap is required to ensure uniform application. The wide spray angle, ranging from 110° to 130°, allows wider spacings and lower boom heights. Tilting the nozzles 15° to 45° can also improve the uniformity of coverage. Flooding nozzles are especially useful for soil-applied herbicides or for combined application of fertilizers and herbicides. Large droplets are produced at operating pressures between 10 and 25 lb/in^2 (69 to 173 kPa). A modification of the flooding nozzle is a version that employs a preorifice to produce a more uniform tapered-edge pattern. The droplets produced by this nozzle are very large, making it a good choice for reducing drift potential at operating pressures of 15 to 30 lb/in^2 (104 to 207 kPa).

Solid-cone nozzles produce large droplets that are distributed uniformly within the conical spray pattern (Figure 5.4g). While these nozzles can be effectively used by orienting them perpendicular to the ground with a 10° to 30° spray pattern overlap, optimum uniformity along the boom is achieved by overlapping the patterns 30 to 50 percent with the nozzles oriented 30° to 45° from vertical. The solid-cone nozzle is well suited for soil-applied and systemic herbicides at operating pressures of 10 to 40 lb/in^2 (69 to 276 kPa); however, maximum reduction of drift potential occurs at 20 lb/in^2 (138 kPa) or lower.

Hollow-cone nozzles (Figure 5.4h) produce spray patterns with the herbicide mixture concentrated along the periphery of the cone, resulting in a saddle-shaped pattern of distribution in which there is less liquid in the center than along the edges. As a consequence, uniform application along the boom is difficult to achieve with these nozzles. Generally, hollow-cone nozzles produce the smallest droplet sizes and are best suited for applying materials requiring maximum foliar coverage and penetration, and where spray drift potential is not an important factor. Typical operating pressures for these nozzles is 40 lb/in^2 (276 kPa) or higher. The "Rain Drop" nozzle is a drift-reducing hollow-cone nozzle that produces fewer small droplets. Its operating pressure should be between 20

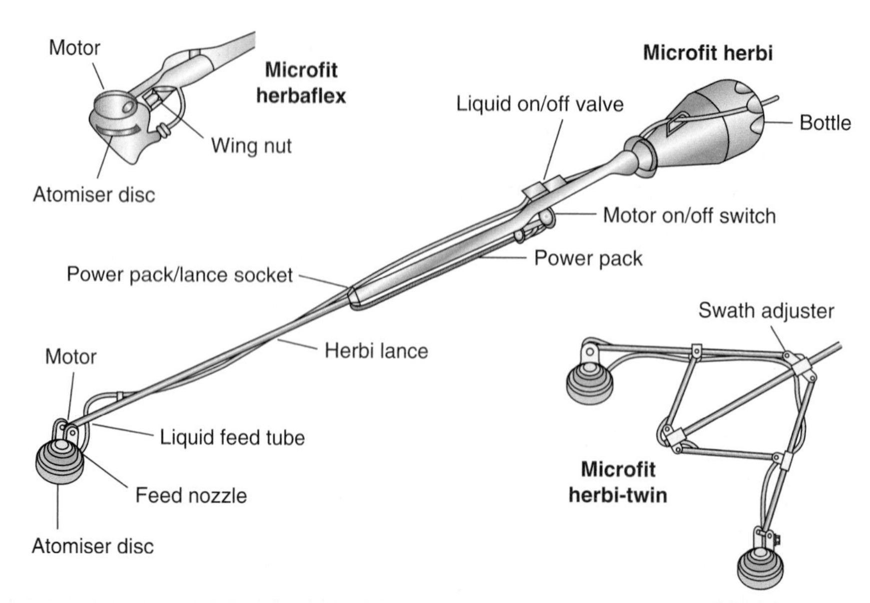

Figure 5.5 *The Microfit System, including the Herbaflex, Herbi, and Herbi-Twin. These applicators employ spinning nozzles to produce uniform droplets at ultra-low spray volumes.* (Courtesy Micron Sprayer Limited, Herfordshire, UK.)

and 60 lb/in^2 (138 to 414 kPa). To achieve optimum uniformity, these nozzles should be tilted 15° to 45°, with spray patterns overlapping 100 percent.

Specialized Applicators In addition to the nozzle types already discussed, an additional one worth mentioning is the spinning nozzle (Figure 5.5). This nozzle utilizes the centrifugal force of a spinning disk, driven by a variable-speed electric motor, to atomize the spray. The concentrated herbicide mixture—usually in an oil carrier—enters the spinning disk through a metered orifice, from which it is channeled through grooves to form individual droplets. Droplet size is determined by several factors: the rotational velocity of the disk, the flow rate and surface tension of the herbicide mixture, and the physical design of the disk. Because the resulting droplets are fairly uniform in size, measuring approximately 250 μm, devices using rotary nozzles are often called controlled droplet applicators, or CDAs. Adequate foliar coverage is obtainable at spray volumes as low as 5 gal/acre (47 L/ha) or lower. These devices are used mostly for postemergence broadcast or spot treatment with systemic herbicides for controlling weeds in turf (Figure 5.5).

Figure 5.6
Photograph of the Lawn Gun with interchangeable 45° nozzle tips. (Courtesy TeeJet® Image Library, www.teejet.com.)

Lawn guns employ full-cone shower-head nozzles that produce multiple streams of spray (Figure 5.6). They are used mostly by lawn-care operators for applying fertilizers and pesticides, including herbicides, to residential lawns. The gun is connected by long lengths of flexible hose to the spray truck, which houses the spray tank and pumping system. The spray mixture is applied by side-to-side movement of the gun by the operator while walking across sections of a lawn. Typical spray volumes employed with lawn guns range from 1 to 4 gal/1000 ft^2 (4.1 to 16.4 L/100 m^2) or 44 to 175 gal/acre (412 to 1637 L/ha). This is substantially higher than the spray volumes—typically, 20 gal/acre (187 L/ha)—employed with conventional sprayers. At such high spray volumes, the potential for spray drift onto adjacent ornamental plants is minimized because of the large droplets produced. Given the irregular distribution of turfgrasses and ornamental plants in residential landscapes, a properly operated lawn gun can be useful in providing accurate placement of the spray to lawn areas while avoiding ornamental plantings.

System Monitors and Controllers Historically, sprayer-mounted vehicles were operated mostly by feel, with the pressure regulator set and the throttle or accelerator positioned to maintain the proper traveling speed. Some spraying systems have employed ground-speed governors to maintain constant ground speed over variable terrain in an attempt to provide consistent chemical application rates. Today, highly sophisticated systems can sense and continuously display on a monitor various operating conditions, including pressure, travel speed, spray volume, field capacity (in acres or hectares per hour), amount applied, amount remaining in the tank, area covered, and distance traveled. Accurate monitoring offers the possibility of achieving greater effectiveness and efficiency (e.g., better weed control and reduced herbicide consumption) from spraying operations. Some systems employ speed-compensating controllers that sense the ground speed of the vehicle and adjust the flow rate from the sprayer in order to ensure a consistent application rate. Geographic positioning systems (GPS) can be used to acquire data from satellites to accurately determine the vehicle's locations on the ground and then plot the areas treated in spatial displays and relay application data to a digital storage unit for automated record-keeping. Where GPS is linked to a geographic information system (GIS) containing site-specific information from soil maps and/or soil test results that can be spatially displayed on a monitor, an option available is variable rate technology (VRT), in which the application rate of specific materials (e.g., preemergence herbicides, fertilizer

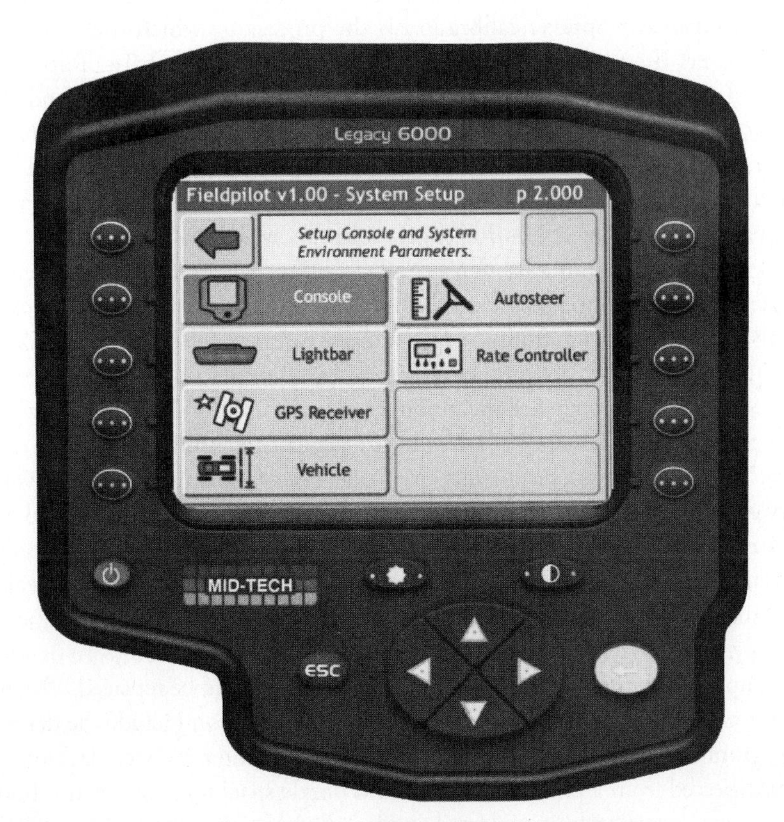

Figure 5.7 *Legacy 6000 Controller.* (Courtesy TeeJet® Image Library, www.teejet.com.)

nutrients) is based, in part, on variable field conditions. Another technology that shows promise is direct injection (DI) of chemicals into the water stream moving toward the nozzles, thus providing the capability of choosing materials on a site-specific basis while avoiding the problem of spray-tank chemical residues afterwards. Linking GPS, GIS, and DI, along with specialized sensors, could provide a truly site-specific approach to management, in that the materials applied could be based not only on historical records, but on direct and remote (satellite-based) observations of field conditions as well. In choosing from an increasing array of rapidly evolving technologies, one must balance the presumed advantages against the acquisition and maintenance costs of these technologies, along with the technical complexities of their use. Finally, no technology, no matter how sophisticated, can fully replace the proper selection and timing of herbicide application, accurate sprayer calibration, and common sense.

Spray Markers Important concerns in sprayer operations are skips and excessive overlaps. Skips occur where no spray has been received, often resulting in poor weed control and the necessity to make supplemental applications. Some overlap is necessary for ensuring that the entire treated area receives the proper rate of the herbicide. Depending on nozzle selection, this may involve overlapping by 15 to 50 percent of the last nozzle's spray width, but overlapping by more than these percentages means that the herbicide application rate will be greater than the intended rate, possibly resulting in phytotoxicity to nontarget turfgrasses or ornamental plants. A sprayer accessory that can be helpful for minimizing these problems is a foam or dye marker system that clearly shows the edge of the spray swath. The spray operator is then aided in determining exactly where the next pass should be made in order to avoid skips and excessive overlaps. An alternative would be to include a turf colorant to the spray mixture to show previously treated areas.

Sprayer Calibration Sprayer calibration is the process by which the rate at which a sprayer delivers the spray mixture, or its liquid carrier, to the site of application is determined. Once the calibration process has been performed properly and the spray volume accurately determined, the appropriate dilution of the herbicide formulation in the carrier liquid can be calculated. Before the calibration process is attempted, however, a spray distribution test should be performed to ensure that the nozzles are delivering the spray uniformly along the entire length of the spray boom.

A spray distribution test has three components: nozzle uniformity, nozzle flow, and boom height. To determine nozzle uniformity, check to ensure that all of the nozzles are identical; that is, they all are from a single manufacturer and bear the same part number. To determine nozzle flow, collect the output from all nozzles simultaneously in containers; then, measure and compare the amounts collected to ensure that the output from each nozzle is within 5 percent of the average. Finally, after adjusting the boom to the height suggested by the manufacturer for that particular nozzle, operate the sprayer with water on a dry concrete or asphalt surface and observe the drying pattern. Differential drying of the sprayed area indicates nonuniform application. If wide streaks appear directly under the nozzles, the boom may be too low and should be raised. If narrow streaks occur between the nozzles, the boom height may be too high or too low and should be adjusted accordingly. These patterns may also indicate that the nozzles are too worn or were damaged by improper cleaning, in either case of which they should be replaced. When choosing new nozzle tips, factors to consider in making the selection include the desired spray volume, ground speed, and spray width. Nozzle manufacturers provide data on the spray volumes expected from a particular nozzle (and nozzle spacing) and at various operating pressures. Once new nozzles are selected and installed, the sprayer should be recalibrated.

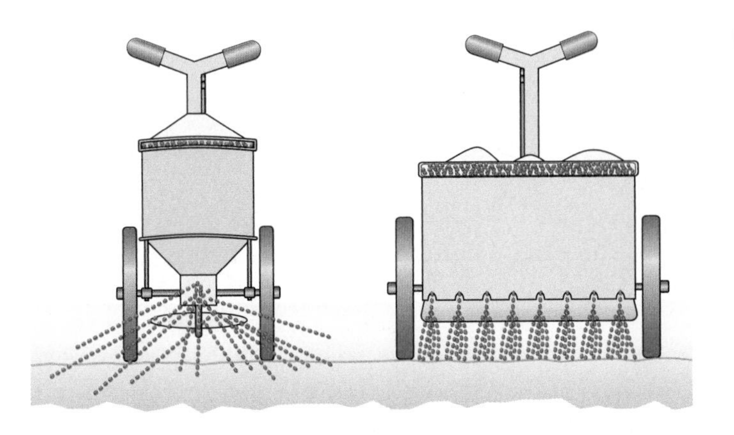

Figure 5.8
Illustration of two spreader types: broadcast (left) and drop (right).

While there are many ways of calibrating a sprayer, the following procedure is suitable:

- While operating the sprayer at the desired operating pressure, collect the output from a single nozzle for one minute, using a container with graduated markings in fluid ounces (fl oz) or milliliters (mL). The resulting measurement is the flow rate in fl oz/min (or mL/min).
- Convert this flow rate from fl oz/min to gal/min by multiplying by 128 (or from mL/min to L/m by multiplying by 1,000).
- Determine the spraying speed by timing the movement of the sprayer from one point to another on the ground. The speed in mi/hr can be calculated by dividing the distance (ft) by the number of seconds required to travel that distance, then multiplying by 0.67. For example, if 17 seconds is required to cover 100 ft, then the speed of the sprayer equals 100/17 * 0.68 = 4 mi/hr (where 0.68 is calculated by dividing the number of seconds in an hour [3,600 s/hr] by the number of feet in a mile [5,280 ft/mi]). The speed in m/s can easily be calculated by dividing the distance traveled in meters by the seconds required to cover the distance; thus, if a distance of 30.4 m (100 ft) is covered in 17 seconds, the speed of the sprayer is 1.79 m/s, which is equivalent to 4 mi/hr.
- Measure the nozzle spacing (W) on a boom sprayer (or the sprayed width of a single-nozzle sprayer) in inches (in) or centimeters (cm).
- Calculate the sprayer application rate or spray volume in units of gal/acre by the formula gal/acre = gal/min * 5,940 ÷ mi/hr * W (in). To calculate this in units of L/ha, use the formula L/ha = L/min * 16,660 ÷ (m/s * W [cm]).

Spreaders

Spreaders are designed for applying dry granular formulations of herbicides, fertilizers, and other materials. The two basic types of spreaders are the broadcast and drop spreaders (Figure 5.8). The basic components of a spreader include a hopper for holding the granular formulation, one or more openings (feeder gate) at the bottom of the hopper, an on-off lever or other device for opening and closing the feeder gate, and a feeder-gate control for setting the sizes of the openings and determining the flow rate. Some spreaders may also have an agitator to generate a consistent flow of granules through the feeder gate.

Broadcast Spreaders Broadcast spreaders employ one of two mechanisms for distributing granular materials: a rotary impeller or an oscillating spout. The granules fall through the feeder gate at the bottom of the hopper onto the impeller or spout, which distributes them across a wide swath. As the throw distance is a function of both the throw force generated by the impeller or spout and the weight of the granules being

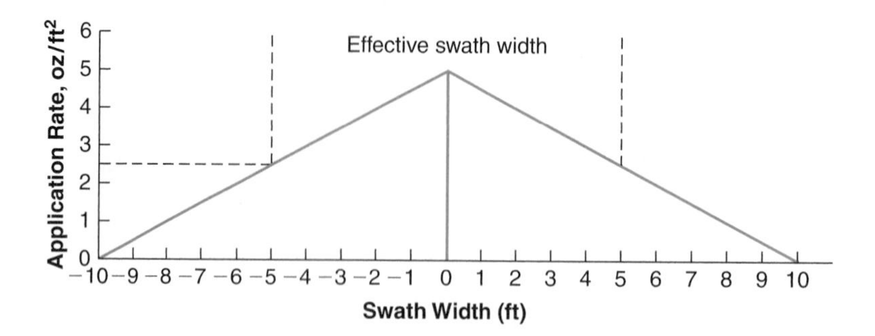

Figure 5.9 *Distribution plot of the application rate of a granular formulation across the entire swath width. The effective swath width is the region in which the application rate is half of the maximum rate.*

thrown, a formulation in which the granule size varies may be differentially distributed, with the larger, heavier granules thrown longer distances. Even with uniform granules, the distribution pattern is not likely to be uniform, as the application rate is generally highest at the center of the swath and lowest near the edges; however, the distribution pattern of a particular broadcast spreader is dependent upon many factors, including the characteristics of the impeller or spout (height, angle, speed, and shape), ground speed, physical characteristics of the granular formulation (density, shape, and size), and atmospheric conditions (temperature and humidity).

The distribution pattern of a broadcast spreader can be determined by the following procedure:

- Place a series of shallow containers in a paved test area along a line perpendicular to the direction of travel and covering the entire swath width of the spreader.
- Operate the spreader several times in the same direction of travel to apply the granular herbicide formulation at the recommended spreader setting.
- Measure the volume or weight of the material in the containers.
- Plot the data to develop a visual picture of the distribution pattern.

The *effective* swath width is twice the distance from the center of the swath out to a point at which the application rate is half that measured at the center (Figure 5.9). Operating the spreader using the effective swath width can provide up to 100 percent overlap, which should be sufficient to ensure reasonably uniform distribution of the granular formulation over most of the treatment area. Some broadcast spreaders have a feature that enables the operator to cut off one side of the distribution pattern. This is desirable when edging along the borders of the treatment area. If the spreader's application rate and pattern are not changed when this feature is employed, then the uniformity of application can be extended to the entire treatment area, including the borders, by overlapping the last effective swath width with this feature employed. If the spreader's application rate and pattern are different when this feature is employed, make successive adjustments to the feeder gate until they are the same or as close as possible. This feature is also useful in providing a *header strip* at opposite ends of the treatment area. The spreader should then be operated back and forth between header strips, with the spreader's on-off lever turned off once a header strip has been reached.

Drop Spreaders All drop spreaders are wheel-mounted and have a hopper that is as wide as the spreading width. The agitator runs the full length of the hopper. Drop spreaders provide a more accurate and uniform distribution of granular material than that obtainable from broadcast spreaders, but they are less efficient and, with respect to push-type units, require more effort to operate. As there is no pattern feathering, the effects of skips and overlaps may be especially noticeable. And, because the feeder gate is close to the ground, turf clearance is sometimes a problem, especially at high mowing heights.

Spreader Calibration Spreader calibration is the process of determining the rate at which a spreader delivers a granular formulation to the site of application. Unlike sprayers, which can be calibrated for the water carrier containing a large variety of herbicides and herbicide formulations, a spreader must be calibrated for each granular herbicide formulation used. This is because each granular formulation is unique with respect to the size, shape, and density of the individual granules constituting the formulation. Also, a spreader's calibration should be rechecked on a regular basis to ensure that the feeder-gate setting established earlier for a particular granular formulation continues to provide the desired application rate. This is especially important where spreaders are subjected to abuse in their handling and/or operation.

The calibration process must be conducted in a fashion that is consistent with the way the spreader will be used to apply the formulation to the site of application. This is especially important with respect to operating speed, as the flow rate of material through the feeder gate of broadcast spreaders is constant (due to *gravimetric* metering), regardless of operating speed. As a consequence, the faster the operating speed, the less granular formulation will be applied per unit area. While the flow rate from drop spreaders does increase with faster operating speed, the increase is not entirely volumetric (i.e., delivering the same rate of formulation per unit area, regardless of operating speed); therefore, it will vary depending on the operating speed, but not as much as with broadcast spreaders.

A procedure that can be used for spreader calibration is as follows:

- Place a weighed amount of a particular granular formulation in the hopper.
- Select the feeder-gate control to a setting suggested by the manufacturer. (If there is no suggested setting, select one for testing.)
- Apply the material to a marked-off area (e.g., 1,000 ft^2 or 100 m^2).
- Weigh the amount of granular formulation remaining in the hopper.
- The difference between the initial and remaining weights is the amount of formulation applied to the measured area (1.84 lb/1,000 ft^2) .

For example, a 2.5 G herbicide formulation is to be applied at the rate of 3 lb a.i./acre (3.36 kg a.i./ha). The amount of this formulation applied to a 1,000-ft^2 (93 m^2) test area was measured at 2 lb (0.9 kg); therefore,

$$2 \text{ lb G}/1{,}000 \text{ ft}^2 * 2.5 \text{ lb a.i.}/100 \text{ lb G}^* \ 43{,}560 \text{ ft}^2/\text{acre} = 2.2 \text{ lb a.i./acre}$$

$$0.9 \text{ kg G}/94 \text{ m}^2 * 2.5 \text{ kg a.i.}/100 \text{ kg G}^* \ 10{,}000 \text{ m}^2/\text{ha} = 2.4 \text{ kg a.i./ha}$$

Because the application rate of this granular formulation was less than desired (i.e., 2.2 instead of 3 lb a.i./acre, 2.4 kg a.i./ha), a higher feeder-gate control setting will be needed to get to the desired application rate. The tests should be continued until the proper setting has been determined.

With some drop spreaders, a catch device can be installed just below the feeder gate to intercept the granular formulation as it exits the spreader. If such a device is available, the calibration procedure can be simplified by simply intercepting the granular formulation as the spreader is operated over the test area and weighing it directly to determine the amount that *would* have been applied.

The reader may wish to access Christians and Agnew (2000) for a more detailed discussion of calibration procedures.

Spreader Maintenance The usable life of a spreader can be extended by implementing a proper maintenance program. Corrosion of metal parts and caking of formulation material are the principal concerns. A spreader should be washed and hand dried (or allowed to dry in the sun) after each use. Care should be taken to remove all caked material, perhaps with hot water and a brush, or a power washer, to break it loose. Afterwards, all moving parts should be oiled in accordance with manufacturer's instructions. Dry graphite lubricants can be substituted where oil might attract grit, causing accelerated wear of some gears and other moving parts. Spreaders should be checked on a regular basis, and an inventory of replacement parts should be maintained so that worn or damaged parts can be replaced as needed to ensure proper operation. Spreaders should be stored in protected areas where they can be kept clean and dry, preferably away from direct sunlight or other closely positioned equipment, to minimize the potential for damage.

unit three
Weed Control

CHAPTER 6 **Control of Turfgrass Weeds** CHAPTER 7 **Turfgrass Plant Growth Regulators**
CHAPTER 8 **Weed Control in Landscapes and Nonturf Areas**

Weed control is any practice designed to eliminate weed species from a population or community of turfgrasses or ornamental plants. While weed control practices could certainly involve the use of one or more herbicides, they also could include mulching an ornamental planting bed, selecting weed-free turfgrass seed, and modifying environmental conditions so that they are more favorable for the growth of a particular turfgrass or ornamental plant. Some weeds are indicators of unfavorable growing conditions. For example, large populations of knotweed (*Polygonum aviculare* L.) may occur where severe soil compaction limits turfgrass growth. Yellow nutsedge (*Cyperus esculentus L.*) often occurs where soil drainage is poor. Broadleaf plantain (*Plantago major* L.) and buckhorn plantain (*Plantago lanceolata* L.) usually reflect low soil fertility or inadequate fertilization. And the presence of red sorrel is indicative of acid soil conditions. Effective measures taken to correct these conditions can go a long way toward preventing these weeds from becoming established and competing with landscape plants.

Reasons for classifying a plant as a weed are numerous. In addition to being unsightly, weeds compete with desirable plants for light, oxygen, soil nutrients, soil moisture, carbon dioxide, and space. Weeds also act as hosts for pests such as plant pathogens, nematodes, and insects. Certain weeds cause allergic reactions in humans due to their pollen or their volatile chemicals. When seed is sold, the definition of a weed can become a legal matter.

Probably the most undesirable characteristic of weeds in turf and ornamentals is the disruption of plant-stand uniformity. Different leaf width and/or shape, variations in growth habit, and/or different color contribute to unsightliness. For example, many broadleaf weeds such as dandelion, plantains, and pennywort have leaf widths different from turf or ornamentals. Goosegrass, pathrush, smutgrass, and dallisgrass tend to form clumps or patches that also disrupt plant-stand uniformity. In addition, large clumps are difficult to mow effectively, and they increase maintenance problems. Plant stand uniformity also is

Figure 1
Weed seedheads disrupt the smoothness and trueness of a turf's surface.

disrupted by weed seedheads. Annual bluegrass, for example, is largely unnoticed on putting surfaces until prolific seedheads appear in late winter and spring. Seedheads also disrupt the smoothness and trueness of the turf's playing surface (Figure 1).

Plant color is another factor in determining the potential of a weed problem. The lighter green color typically associated with annual sedge often distracts from the playing surface. Bahiagrass and *Poa trivialis* often have different color when grown in combination with other turf species.

The next three chapters provide a comprehensive coverage of weed-control practices in turf (Chapter 7) and ornamentals (Chapter 9). And because nonauxinic plant-growth regulators have become so popular for selectively suppressing weeds—especially annual bluegrass (*Poa annua* [L.])—in turfgrass communities, Chapter 7 was included to cover these materials as well.

*6
Control of Turfgrass Weeds

Weeds often are the result of a weakened plant stand, not the cause of it. Knowing a plant's weaknesses often helps to illuminate the reason for weed encroachment within an area. Lack of stand density and areas devoid of growth allow light to the soil surface, promoting weed encroachment (Figure 6.1). Weak turf and bare areas exist because of (a) improper turf species for a given area; (b) damage from pests such as diseases, insects, nematodes, and animals; (c) environmental stresses such as excessive shade, drought, heat, cold, and poor drainage; (d) improper turf management practices such as misuse of fertilizer and chemicals, improper mowing height or frequency, and lack of proper soil aeration; and (e) physical damage and compaction from concentrated or constant traffic. Unless the factors contributing to turf decline are corrected, continued weed encroachment should be expected.

Controlling weeds in landscape plantings is much more complex than controlling weeds in turfgrass. Stands of turfgrass are generally made up of one or two perennial grass species. These species can be intensively managed to promote their growth, to the competitive disadvantage of weeds. When weeds do encroach, calibrated herbicide applications are relatively easy to make in the open areas in which turfgrasses grow. Many herbicides are available to selectively remove weeds from turfgrasses.

Landscape plantings often include a much greater variety of plant materials. Annual and perennial flowers, ornamental grasses, low-growing ground covers, and a variety of shrubs and trees often grow in the same landscape bed. It is difficult, if not impossible, to find an herbicide labeled for use in some landscape beds containing a wide variety of plant material. For example, selectively removing perennial broadleaf weeds from herbaceous perennial flowers is extremely difficult.

Another problem in landscape beds is the difficulty in making uniform, calibrated herbicide applications. Most landscape beds have irregular shapes and contain plants of various sizes spaced irregularly throughout the beds. It is not possible to walk a straight line, maintaining a uniform speed, through most landscape beds. Uniform broadcast granular applications of herbicide are difficult because of the canopies created by shrubs such as rhododendrons and viburnums. The herbicide granules tend to accumulate at the drip line of the shrubs.

Because of all of the problems associated with controlling weeds in landscape plantings, it is especially important that detailed weed control programs be developed and strictly followed for each landscape bed.

Weed control techniques include prevention, as well as cultural, mechanical (or physical), biological, and chemical control.

Figure 6.1
The first line of defense against most weeds is to provide a dense thick turf stand to minimize the amount of sunlight reaching the soil surface.

PREVENTION

Weeds are prolific seed producers. A single plant can produce from 1,000 to 500,000 seeds (Table 6.1). Seed can easily be transported to nearby turf and ornamental areas by way of wind, water, machinery, and animals. Prevention is avoiding the introduction of weeds within an area. There are national, state, and local prevention efforts against the

Table 6.1
APPROXIMATE NUMBER OF SEEDS PRODUCED PER PLANT (COMPILED FROM ROSS AND LEMBI, 1999; AND EMMONS, 1984)

Weed	Seeds Produced per Plant
Annual bluegrass	2,000
Black medic	2,400
Broadleaf plantain	36,000
Common lambsquarters	72,000
Common purslane	52,000
Curly dock	40,000
Dandelion	12,000
Giant foxtail	10,000
Goosegrass	1,35,000
Knotweed	6,000
Redroot pigweed	1,17,000
Sandbur	1,000
Speedwell	2,900
Witchweed	5,00,000
Yellow nutsedge	2,400

introduction and spread of weeds. A local preventive program is one of the best methods of avoiding future weed problems. Many of these methods are commonsense approaches that ensure sanitary conditions and minimize weed introduction. Some of these methods include the use of weed-free turf seeds, stolons, sprigs, plugs, or sod. The use of clean mulch, container soil mixes, landscape plants, and topdressing material and the avoidance of manure are also examples of preventative methods.

Clean Seed

Plant only cleaned seed. Cleaned seeds have most of the weed seeds removed. The Federal Seed Act of 1939 regulates the seed trade in order to protect consumers from mislabeled or contaminated crop seed. The following information is required by this act to be listed on seed labels in interstate commerce:

1. Percentage of pure seed of the named crop
2. Percentage of other crop seed
3. Percentage of weed seed
4. The name and occurrence of noxious weed seed

The following is an example of a typical turf seed label:

Brand name: 'XYZ' Tall Fescue Seed
98.75% Pure Seed
Other Ingredients:
0.00% Other Crop Seed
1.00% Inert Matter
0.25% Weed Seed
85.00% Germination
Tested: 1/09
Lot 0001-A
Net Wt. 50 lbs. (23 kg)

Noxious weed seeds are prohibited from entering the United States by the Federal Seed Act. Also prohibited from entering the country are crop seeds containing in excess of 2 percent weed seeds of all kinds.

Vegetative Materials

It is important to purchase weed-free vegetative materials (Figure 6.2). Weeds are not always evident in a commercial production field, but become more obvious when cultural practices change after planting.

Figure 6.2
Using clean planting stock prevents weed seed introduction.

The Turfgrass Producers International has proposed maximum weed infestation numbers in nursery and field-grown grass. Nursery-grown sod is considered weed free if fewer than five weeds are found per 100 sq. ft. (9.3 m^2). No more than 10 weeds are allowed per 100 sq. ft. (9.3 m^2) for field sod to be considered a weed-free product. Individual states often have specific limits of hard-to-control weeds. For example, limits of "off-type" bermudagrass occurrence are in place for hybrid bermudagrass production fields. In addition, sod is considered unacceptable if common bermudagrass, quackgrass, or bromegrass is present.

Once established, certain weeds cannot selectively be controlled with herbicides. For example, common bermudagrass and tropical signalgrass are primary

Figure 6.3 *Sanitation practices, such as washing maintenance equipment before entering a weed-free turf area, help minimize weed dispersal.*

weed problems in St. Augustinegrass sod production. This is the result of the unavailability of selective herbicides and unsatisfactory control from nonselective herbicides. Inspection of planting material for weeds should occur before purchase.

Sanitation Practices

Sanitation practices include the prevention of weed seed movement by mechanical or human means. Mechanical devices, such as mowers or cultivators, pick up weed seeds easily and transport them to adjacent areas. Annual bluegrass, crabgrass, chickweed, and goosegrass are weed seeds transported and deposited by these devices, especially when shoes and maintenance machinery are wet (Figure 6.3). Mowers should be rinsed before being transported from infested to weed-free areas during weed-seed production season. Weeds also should not be allowed to go to seed. Seeds are blown, washed, or carried into high-maintenance areas.

Frequently, weeds are introduced through the addition of contaminated topsoil. Most weed seeds are located near the soil surface and can remain viable for an extended period (Table 6.2). When contaminated topsoil is used for topdressing, prolific weed-seed germination and subsequent emergence often results. Therefore, materials used for soil modification or topdressing should be free of viable weed seeds

Table 6.2
LONGEVITY OF WEED SEEDS BURIED IN THE SOIL (MODIFIED FROM ROSS AND LEMBI, 1999)

Weeds	Years Viable
Annual bluegrass	6+
Quackgrass	1 to 6
Cocklebur	16
Foxtail	20
Johnsongrass	20
Canada thistle	21
Common lambsquarters	40
Redroot pigweed	40

Figure 6.4
An often unrecognized source of weed seed is open irrigation ditches or water storage sites.

or vegetative propagules through fumigation or heat treatment prior to use. These tasks require time and attention to matters that seem to have little short-term impact but, in the long run, are very worthwhile.

In ornamentals, selecting a growing medium that is naturally and consistently weed free avoids many weed problems in container-grown plants. Vermiculite and perlite are weed free because of their production methods. Most peats contain very few, if any, weed seeds. Properly composted organic matter will be weed free because the high temperatures reached during composting kill weeds and their seeds. Composted or aged bark, sawdust, leaves, or sewage sludge may also be used.

Steam pasteurization is an effective way of controlling weeds and most soilborne insects and pathogens. Media should be heated between 160° and 180°F (71° and 82°C) and held for 30 minutes. Steamed media should be covered or used within about a week to prevent reinfestation while stored. Portable steam generators are available for steaming soil outdoors.

Irrigation systems fed by open ditches, canals, and ponds also contribute to the spread of weeds. One study found more than 130 weed species in irrigation water. The researchers concluded that irrigation deposits more than 35,000 seeds per acre (14,165 ha^{-1}) (Kelley and Burns, 1975). Therefore, production managers should strive to keep irrigation sources and embankment areas weed free (Figure 6.4). Areas periodically flooded by streams, canals, or rivers also are subject to weed infestation.

Another source of weed seed is from unmowed adjacent areas such as fence rows, roadsides, and open fields (Figure 6.5). Cultural practices minimizing seed production, or the use of herbicides, should be a regular part of a maintenance program.

Figure 6.5
Not allowing weeds to produce seedheads also helps reduce weed spread.

CULTURAL

Cultural practices promoting vigorous, dense turf are the most important and least recognized means of preventing weed establishment and encroachment (Table 6.3). Since high light intensity is required for germination of some weeds such as crabgrass and goosegrass, cultural practices increasing turf density (competition) will prevent

Table 6.3
PRE-PLANT NONSELECTIVE WEED CONTROL *(REFER TO HERBICIDE LABEL FOR SPECIFIC USE LISTING)*

Common Name	Trade Name(s)	Soil Fumigant	Soil Residual/ Root Uptake	Foliar Uptake	Contact Activity
Ammoniated soaps of fatty acids	Quick Fire	—	—	—	Y
Bromacil	Acti-Cil, Hyvar, Opti-Kill,	—	Y	Y	—
Bromacil + diuron	Krovar	—	Y	Y	—
Dazomet	Basamid	Y	—	—	—
Diquat	Reward, Aquatrim II	—	—	—	Y
Glufosinate-ammonium	Finale, Derringer	—	—	Y	Y
Glyphosate	Gly-Flo, Prosecutor, Razor, Roundup Pro & Pro Dry, Trailblazer, Touchdown Pro, + others	—	—	Y	—
Glyphosate + diquat dibromide	QuickPRO, Prosecutor Swift Acting			Y	Y
Imazapyr	Arsenal	—	Y	Y	—
Imazapyr + diuron	Sahara	—	Y	Y	—
Metam sodium	Metam CLR, Vapam HL, Soil Prep	Y	—	—	—
Methyl bromide	MB 98, MBC, Dowfume MC-2, Brom-o-gas, Profume, Terr-o-gas	Y	—	—	—
Pelargonic acid	Quik, Scythe	—	—	—	Y
Prometon	Pramitol, Spot	—	Y	—	—
Prometon + 2,4-D	Vegemec	—	Y	Y	Y
Tebuthiuron	Spike	—	Y	—	—

light from reaching the soil surface. Exclusion of light from the soil surface also delays germination of weed seeds in spring, since the soil surface is better insulated and remains cooler. Soil fertility, soil aeration, and soil moisture should therefore be maintained at optimal levels for turfgrasses.

Mowing and Fertilization

Mowing is not used as a method of weed control in landscape plantings, but is a form of weed management in turf. It kills weeds, such as tree seedlings, that cannot survive short mowing heights. Some weeds, like redroot pigweed that normally grow several feet tall, can adapt to low mowing heights and still produce seed.

An example of the cultural practices, mowing height, and level of fertilization that influence the occurrence of weeds involves smooth crabgrass in a red fescue lawn. Red fescue mowed at 2.2 in (5.6 cm) has better turf coverage and less crabgrass occurrence (4 to 14 percent) than turf mowed at 1.25 in (3.2 cm) (18 to 39 percent). The yearly addition of nitrogen fertilization at 1 or 2 lb N/a (49 or 98 kg N/ha) further strengthened the turf stand and thus reduced crabgrass occurrence. Heavy weed infestation and poor turf quality resulted at a low fertility rate and an excessively low 0.75 in (1.9 cm) of mowing height.

The same principles apply to landscape beds containing ornamental groundcovers such as ivy, vinca, and pachysandra. These resist weed invasion if properly maintained. During establishment is the most critical time to initiate a weed control program. The site should be prepared properly by eliminating weeds and loosening the soil prior to planting. Only healthy, actively growing stock should be used and then planted at recommended spacings. Organic mulch, fabrics, and preemergence herbicides should be used at this point to prevent weeds from germinating in the groundcover.

Soil nutrient levels also may favor one plant species over another. For example, phosphorus fertilization repeatedly used at high rates increased annual bluegrass populations in bentgrass (Goss et al., 1975). Bentgrass apparently is better able to absorb P in low P-containing acidic soils and can out-compete annual bluegrass under these conditions (Kuo et al., 1992). The application of sulfur to acidify soil, or limiting P application, is suggested to decrease annual bluegrass encroachment (Goss et al., 1975). Bentgrass also has a competitive advantage over annual bluegrass when nitrogen fertilization is withheld (Dest and Guillard, 1987). A range of soil pH between 5.8 and 7.2 is recommended to favor a dense and competitive turf (Turgeon, 2005), although weed growth varies with different soil pH.

Irrigation Practices

Maintaining proper soil moisture through irrigation and soil drainage encourages desirable vigorous plant growth. Over-irrigation and poor surface and subsurface soil drainage result in low soil-oxygen levels. Soil compaction also reduces oxygen diffusion and restricts rooting. Turf density decreases with compaction, and weeds such as annual bluegrass, goosegrass, prostrate knotweed, and various sedges often invade because they can tolerate these conditions (Figure 6.6).

Figure 6.6
Improper turf growing conditions often favor weed growth. Shown is goosegrass infestation in compacted turf adjacent to a cart path.

Pest Damage

Pest damage also decreases stand density and allows weed encroachment. Insects, diseases, and nematodes are common pests that affect stand density and should

Figure 6.7
Pests that thin or damage turf allow weeds to invade. Shown is an infestation of crabgrass in damaged bermudagrass from the disease called spring dead spot.

be controlled when possible. If other pests are not controlled, weeds easily infest damaged areas (Figure 6.7).

Specific weed occurrence may provide insight to specific pest presence. For example, weeds commonly associated with nematode-thinned turf include prostrate spurge, prostrate knotweed, and Florida pusley (McCarty, 2005).

Mulches and Synthetic Covers

Mulches are an integral part of almost all landscape plantings. They are used for aesthetic purposes, serving as a unifying element in landscape design, and they provide several cultural benefits to plants. Mulches exclude light and thus moderate soil temperature, conserve soil moisture, provide nutrients as they decompose, and limit weed growth (Figure 6.8). Commonly used mulching materials are straw, wood chips, sawdust, grass clippings and various synthetic covers. These materials generally are useful only on small, specific areas. If organic mulches are used, a minimum depth of 2 to 3 inches (5 to 7.6 cm) is required to prevent weed germination. Care must be taken to prevent these materials from moving into mowed areas, as they may become a hazard to equipment operation and personnel.

Figure 6.8
Mulch in flower and ornamental beds helps retain moisture and prevent weeds.

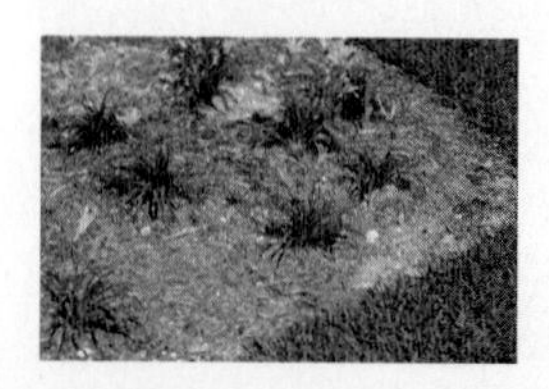

Fire

Fire also is used to remove undesirable brush and weeds. For example, bermudagrass sod production fields can be burned in late winter or early spring just prior to green-up. This burns the weed tops and reduces thatch and mat accumulation. It also controls some insects and disease. Heat generated from burning, however, may break seed dormancy of certain weeds. This may be true because (1) less competition from dominate vegetation results in greater fluctuations in day and night soil temperatures while more light reaches the soil surface; and (2) the removal of plants likely has allopathic effects (Klingman and Ashton, 1982). In most areas, open burning now is regulated or is restricted by local governmental agencies. For example, burning was a long-standing practice for many seed production fields in the northwestern United States. Since the ban on burning, certain weeds, such as annual bluegrass and *Poa trivialis*, have become more problematic.

Miscellaneous Control Techniques

Numerous miscellaneous products and techniques have been proposed for control of very specific weeds or in unusual situations. Examples include using rock salt in seashore paspalum for selective broadleaf and grass weed control (suppression), soaps and fatty acids, desiccates such as lime and baking soda, heavy metals, and acetic acid (vinegar) for broadleaf weed control in cool-season grasses (McCarty and Tucker 2005).

Perhaps the most widely available and currently used of such products is corn gluten meal, a by-product of the wet-milling process, which has long been used in animal feeds. Over the last decade, it has been discovered to inhibit root development in certain germinating weed seeds (Polomski, 1998). Corn residues and extracts can inhibit shoot and root elongation of selected plant species including wheat, sorghum, and lettuce (Chou and Patrick, 1976). The protein fraction of the corn grain has been linked to the inhibition of root growth and has potential to be a natural herbicide in turfgrass (Larson, 1997; Christians, 1993b). Corn gluten meal provides only preemergence activity (Christians, 1993a). Field work indicates that 20 lbs per 1,000 sq. ft. (975 kg/ha) will reduce crabgrass in Kentucky bluegrass turf by 50 to 60 percent the first year (Christians, 1993a). Timing is important for best control, since microbes will degrade the material very rapidly. Corn gluten meal needs to be applied as close to predicted germination as possible. Moisture is needed to activate the material, but control following activation has been best under extended dry conditions (drought). The effectiveness of this treatment is due primarily to plants being unable to survive with a root system stunted by the corn gluten. A reduction in root development by 275 percent and a decrease in shoot length by over 50 percent were reported on black nightshade, lambsquarters, creeping bentgrass, curly dock, purslane, and pigweed when preplant-incorporated treatments of corn gluten meal at 3 lbs per acre (3.36 kg/ha) were applied (Bingaman and Christians, 1995). If wet conditions persist after application, damaged plants can recover and resume rooting (Christians, 1993a). Weeds controlled by corn gluten meal include crabgrass, dandelions, smartweed, pigweed, purslane, foxtail, barnyardgrass and lambsquarter (Larson, 1997).

MECHANICAL (PHYSICAL) CONTROL

Mechanical control of weeds involves hand pulling or various types of tractor-powered tillage operations. The most commonly used mechanical weed control methods in turf and ornamentals are tillage (preplant), hand pulling, hoeing, and mowing.

Tillage

Tillage, or cultivation, usually is practiced before plant establishment. Weeds are destroyed by mechanical disruption, which removes them from the soil, disturbing their root systems, causing desiccation, and smothering or burying tender tissue. Depleting stored food reserves and reducing soil reserves of vegetative propagules also destroy weeds.

Other potential benefits from proper tillage are increased soil aeration and water penetration, breakage of surface crusts and soil clods, surface smoothing for planting, and the incorporation of surface-applied fertilizer, liming material, and soil amendments

Figure 6.9 *Tillage prior to planting helps distribute soil amendments, break up compacted soils, and destroy existing weeds.*

(Figure 6.9). Best results from tillage are obtained when the soil surface is dry, so that the disturbed weeds are subjected to desiccation. When soils are too wet or water is applied shortly after tillage, the disturbed plants are more likely to survive. Tillage of wet soils also increases the incidence of clods, crusts, and compaction layers, which may interfere with subsequent planting and turf-seed germination.

Implements used to work the soil commonly are referred to as *primary* and *secondary* tillage equipment. Primary equipment is used to break and loosen the soil at depths of 6 to 36 inches (15 to 91 cm). Moldboard, disk, rotary, chisel, and subsoil plows are examples of primary tillage equipment. These pieces are designed to break up hardpan layers or compacted zones. Secondary equipment is used to work the top 6 inches (15 cm) of soil. Power-driven tillers, harrows, cultivators, tandem disks, and rotary hoes are examples of secondary tillage equipment. This equipment provides control of weeds, prepares the seedbed, and incorporates soil amendments.

Tillers normally are powered by a tractor's power take-off (PTO) or an auxiliary mounted engine. Nonpowered equipment usually is pulled or dragged through the soil. A disk harrow, or a nonpowered unit, is perhaps the most commonly used tillage implement in preplant turf operations. The disk harrow provides both cutting and burying of weeds as a method of control.

Repeat tillage usually is necessary for control of perennial weeds and the continued emergence of annuals. Multiple growing seasons of intensive tillage normally are required to deplete the vegetative reproductive structures of most perennial weeds. Tillage should be repeated over a two- to-three-week interval and continued through the end of the growing season in order to deplete the underground carbohydrate food reserves of tubers, rhizomes, and bulbs. Tillage should be performed when the effects of water and wind on soil erosion can be avoided.

Though cultivation is an effective way of controlling weeds, there are several problems associated with it. Cultivating is time consuming, and to maintain a satisfactory level of control, this must be repeated six to eight times through the season. Weeds must be cut down before they grow too large, and immediately after a cultivation, additional weed seeds may germinate.

Another problem associated with cultivation is the potential spread of perennial weeds. Cultivation may break apart underground structures of perennial weeds, such as roots (mugwort), rhizomes (quackgrass), or clusters of nutlets (nutsedge), spreading them throughout the landscape bed. It is critical that these weeds be killed with herbicides prior to cultivation.

Disadvantages of tillage include the fact that the exposed soil is subjected to wind and water erosion. Also, if weeds are mature when tilled, their seeds are buried and become a future source of weeds. Other possible disadvantages of tillage are scheduling to coincide with proper soil moisture and the cost of tractors, tillage implements, labor, and fuel.

Hand Pulling, Hoeing, and Rouging

Manual weed control often is performed by hand pulling, hoeing, and rouging. Manual weed control is not widely practiced and is generally impractical on large areas. However, because of herbicide-sensitive constraints on certain small turf areas and on landscapes with sensitive ornamental species, manual weed control practices still are used.

Hand pulling and hoeing effectively control annual and biennial seedling weeds. These practices are less effective on established perennial weeds, because underground reproductive parts often remain in the soil and can regenerate. Hand pulling, hoeing, and rouging are methods everyone involved in landscape management would like to see become obsolete, but they probably never will. Though mulch and herbicide combinations can provide excellent control of most weeds, total weed control is difficult to achieve. The only way to achieve the level of weed control required in highly visible landscape sites is to finish the job with some hand-pulling to eliminate the weeds that escape the preventive measures.

After a landscape bed is established, cultivation with heavy equipment is almost impossible because the density of plants leaves insufficient room to operate equipment. Handheld hoes are often used to eliminate weeds from landscape plantings. Specialized hoes, called scuffle hoes, are made of a ring of steel strap rather than the steel plate found on the traditional hoes. The steel strap on the scuffle hoes is sharpened, and it slides along the ground right at ground level or slightly under it, cutting off the weeds at the ground line with minimal effort and minimal disturbance of soil around the plants.

Rouging often is used in combination with hand pulling and involves the use of a special implement that has a hooked and sharpened metal blade end. The implement is pushed into the soil to sever the roots, and the plant then is pulled from the soil. Rouging still is widely used to remove weeds such as goosegrass from creeping bentgrass golf greens, because the turfgrass species has low tolerance to postemergence herbicides (Figure 6.10).

Despite its drawbacks, cultivation is a useful way to control weeds in herbicide-sensitive crops or to clean up a bed prior to applying a preemergence herbicide. Cultivation may be used in landscape plantings when a large variety of plants growing in a small area limits the use of herbicides.

Mowing

Proper mowing practices are a valuable weed control method. When frequently repeated at the appropriate height (McCarty, 2005; Turgeon, 2005), mowing depletes underground food reserves, prevents seed maturation, and favors the growth of turfgrass

Figure 6.10
In some instances, hand removal is the only safe means to achieve a weed-free surface. (Courtesy of Todd Lowe.)

species. Usually, tall annual broadleaf weeds are the ones most weakened or eliminated by mowing. Stem tips of these weeds produce growth-inhibiting substances. These substances, when present, inhibit bud growth on the lower and underground stems, as well as on roots. This suppression is known as *apical dominance* and is reduced with the removal of the stem tips by mowing. Buds may then grow, resulting in a bushier appearance. Once these stems are removed by mowing, new stem growth occurs at the expense of below-ground food reserves. Over time, repeated clipping essentially causes the plant to become weakened through starvation of these below-ground root reserves.

Common lambsquarters, dogfennel, burnweed, sicklepod, horseweed, pigweed, and common ragweed are common weeds in newly established areas intolerant of frequent mowing. These are weakened as mowing commences. Weeds should be mowed in the bud stage or earlier in order to prevent seed development. Prostrate or rosette type broadleaf weed species, as well as grasses, have their primary growing point, or crown, located at or just below the soil surface. Therefore, mowing does not injure the growing point, but does remove the oldest portion of the leaf. New growth resumes because the growing point is unaffected. If the crown is injured by other means, the plant may not recover. Examples of weeds tolerant to mowing include goosegrass, annual bluegrass, common purslane, spotted and prostrate spurge, Virginia buttonweed, and prostrate knotweed.

BIOLOGICAL

Biological methods use weeds' natural antagonists as control agents. The objective of biological control is not weed eradication, but rather the reduction of the population to below a level of economic or aesthetic injury. Unfortunately for turf and ornamental managers, a high percentage of control of a weed species is necessary to satisfy their clientele.

A successful biological weed control agent should (Klingman and Ashton, 1982)

1. Satisfactorily weaken or kill the weed;
2. injure only the intended species and no others;

3. be mobile enough to reach the weed;
4. reproduce desired plants faster than the weed;
5. be adapted to the weeds' environment; and
6. keep the desired plants free of their own predators or pathogens.

Biological control agents rarely have been used in turf, because they often are pathogenic to the turf or do not provide sufficient control. An example involves the use of *Bipolaris* and *Pyricularia* pathogens as potential biological control agents of goosegrass (Figliola et al., 1988). However, with either pathogen, intricate and exact conditions must exist for disease infection, or control is lost. In addition, various species of these pathogens are disease-causal agents of turfgrasses, including grey-leaf spot of perennial ryegrass, St. Augustinegrass, and various leaf spots of bermudagrass. However, progress in using pathogenic fungi for weed control in turfgrasses has been achieved. Riddle et al. (1991) reported that selected strains of *Sclerotinia sclerotiorum* (Lib.) de Bary effectively controlled dandelion without injuring Kentucky bluegrass, creeping bentgrass, or annual bluegrass. Integrations of 2,4-D and *Phoma herbarum* have produced enhanced control of dandelion (Schnick and Boland, 2004).

Interest has increased on the potential use of bacterium as a biological means of weed control. Various members of the *Xanthomonas* genera have been isolated and screened for control of annual bluegrass (McCarty and Tucker, 2005). Much work on using various rust (*Puccinea* and other species) organisms for nutsedge control also has been performed. For example, the fungus *Dactylaria higginsii* has been used to infect leaves and emerging shoots of purple nutsedge. Significant reductions in purple nutsedge shoots, shoot dry weights, and tuber dry weights can occur after inoculation (Kadir and Charudattan, 2000).

These biological controls agents typically do not provide long-term control of weeds and, due to the high concentrations needed for control, can lead to sensitivity to environmental conditions (Busey, 2003). Future work on these and other biological control agents will continue, resulting in alternatives to current methods.

A successful and commercially available biological control agent for aquatic weeds involves the white amur, or grass carp (*Ctenophryngodon idella*). This fish feeds mostly on filamentous algae, chara, submersed weeds, and duckweed. It does not, however, feed extensively on emergent vegetation or large free-floating weeds such as water hyacinth. Stocking rates range from 5 to 20 fish per acre (2 to 8 per ha), and white amur has a life span of about 16 years (Ross and Lembi, 1999).

Although some outstanding successes have been achieved with biological agents in other commodities, additional research is needed to identify biological weed-control agents for turfgrasses and ornamentals. Through the wide publicity of misinformation, the general public has incorrectly viewed this as a viable and immediately available alternative for the control of all types of turfgrass and ornamental pests.

CHEMICAL WEED CONTROL

One of the major contributing factors to the advancement of man's way of life has been the development of pest-control compounds. The first major selective pest controlling compound used was a lime-copper-sulfur mixture known as the Bordeaux mixture. It was discovered to have fungicidal properties on powdery mildew of grapes in 1896. Selective broadleaf weed control also was investigated with this mixture.

Attempts in developing selective herbicides in cereal crops during the period from 1900 to 1915 centered on solutions of copper nitrate, ammonium salts, sulfuric acid, iron sulfate, and potassium salts. These were investigated for selective weed control in cereal crops. Compounds developed between 1900 and 1940 included the arsenicals, chlorates, borates, ammonium sulfamate, and the dinitrophenols. The organic arsenicals (MSMA, DSMA, AMA, and CMA) were developed during this time and still are used for selective grass control in turf. However, the trigger for the development of modern herbicide technology did not occur until World War II.

The discovery of the herbicidal properties of 2,4-D (2,4-dichlorophenoxyacetic acid) during World War II began the era of modern herbicide technology. Postemergence perennial broadleaf weed control was provided by 2,4-D. These weeds previously were controlled by hand labor or land fallow. The first use of 2,4-D was for dandelion control in Kentucky bluegrass. The compound proved to be economical, reasonably predictable and consistent, and highly efficacious. Most importantly, it was safe and relatively easy to use. It still is one of the most widely used herbicides.

NON-SELECTIVE WEED CONTROL

PREPLANT WEED CONTROL

If perennial weeds such as bromegrass, quackgrass, torpedograss, kikuyugrass, nutsedge, and bermudagrass are present, it is advisable that weed control be implemented before planting or establishing a planting area (Table 6.3). Once the plants are established, control of these weeds becomes more difficult, costlier, and, oftentimes, not achievable.

Soil Fumigation

Soil fumigants are volatile liquids or gases that control a wide range of soil-borne pests. Soil fumigants such as methyl bromide are expensive and extremely toxic. Their use is limited to small high-volume cash crops such as tobacco, certain vegetables, fruits, bedding plants, and turf. The expense is due to the need to purchase a plastic cover designed to trap the fumigant vapors in the soil. Fumigants control both weeds and many nematodes, fungi, and insects. However, weed species that have a hard, water-impermeable seed coat, such as sicklepod, white clover, redstem filaree, and morning glory, are not effectively controlled with soil fumigants. Among the important considerations involved in choosing a particular soil fumigant are its expense, soil moisture level, soil temperature, and the time elapsed before planting. The two most commonly used materials as fumigants in turf are *methyl bromide* and *metham* (also referred to as *metam-sodium*).

Methyl Bromide Methyl bromide is a colorless, nearly odorless liquid or gas. At 38°F (3.4°C), the liquid turns into a gas and is 3.2 times heavier than air at 68°F (20°C). These properties require that a gas-impermeable cover be used with the herbicide, or else the material will escape (Figure 6.11). Methyl bromide is extremely toxic and commonly is combined with an odor-detectable warning agent such as chloropicrin (teargas).

Figure 6.11
Fumigating destroys existing weed seeds in the soil. The most effective fumigant, methyl bromide, is slated to be phased out of because of its potential ozone-depleting properties.

When a fumigant is used, the soil should be prepared so that it is suitable for planting. Weed control normally is achieved only as deep as the soil is properly prepared. The soil should be moist for adequate fumigant penetration and dispersion. Moisture-saturated or extremely dry soils limit penetration and dispersion, which subsequently affects weed-seed absorption. Soil temperatures at 4 inches (10 cm) should be a minimum of 60°F (15.6°C). Fumigation is not effective if soil temperatures are below 50°F (10°C). A plastic or polyethylene cover should be placed with ends properly secured prior to application in order to prevent gas leakage. Once the area is treated, it should remain covered for 24 to 48 hours. The cover then may be removed, and the soil aerated for 24 to 72 hours, before planting.

Methyl bromide has been noted as a potential ozone-depleting substance (Unruh and Brecke, 2001). As a result, methyl bromide is under a mandatory phaseout based on the EPA's ozone depletion potential (ODP) data. The EPA states that any product having an ODP greater than 0.2 must be phased out in seven years (Unruh, 1998). The ODP for methyl bromide has been estimated at 0.45 to 0.49 (Unruh, 1998).

Metham or Metam-Sodium Metham (sodium methyl-dithiocarbamate) is a member of the thiocarbamate herbicide family. Metham is water soluble and, upon contact with moist soils, breaks down to form the highly toxic and volatile chemical *methyl isothiocyanate.* Like methyl bromide, metham should be applied to moist soils with temperatures of at least 60°F (15.6°C). It is most effective when its vapors are confined by a tarp. A water-based soil-seal method may also be used. With this method, the soil is cultivated and kept moist with water at 15 gallons per 100 square feet (61 L per 10 square meter) for a week before treatment. The material is applied, rototilled, and watered-in immediately to the depth of desired control (approximately 4 to 6 inches [10 to 15 cm]). Approximately seven days after treatment, the area should be cultivated to help release any residual gases. One to two weeks later (two to three weeks after initial application), the treated area may be planted. The longer preplanting waiting period and lowered effectiveness if a tarp is not used are the primary disadvantages of metham.

Dazomet Dazomet, unlike methyl bromide and metham, is a granular product and is not a restricted-use product. Being a granular, dazomet must be evenly applied and incorporated for maximum effectiveness. Its breakdown characteristics, application preparation, and effectiveness are closely associated to metham, as are its advantages and disadvantages.

Nonselective Herbicides

Nonselective herbicides (e.g., glyphosate) also are used for preplant weed control. Normally, multiple applications spaced two to four weeks apart are needed for control of existing weeds. For example, a minimum of three applications of glyphosate are necessary to completely control bermudagrass (Johnson, 1988). Control of subsequent germinating weeds, as well as of other soil-borne pests, is not achieved with this method.

Two or three applications of glyphosate plus fluazifop-P-butyl have provided control similar to three applications of glyphosate alone (Boyd, 2000). Glyphosate 4L is applied at 0.5 gal per acre (4.7 L/ha or 2 lb ai/a or 2.24 kg ai/ha) plus fluzaifop-P-butyl 2L at 24 ounce per acre (1.8 L/ha or 0.4 lb ai/a or 0.45 kg ai/ha). Applications are spaced three to four weeks apart or done as often as sufficient bermudagrass regrowth occurs. Applying ammonium nitrate between applications increases bermudagrass regrowth and makes it more succulent, increasing control.

SELECTIVE WEED CONTROL

Maintaining turf and ornamental complexes at the desired level of aesthetics requires knowledge of specific weeds, weed biology, and available weed-control measures. Weed identification and biology are discussed in other chapters of this textbook. The next subsection discusses selective weed-control options.

DEVELOPING A WEED-CONTROL PROGRAM

Weed control should be a carefully planned and coordinated program instead of being a hit-or-miss operation. Understanding why weeds are present on a site is just as important as the chemical control options. The following is a suggested list for developing a logical, long-term effective weed control program in turf and ornamentals.

Maintain Healthy Plants

As mentioned earlier, the best method for weed control is to encourage a strong, vigorous growing ground cover that prevents or out-competes the weeds. Proper cultural practices and other pest-control techniques must be provided in conjunction with herbicide use. Unless such a competitive plant material fully covers the soil surface, continued weed problems will occur regardless of the herbicide program employed. For example, sedges are often associated with perennially wet areas. Unless the moisture is moderated in these areas, sedges will persist regardless of the herbicides used.

Weed Identification, Biology, and Habitat

Weed identification is the first step in understanding why weeds occur and how to control them. Each plant should be identified by species, if possible. The identification characteristics of common turf weeds are listed in an accompanying chapter.

Understanding the biology or growth and reproductive characteristics of a weed is the second most important step in developing a weed control strategy. Supervisors should understand environmental and cultural factors influencing weed infestations and how these can be manipulated in order to favor desirable plant growth over weed establishment.

Soil

Understanding and knowing the chemical and physical composition of soils becomes important when trying to disrupt weed habitat by using preemergence herbicides. Soil texture, colloidal components, organic-matter content, and pH affect the efficacy, availability, and duration of control of most organic pesticides. Knowing the following factors about soil will enable better decision making on herbicide products and their control levels:

a. *Texture.* Soil texture involves its percentages of sand, silt, and clay present. The proportions of these elements determine properties such as water-holding capacity, tilth, and potential herbicide absorption and movement.
b. *Soil condition and colloidal components.* Surface moisture, plant residue, surface cracking, and evenness are soil conditions impairing the effectiveness of soil-applied herbicides. Organic matter and clay content are colloidal components that influence soil-applied herbicides. For example, soils high in organic matter (e.g., muck or peat soils) may bind certain herbicides so tightly that they become ineffective.
c. *pH.* The soil pH determines, in part, the ionization (ability to attract or release ions) of some herbicides and thus their binding affinities and availability for weed control.

Proper Herbicide Selection, Application, and Record Keeping

The safety, effectiveness at certain weed-growth stages, stage of plant establishment, tolerance or susceptibility of treated plant species, length of time required for control, and economics all play a role in an herbicide's effectiveness. Knowledge of these parameters is thus an important consideration when one is choosing among herbicides. Enormous amounts of information are available from local county extension offices, state horticulture specialists, colleagues, and representatives of the chemical company. However, even the best herbicide is only as good as its application. Many variables influence successful herbicide application, including pesticide formulation, proper equipment, environmental factors at the time of application, proper and constant calibration, and adequate agitation. Most herbicide failures involve applying a nonlabeled chemical or applying the chemical at an improper time, at an incorrect rate, or in an inappropriate manner.

Once a particular herbicide is chosen and applied, accurate, detailed information should be kept. In addition to a listing of what material was used and where it was applied, information on how and at what rate it was applied, who made the application, and the environmental conditions at the time of application are needed. Environmental information should include soil and air temperatures, soil moisture and pH, relative humidity, wind speed and direction, water pH, cloudiness, weed-growth stage and turfgrass condition at the time of application, time and amount of irrigation or rainfall following application, and whether dew was present (Table 6.4). By having this information, supervisors can determine the effectiveness of a material when applied under specific environmental conditions and can more accurately pinpoint contributing factors for either satisfactory or unsatisfactory results.

Table 6.4
INFORMATION THAT SHOULD BE RECORDED WHEN PESTICIDES ARE APPLIED (MCCARTY, 2005)

Pesticide Application Record

Company Name ____ Commercial Applicator ____
Application Date & Time ____ Site Location ____
Pesticide License Category ____ Number ____
Pesticide Name(s) ____ Manufacturer ____
EPA Registration No. ____ Restricted-entry Interval (REI) ____
Active Material & Formulation ____
Lot No. ____ % Concentration ____
Safety Equipment Needed ____

Application Information

Type of Area Treated ____ Target Site ____
Target Pest(s) ____ Total Treated Area ____
Application Rate (e.g., per acre or per 1000 sq. ft.) ____ Application Timing ____
Amount of Pesticide Product Mixed ____ Per ____ Gallons of Water
Additives (Surfactant/Wetting Agent/Crop Oil, etc.) ____ Rate ____

Weather Conditions

Air Temperature (EF) ____ % Relative Humidity ____ Dew Presence (Y/N) ____
Initial Wind Velocity (MPH) ____ Wind Direction ____
First Hour ____ Second Hour ____ Third Hour ____
Soil Temperature at 4 inches (F) ____ Soil Moisture ____ % Cloud Cover____

Application Equipment

Method of Application ____ Speed (mph) ____ Motor Speed (RPM) ____ Nozzle Type ____ Number ____
Nozzle Height ____ Spacing ____ Boom Width ____ Gallon Per Acre (GPA) ____ Spray Pressure (PSI) ____
Nontarget Plant, Animal, or Human Exposure: Yes ____ No ____ (If yes, list corrective or emergency action taken)

____

Other Comments:

Signature ____ **Date** ____

PREEMERGENCE HERBICIDES

Some of the first chemicals evaluated for preemergence weed control included calcium cyanide (Sturkie, 1933), arsenate, and naphthylacetic acid. The first true and consistent preemergence herbicide became available for turf and ornamental producers around 1959. Dimethyl Tetrachloroterephthalate (DCPA) provided more consistent weed control with less plant damage than was previously available (Engel and Ilnicki, 1969). With subsequent release of dinitroaniline chemistry, the widespread acceptance of preemergence weed control was established.

Preemergence herbicides are applied to the plant site prior to weed-seed germination. The mode of action for most preemergence herbicides (e.g., bensulide, benefin, dithipyr, oryzalin, pendimethalin, prodiamine) is the inhibition of certain phases of cell division (mitosis) or inhibition of cell elongation. As the weed seedling germinates, its root and shoot absorb the herbicide and the seedling stops growth and eventually dies.

Turfgrass Tolerance

With any herbicide, the first consideration is the tolerance of the desirable turfgrass species to the chemical in question. As a general rule, preemergence herbicides are not as phytotoxic to established turfgrass species as postemergence herbicides are. Notable exceptions are atrazine, simazine, and pronamide on cool-season grasses. Table 6.5 lists the most widely used turfgrass species and their tolerance to herbicides, while Table 6.6 lists preemergence herbicides currently labeled for golf greens.

Effectiveness of Preemergence Herbicides

The effectiveness of preemergence herbicides varies. Reasons for this variation include timing in relation to weed-seed germination, soil types, and environmental conditions (e.g., rainfall and temperature) during and immediately following herbicide application. Other factors are the target weed species and biotypes, and the cultural factors (e.g., aerification) following application. Preemergence herbicides generally are more effective for annual grass control, although some annual broadleaf weeds also are suppressed.

Timing

An important consideration in preemergence herbicide use is its application timing. Most preemergence herbicides are ineffective on visible weeds. Applications should occur prior to seed germination. However, if preemergence herbicides are applied too soon, the degradation of the herbicide in the soil may reduce its concentration to an ineffective level by the time the weed germinates.

Crabgrass and goosegrass are two of the most troublesome annual grass weeds in turf. If preemergence herbicides are used for control, application timing is critical. Crabgrass initiates spring germination when soil temperatures at a 4-inch (10 cm) depth reach 53° to 58°F (11.7° to 14.4°C) for 24 continuous hours (Lewis, 1985). Alternating dry and wet conditions at the soil surface, as well as high light intensity, also encourage crabgrass germination. Goosegrass, meanwhile, germinates when soil temperatures are 60° to 65°F (15.6° to 18.3°C) for 24 continuous hours. Goosegrass also requires high light intensity for seed germination (Fulwider and Engel, 1959). Because of higher temperature requirements for germination, goosegrass normally germinates two to eight weeks later in spring than crabgrass (Bingham, 1985; Lewis, 1985). Therefore, when goosegrass weed-control program is being developed, the spring application of preemergence herbicides should be delayed approximately three to four weeks longer than its spring application in a crabgrass program (Nishimoto and McCarty, 1997).

Table 6.5

Established Turfgrass Tolerance to Preemergence Herbicides (McCarty, 2008) *(Refer to Herbicide Label for Specific Use Listing)*

These are relative rankings that depend on factors such as environmental conditions, turfgrass vigor or health, application timing, etc., and are intended only as a guide.

Herbicides (Trade Name)	Bahiagrass	Bentgrass[1]	Bermudagrass[1]	Buffalograss	Centipedegrass	Kentucky Bluegrass	Kikuyugrass	Overseeded Ryegrass	Perennial Ryegrass	Red Fescue	Seashore Paspalum	St. Augustinegrass	Tall Fescue	Zoysiagrass
atrazine (Aatrex)	NR[2]	NR	I (D)	I (D)	S	NR	NR	NR	NR	NR	NR	S	NR	I-S
benefin (Balan)	S	NR	S	NR	S	S	NR	NR	S	S	NR	S	S	S
benefin + oryzalin (XL)	S	NR	S	I (D)	S	NR	NR	NR	NR	NR	NR	S	S	S
benefin + trifluralin (Team)	S	NR	S	NR	S	S	NR	NR	S	S	NR	S	S	S
bensulide (Betasan, PreSan)	S	S	S	NR	S	S	NR	I-S	S	S	NR	S	S	S
bensulide + oxadiazon	NR	S	S	NR	NR	S	NR	NR	S	S	NR	NR	S	S
dithiopyr (Dimension)	S	S	S	S	S	S	S	I	S	I	S	S	S	S
ethofumesate (Prograss)[3]	NR	S	S(D)	NR	NR	I	NR	S(D)	S	I	NR	I	I	NR
isoxaben (Gallery)	S	NR	S	S	S	S	NR	I-S	S	S	NR	S	S	S
fenarimol (Rubigan)	NR	NR	S	NR	NR	S	NR	S	NR	S	NR	NR	S	NR
mesotrione (Tenacity)	NR	NR	NR	NR	S	S	NR	NR	S-I	S-I	NR	S-I	S-I	NR
metolachlor (Pennant)	S	NR	I	NR	S	S	NR	NR	NR	S	NR	S	S	S
napropamide (Devrinol)	S	NR	S	NR	S	NR	NR	NR	NR	NR	NR	S	S	NR
oryzalin (Surflan)	S	NR	S	S	S	NR	NR	NR	NR	NR	NR	S	I	S
oxadiazon (Ronstar)	NR	NR	S	S	NR	S	NR	I	S	S	S	S	S	S
pendimethalin (Pre-M)	S	NR	S	S	S	S	NR	NR	S	S	NR	S	S	S
prodiamine (Barricade)	S	NR	S	S	S	S	NR	I	S	S	S	S	S	S
pronamide (Kerb)	S	NR	S	S	S	NR	NR	NR	NR	NR	NR	S	NR	S
siduron (Tupersan)	NR	I	NR	NR	NR	S	NR	NR	S	S	NR	NR	S	S
simazine (Princep)	NR	NR	I (D)	NR	S	NR	NR	NR	NR	NR	NR	S	NR	S

[1]Check herbicide label to determine whether product can be used on golf course putting greens.
[2]**S** = Safe at labeled rates on mature, healthy turf; **I** = Intermediate safety—may cause slight damage to mature, healthy turf. Use only one-half the normal rate when temperatures are hot (>85°F) or if the turf is under water stress; **NR** = Not registered for use on, and/or damages, this turf species.
[3]Ethofumesate is labeled only for Dormant (**D**) bermudagrass overseeded with perennial ryegrass.

Table 6.6
PREEMERGENCE HERBICIDES FOR PUTTING GREENS *(REFER TO HERBICIDE LABEL FOR SPECIFIC USE LISTING)*

Trade Names	Ingredients	Bentgrass	Bermudagrass	Bermudagrass to be Overseeded (Refer to Label for Specific Timing)
Weedgrass Preventer	bensulide	Y	Y	Y
Goosegrass/Crabgrass Control	bensulide + oxadiazon	Y	Y	—
Southern Weedgrass Control	pendimethalin	—	Y	—
Devrinol	napronamide	—	Y	—
Betasan	bensulide	Y	Y	Y
Kerb	pronamide	—	Y	Y
Dimension	dithiopyr	Y*	Y	—
Monument (early postemergence control)	trifloxysulfuron	—	Y	Y
Revolver (early postemergence control)	foramsulfuron	—	Y	Y
Rubigan	fenarimol	—	Y	Y
TranXit (early postemergence control)	rimsulfuron	—	Y	Y
Tupersan	siduron	Y	–	—

*Refer to label for proper rates, timings, and any restrictions. Not all creeping bentgrass varieties have been tested; Y = Yes.

Annual bluegrass is a winter annual grass weed that starts germination in late summer and continues germination through the fall and winter when soil temperatures at the 4-inch (10 cm) level drop to the low to mid-70s Fahrenheit (21.1°C) or lower. Preemergence herbicide application should be timed just prior to the expected period of peak germination. Annual bluegrass often has a second germination flush during warm days in mid- to late winter. This is important for turf managers to recognize, as fall herbicide applications normally do not provide season-long control and repeat applications may be necessary.

Efficacy

Table 6.7 lists the expected control of common turf weeds for various preemergence herbicides. Continued use of a particular preemergence herbicide may control the intended species, but also may increase the cover of other weed species. For example, bensulide provides good crabgrass control, but may increase the cover of clover and speedwell (Johnson, 1982a). In this same study, benefin and DCPA increased lawn burweed, wild parsnip, and clover, while oxadiazon increased wild parsnip and sandwort populations.

Sequential Applications

For season-long control, repeat applications of most preemergence herbicides are necessary, especially in tropical and subtropical growing regions. As a general rule of thumb, if weed control is desired past 100 days, a repeat application will be needed. Most herbicides begin to degrade when exposed to the environment. Degradation generally is enhanced by increasing soil moisture and warm weather, and with soils having low cation exchange and water-holding capacities. Most herbicides are effective in preventing subsequent weed-seed germination over a two-to-five-month period after application. Due to degradation losses, repeat applications become necessary for continued preemergence weed control. Turf managers must recognize and understand that most preemergence herbicides will prevent the germination and establishment of desirable turfgrasses as well as weeds for up to two to five months after application. Timing, therefore, is important on those areas to be turf established.*

Some evidence suggests that after initial herbicide application at normal rates over one or two years, subsequent yearly rates may be reduced. When proper mowing height (2.25 in. [5.7 cm]) and nitrogen fertilizer (2 lb N/M [98 kg N/ha]) are maintained, the application of a preemergence herbicide in a red fescue lawn, in the first year, maintained low crabgrass populations (1 to 3 percent) during the next four years. Similarly, herbicide rates required to control crabgrass or goosegrass could be halved or eliminated in subsequent years when a normal rate was applied the first year (Johnson, 1982a).

Aerification

Core aeration generally has not been recommended or practiced following a preemergence herbicide application. Core aeration was believed to disrupt the herbicide barrier in the soil and stimulate weed emergence. Johnson (1987a) reported that core aeration immediately prior to or one, two, three, or four months after applications of benefin, bensulide, DCPA, and bensulide + oxadiazon to common bermudagrass did not stimulate large crabgrass emergence. Aeration at one or two months after application increased large crabgrass cover 5 percent for oxadiazon at 2 lb ai/a (2.2 kg ha^{-1}), but not at 4 lb ai/a (4.4 kg ha^{-1}). In a related study, it was shown that core aeration at

Note: On those areas to be turf established, most preemergence herbicides should not be used two to four months prior to planting. Severe turf root damage and turf germination reduction may result.

Table 6.7

Preemergence Herbicide Efficacy Ratings on Common Turfgrass Weeds *(Refer to Herbicide Label for Specific Use Listing)*

These are relative ratings that depend on many factors such as environmental conditions, turfgrass vigor or health, application timing, etc., and are intended only as a guide.

Herbicide (Trade Name)	Crabgrass	Goosegrass	Annual Bluegrass	Common Chickweed	FL Pusley	Henbit	Lawn Burweed	Purslane	*Phyllanthus* sp.	Speedwell spp.	Spurges	Woodsorrel (*Oxalis*)
atrazine (Aatrex)	F[1]	P	E	E	G	E	G	G	—	E	G	F
benefin (Balan)	G-E	F	G-E	G	—	G	P	—	—	P	P	—
benefin + oryzalin (XL)	E	G	G	G	G	G	—	G	—	—	F	F-G
benefin+trifluralin (Team)	F-G	F	G	G	—	G	—	—	—	—	F	F
bensulide (Betasan, PreSan)	G-E	P-F	F	P	—	P	P	F	—	P	—	—
bensulide + oxadiazon	E	G-E	G-E	G	—	—	—	—	—	—	G	—
dithiopyr (Dimension)	E	G-E	G-E	G	—	G	F	F	—	G	G	G
fenarimol (Rubigan)	P	P	G-E	P	P	P	P	—	—	P	P	P
isoxaben (Gallery)	P-F	P	P-F	E	F-G	G	E	G	—	G-E	G	G
mesotrione (Tenacity)	G	F-G	F	G	G	G	G	F	—	G	—	G
metolachlor (Pennant)	F-G	P-F	G	F	G	—	—	F	P	—	F	P
napropamide (Devrinol)	G-E	G	G	E	P	P	E	G	—	E	P	G
oryzalin (Surflan)	E	G	G-E	G	G	G	F	G	—	P	F-G	G
oxadiazon + prodiamine	E	G-E	G-E	G	G	G	F	G	F-G	G	G	G
oxadiazon (Ronstar)	G-E	E	G-E	P	G	P	P	G	F-G	G	G	G
pendimethalin (Pendulum)	E	G-E	G-E	E	G	G	G	G	F-G	G-E	G	G
prodiamine (Barricade)	E	G-E	G-E	G	G	G	F-G	G	F-G	F-G	G	G
pronamide (Kerb)	P-F	P	G-E	E	—	F-G	P	G	—	E	P	P
simazine (Princep T&O)	P-F	P	E	E	G	E	G-E	G	—	E	F-G	F

[1]**E** = Excellent, >89% control; **G** = Good, 80 to 89% control; **F** = Fair, 70 to 79% control; **P** = Poor, <70% control; — = Data not available.

one, two, or three months after an application of oxadiazon did not decrease goosegrass control on a Tifgreen bermudagrass putting green (Johnson, 1982b). Branham and Rieke (1986) reported that core aeration, or vertical mowing, immediately or one month after an application of benefin, bensulide, or DCPA did not affect large crabgrass control in annual bluegrass. Monroe et al. (1990) noted that aeration did not affect the activity of several preemergence herbicides in controlling crabgrass species in either Tifgreen or common bermudagrass. However, in creeping bentgrass, greater amounts of crabgrass occurred in aerified plots with the cores returned than in plots that were not aerified or in aerified plots with the cores removed.

Newly Established Turf

Preemergence herbicides such as bensulide, dithiopyr, and members of the dinitroaniline family (e.g., benefin, oryzalin, pendimethalin, prodiamine, trifluralin) should be used only on well-established turfgrasses. The previously mentioned herbicides may inhibit rooting of immature or newly established turfgrasses. A waiting period of three to five months normally is required between the last herbicide application and establishment of the turfgrass.

Oxadiazon (Ronstar) generally is the safest preemergence herbicide when grass is vegetatively established (McCarty, 2005), while siduron, mesotrione, or quinclorac are available for seeded grasses. Oxadiazon, a preemergence herbicide that inhibits protoporphyrinogen oxidase (PPG, or Protox), blocking chlorophyll, and heme biosynthesis, does not negatively affect rooting to the extent that other mitotic inhibiting preemergence herbicides do. Oxadizaon, therefore, can be used prior to or immediately following sprigging or sodding, with much less turf injury or delayed establishment (McCarty and Weinbrecht, 1997). Oxadiazon functions by inhibiting shoot emergence of susceptible weeds and does not prevent cell division of roots as do most other preemergence herbicides. Minimal effects, therefore, occur on the rooting of vegetatively established turf species, but selective weed control of goosegrass occurs. Only the granular formulation of oxadiazon should be used for this purpose.

Siduron (Tupersan) is a member of the substituted urea herbicide family. It appears to interfere with photosynthesis of the developing weed. Since siduron does not appreciably inhibit rooting, it safely can be applied to newly seeded turf that has established a secondary roots system. Siduron effectively controls annual grass weeds such as crabgrass and foxtail, but does not effectively control annual bluegrass, goosegrass, or most of the broadleaf weeds. Siduron should not be used on bermudagrass.

Quinclorac (Drive) also can be used as a preemergence or early postemergence herbicide on newly seeded annual bluegrass, ryegrass, bentgrass fairways, common bermudagrass, Kentucky bluegrass, tall fescue, and zoysiagrass grasses, as well as on sprigged areas, to control crabgrass and numerous broadleaf weeds. Creeping bentgrass, hybrid bermudagrass, and fine fescue have intermediate tolerance. Weeds controlled include crabgrass, signalgrass, barnyardgrass, foxtail, and selective broadleaf weeds such as pennywort, speedwells, dandelion, black medic, white clover, and violets. Good soil moisture should be present before treatment. Do not apply to desirable bahiagrass, centipedegrass, St. Augustinegrass, or dichondra, or on golf greens and collars. Tank mixing with N or Fe may lessen turf discoloration. Add a crop oil concentrate (2 pt/a [2.34 L/ha]) or methylated seed oil (1.5 pt/a [1.75 L/ha]) to increase performance and avoid drift onto ornamentals.

Mesotrione (Tenacity 4L) is a recent product used for pre- and postemergence annual grass control. For preemergence crabgrass, goosegrass, barnyardgrass, foxtail, and certain broadleaf weed control, apply at seeding of the cool-season grasses tall and fine fescue, Kentucky bluegrass, and perennial ryegrass at 4 to 8 fl. oz per acre (0.125 to 0.25 lb ai/a, 0.14 to 0.28 kg ai/ha) in 40 gpa (375 L/ha). Repeat this in three weeks. Do not apply mesotrione to desirable bentgrass, bermudagrass, or zoysiagrass. Add a nonionic surfactant at 0.25% v/v.

Figure 6.12
Young weeds are the easiest to control with the least amount of herbicide. Shown is young goosegrass.

POSTEMERGENCE HERBICIDES

Postemergence herbicides are effective only on visible weeds. The timing of the application should be when weeds are young, preferably during the two-to-four-leaf stage (Figure 6.12). Younger weeds are more favorable to having greater herbicide uptake, thus allowing translocation of the herbicide to underground storage areas of the plant. Turfgrasses also are then better able to fill in voids left by the dying weeds.

Turfgrass Tolerance

Generally, postemergence herbicides have a much greater likelihood of injuring a turf species than preemergence herbicides do. Turf managers must be careful about the herbicide they select and how they use it. Turf injury usually is more likely when the turf is under stress from temperature, moisture, and other pests such as nematodes. Turfgrass species tolerances to postemergence herbicides are listed in Table 6.8.

Broadleaf Weed Control

Broadleaf weed control in turf traditionally has been with members of the phenoxy herbicide family (e.g., 2,4-D, dichlorprop [2,4-DP], MCPA, and mecoprop) and with the benzoic acid herbicide family (e.g., dicamba). All are selective, systemic, and foliar-applied herbicides. Broadleaf weeds, especially perennials, often are not adequately controlled with just one of these materials. Usually, two- or three-way combinations of these herbicides and repeat applications 7 to 14 days apart are necessary for satisfactory control of, especially, mature weeds. Formulations of 2,4-D, dichlorprop [2,4-DP], MCPA, MCPA, and/or dicamba are available in three-way mixtures. In combination, these products appear to be synergistic; therefore, less is required of each than if each were applied alone. Turf safety with these combinations depends on the herbicide mixture, its rate, and its formulation, as well as on environmental conditions at the time of treatment.

Many formulations and mixtures of the phenoxy herbicides are available (Table 6.9). These include the parent acid, amine salts, esters, and inorganic salts. The water-soluble amine formulations are the most common because of their lack of volatility, high water solubility, ease of handling, and overall low cost. Although more effective, high-volatile esters normally are not used in turf. The likelihood that these will volatilize and drift to nearby desirable plant species is high.

Until recently, these various herbicide combinations were the main chemicals for broadleaf-weed control. Several new chemistries have been introduced as alternatives or additives to phenoxy herbicides for broadleaf control over the last couple of decades. Chlorsulfuron 75 DF (Corsair and TFC) and metsulfuron 60 DF (Manor and Blade) belong to the sulfonylurea herbicide family and are noted for their broadleaf-weed control and selective control of certain cool-season grasses like perennial ryegrass and tall fescue at extremely low rates (e.g., 0.25 to 5 oz product/acre [0.0175 to 0.35 kg/ha]) in warm-season turfgrasses. Metsulfuron also acts on certain bahiagrass cultivars.

Triclopyr (Turflon), fluroxypyr (Spotlight), and clopyralid (Lontrel) belong to the pyridine herbicide family. Compounds in this family have been noted for their high degree of activity. For example, members of this herbicide family are up to 10 times more potent than 2,4-D on some broadleaf-weed species (Klingman and Ashton,

Table 6.8

ESTABLISHED TURFGRASS TOLERANCE TO POSTEMERGENCE BROADLEAF AND GRASS HERBICIDES (MCCARTY, 2008)

These are relative rankings and depend on factors such as environmental conditions, turfgrass vigor or health, application timing, etc., and are intended only as a guide.

Herbicides	Bahiagrass	Bentgrass Fairways	Bentgrass Greens	Bermudagrass	Buffalograss	Carpetgrass	Centipedegrass	Fine Fescue	Kentucky Bluegrass	Kikuyugrass	Overseeded Ryegrass/Blends	Ryegrass	Seashore Paspalum	St. Augustinegrass	Tall Fescue	Zoysiagrass
atrazine (Aatrex)	NR[1]	NR	NR	S-I(D)	I (D)	I[3]	S	NR	NR	NR	NR	NR	NR	S	NR	I
bentazon (Basagran T&O)	S	I	NR-I	S	S	S	S	S	S	NR	S-I	S	S-NR	S	S	S
bromoxynil (Buctril)	S	NR	NR	S	NR	S	S	S	S	NR	S	S	NR	S	S	S
carfentrazone (QuickSilver)	S	S	NR	S	S	NR	S	S	S	NR	S	S	S	I	S	S
carfen. + 2,4-D + MCPP + dicamba (Speed Zone North.)	NR	S	NR	S	NR	NR	NR	S	S	NR	S	S	NR	NR	S	S
carfen. + MCPA + MCPP + dicamba (Power Zone)	NR	NR	NR	S	NR	NR	NR	S	S	NR	S	S	NR	NR	S	S
carfen. + 2,4-D + MCPP + dicamba (Speed Zone So.)	S	S	NR	S	S	NR	S	S	S	NR	S	S	S	S	S	S
chlorsulfuron (Corsair, TFC)	I	I	NR	S	NR	I	I	I-S	S	NR	NR	NR	S	I	NR	I
clopyralid (Lontrel)	S	I	NR	S	S	S	S	S	S	NR	S	S	NR	S	S	S
2,4-D	S	NR	I[1]	S	I	I	S-I	S	S	S	S-I	S	S	I	S	S
MCPP (mecoprop)	S	I	S	S	I	I	I	S	S	NR	I	S	S	I	S	S
dicamba (Vanquish)	S	I	I	S	I-NR	I	I	S	S	NR	I	S	S	I	S	S
2,4-D + dichlorprop (2,4-DP)	S	I	I	S	S	I	I	S	S	S	S	S	S	I	S	S
2,4-D + triclopyr (Turflon)	NR	NR	NR-I	NR	NR	NR	NR	I	S	NR	S	S	NR-P	NR	S	NR
2,4-D + MCPP + dicamba	S	I	I	S	I	I	I	S	S	NR	S	S	NR	I	S	S
2,4-D + MCPP + 2,4-DP	S	I	I	S	NR	I	I	S	S	NR	S	S	NR	I	S	S
MCPA + MCPP + 2,4-DP	S	I	I	S	NR	I	I	S	S	NR	S	S	NR	I	S	I
MCPA + triclopyr + clopyralid	S	S	S	S	S	I	S	S	S	NR	S	S	NR	NR	S	S
fluroxypyr + 2,4-D+dicamba (Escalade)	S	I	NR	S	NR	NR	NR	S	S	NR	NR	S	NR	NR	S	S
fluroxypyr (Spotlight)	S	S	NR	S	S	S	S	S	S	S	S	S	S	S	S	S

(continued)

Table 6.8
ESTABLISHED TURFGRASS TOLERANCE TO POSTEMERGENCE BROADLEAF AND GRASS HERBICIDES (MCCARTY, 2008) *(CONTINUED)*

Herbicides	Bahiagrass	Bentgrass Fairways	Bentgrass Greens	Bermudagrass	Buffalograss	Carpetgrass	Centipedegrass	Fine Fescue	Kentucky Bluegrass	Kikuyugrass	Overseeded Ryegrass/Blends	Ryegrass	Seashore Paspalum	St. Augustinegrass	Tall Fescue	Zoysiagrass
halosulfuron (Sedgehammer)	S	I	NR	S	NR	S	S	S	S	S	S	S	S	S	S	S
imazapic (Plateau)	NR	NR	NR	S	NR	NR	S	NR	NR	NR	NR	NR	NR	NR	NR	NR
imazaquin (Image)	NR	NR	NR	S-I	S-NR	I	S	NR	NR	NR	NR	NR	NR	S	NR	S
mesotrione (Tenacity)	NR	NR	NR	NR	NR	NR	S	S-I	S	NR	NR	S-I	NR	S-I	S-I	NR
metsulfuron (Manor)	NR	NR	NR	S	S	I	S	I	I	NR	NR	NR	NR-S	S-I	NR	S
pyraflufen-ethyl (Octane)	S	S	NR	S	S	NR	S-I	S	S	S	S	S	NR	S	S	S
quinclorac (Drive)	NR	I	NR	S	S	NR	NR	NR	S	NR	S	S	NR-S	NR	S	S
quinclorac + sulfentrazone + 2, 4-D+dicamba (Q4)	NR	NR	NR	NR-I	NR-I	NR	NR	S	S	NR-I	S	S		NR	S	NR-I
simazine (Princep T&O)	NR	NR	NR	S-I(D)	S	I	S-I	NR	NR	NR	NR	NR	NR	S-I	NR	I
sulfentrazone (Dismiss)	S	S	NR	S	S	S	S	I	S	S	NR	S	S	NR	I	S
sulfentrazone + 2,4-D + dicamba + MCPP (Surge)	S	S	NR	S	S	S	S	S	S	S	S	S	NR	S	S	S
triclopyr (Turflon)	NR	NR	NR	NR	NR	NR	NR	S	S	NR	S	S	NR-P	NR	S	NR
triclopyr + clopyralid (Confront)	I	I	NR	I	S	NR	S	I	S	NR	S	S	NR-I	NR	S	S
Grass Weed Control																
asulam (Asulox)	NR	NR	NR[1]	S-I[2]	NR-I	NR	NR	NR	NR	NR	NR	NR	NR	S-I	NR	NR-I
bispyribac-sodium (Velocity)[3]	NR	NR	NR	S[3]	NR	NR	NR	NR	NR	NR	S[4]	S	NR	NR	NR	NR
clethodim (Envoy)	NR	NR	NR	NR	NR	NR	S	NR	NR	NR	NR	NR	NR	NR	NR	NR
diclofop (Illoxan)	NR	NR	NR	S	NR-S	NR	NR	NR	NR	NR	NR	NR	NR	NR	NR	NR
DSMA, MSMA, CMA	NR	I	NR-I	S-I	I	NR	NR	I	I	NR	NR	S-I	NR-P	NR	I	S-I
ethofumesate (Prograss)[4]	NR	I	NR-I	D	NR	NR	NR	I	S	NR	I	S	NR-S	NR	S	NR
fenoxaprop (Acclaim Extra)	NR-I	I	NR-I	NR-I	NR	NR	NR	S	S	NR	I	S	NR	NR	S	I
fluazifop (Fusilade II)	NR	NR	NR	NR	NR	NR	NR	NR	NR	NR	NR	NR	NR-P	NR	S-I	I
foramsulfuron (Revolver)	NR	NR	NR	S	NR	NR	NR	NR	NR	NR	NR	NR	NR	I	NR	S
mesotrione (Tenacity)	NR	NR	NR	NR	NR	NR	S	S-I	S	NR	NR	S-I	NR	S-I	S-I	NR

Table 6.8
(Continued)

Herbicides	Bahiagrass	Bentgrass Fairways	Bentgrass Greens	Bermudagrass	Buffalograss	Carpetgrass	Centipedegrass	Fine Fescue	Kentucky Bluegrass	Kikuyugrass	Overseeded Ryegrass/Blends	Ryegrass	Seashore Paspalum	St. Augustinegrass	Tall Fescue	Zoysiagrass
metribuzin (Sencor Turf)	NR	NR	NR	S-I	NR	NR	NR	NR	NR	NR	NR	NR	NR-I	NR	NR	NR
pronamide (Kerb)	S	NR	NR	S	NR	NR	S	NR	NR	NR	NR	NR	NR-S	S	NR	S
rimsulfuron (TranXit)	NR	NR	NR	S	NR	NR	NR	NR	NR	NR	NR	NR	NR	NR	NR	NR
sethoxydim (Vantage)	NR	NR	NR	NR	NR	NR	S	S	NR	NR	NR	NR	NR-P	NR	NR	NR
sulfosulfuron (Certainty)	I	NR	NR	S	S	NR	S-I	NR	NR	S	NR	NR	NR	S-I	NR	S
trifloxysulfuron (Monument)	NR	NR	NR	S	NR	NR	NR	NR	NR	NR	NR	NR	NR-P	NR	NR	S
quinclorac (Drive)	NR	I	NR	S-I	S	NR	NR	I	S	NR	S	S	NR-S	NR	S	S

[1]**S** = Safe at labeled rates; **I** = Intermediate safety, use at reduced rates; **NR** = Not Registered for use on and/or damages this turfgrass; **D** = Dormant turf only.
[2]Asulam is labeled for 'Tifway' (419) Bermudagrass and St. Augustinegrass.
[3]Used on dormant bermudagrass overseeded with perennial ryegrass.

Table 6.9
GENERAL CHARACTERISTICS OF DIFFERENT FORMS OF 2,4-D (ASHTON AND MONACO, 1991)

Form	Solubility in Water	Solubility in Oil	Color When Mixed with Water	Precipitates Formed in Hard Water	Volatility Potential
Acid	low	low	milky	yes	low
Amine Salts					
water soluble	high	low	clear	yes	none
oil soluble	low	high	milky	yes	none
Esters					
low volatile	low	high	milky	none	medium
high volatile	low	high	milky	none	high
Inorganic Salts	medium	low	clear	yes	none

1982). These herbicides rapidly are absorbed by the roots and foliage of broadleaf plants. They are readily translocated throughout the plants via both xylem and phloem tissues. Problems with this herbicide family include soil mobility and the extreme sensitivity to them by many desirable ornamentals. Triclopyr is used alone and in combination with 2,4-D, or clopyralid, for broadleaf-weed control. These are primarily used on cool-season grasses.

Other additives include carfentrazone (Quicksilver) and sulfentrazone (Dismiss). They provide some broadleaf and nutsedge (especially sulfentrazone) activity and reduced time until herbicide symptom development. Turf tolerance is good for both products, which are available as pre-tank-mix partners with other traditional broadleaf herbicides such as 2,4-D, dicamba, and MCPP.

In 2000, clopyralid was found in compost in Spokane, Washington, from grass clippings collected from lawns (Minter et al., 2003). As a result, clopyralid has been restricted for commercial use only, and clippings are not to be recycled from treated turf. Fluroxypyr recently has been introduced into the turfgrass market and is similar to clopyralid in that it is specifically effective on leguminous plants such as clover, lespedeza, and black medic. Virginia buttonweed is controlled with fluroxypyr plus clopyralid (Taylor et al., 2001; Handly et al., 2001). Fluroxypyr is used alone or in combination with 2,4-D, MCPP, triclopyr, and dicamba. When mixed with triclopyr or MCPP, fluroxypyr controls buckhorn plantain, white clover, black medic, and ground ivy (Loughner et al., 2004). Centipedegrass and zoysiagrass are tolerate at rates less than 0.5 lb ai/a (0.56 kg ai/ha), while bermudagrass and St. Augustinegrass can tolerate rates only of 0.12 lb ai/a [0.13 kg ai/ha] (Handly et al., 2001). Delayed common bermudagrass spring transition can occur when a midwinter application of fluroxypyr is applied at 0.5 lb ai /a [0.56 kg ai/a] (Murphy and Johnson, 1995). Table 6.10 lists the effectiveness of commonly used postemergence herbicides for broadleaf-weed control.

Carfentrazone-ethyl (Quicksilver) and sulfentrazone (Dismiss) belong to the triazolinone family. These compounds are protox inhibitors and generally provide optimum results against young weeds within one to two weeks after application at low dose rates of 0.018 lb ai/a [20 g ai/ha] (Cauchy, 2000). These herbicides cause rapid necrosis and death of broadleaf weeds by disrupting photosynthesis and by subsequent membrane degradation (Weston and Barney, 2004). Due to their quick response, carfentrazone and sulfentrazone are often tank-mixed partners with phenoxy-based herbicides and dicamba. These can be safely applied to newly seeded cool-season grasses, and reseeding can be done in as little as two weeks following an

Table 6.10

Expected Control of Broadleaf Weeds with Turf Herbicides (McCarty, 2008)

Weed	Life cycle	Atrazine/Simazine	2,4-D	MCPP	Dicamba	2,4-D + MCPP	2,4-D + 2,4-DP	2,4-D + MCPP + Dicamba	Bentazon	Bromoxynil	Chlorsulfuron	Clopyralid	Imazaquin	Imazapic	Metsulfuron	Triclopyr	2,4-D + Triclopyr	Triclopyr + Clopyralid	MCPA + Triclopyr + Clopyralid	Carfentrazone + 2,4-D1 MCPP + MCPA &/or Dicamba	Quinclorac	Quinclorac + 2,4-D1 Sulfentrazone + Dicamba	Sulfentrazone + 2,4-D + MCPP + Dicamba	Fluroxypyr	Fluroxypyr + 2,4-D + Dicamba	Sulfentrazone	Carfentrazone	Pyraflufen-Ethyl	Mesotrione
Aster	P[1]	—	G	—	—	F	G	F	P	P	—	G	—	—	G	—	F	G	G	G	—	G	G	—	G	—	—	—	—
Bedstraw, smooth	P	—	P	P-F	G	F	F	G	—	—	G	—	—	G	P	F-G	G	G	G	G	—	G	G	E	G	—	—	G	—
Beggarticks	A	G	G	—	—	—	G	G	G	—	—	—	—	—	—	—	G	G	G	G	—	E	—	—	G	—	—	G	—
Betony, Florida	P	F-G[2]	F	F	F-G	F	F-G	F-G	P	P	—	—	—	—	G	—	G	G	—	G	—	G	G	—	G	—	—	—	G
Bittercress, hairy	WA	—	E	F	E	E	E	E	—	—	—	—	G	—	E	—	—	—	—	—	—	—	—	—	G	G	—	—	—
Bindweed, field	P	—	G	G	G	E-F	G	E	P-F	P	—	—	—	—	—	G	G	—	G	G	E	G	F-G	G	G	—	—	—	—
Burclover	A	—	F-P	E	E	E-F	E	E	—	—	F	G	—	—	G	G	—	—	—	G	—	—	F-G	—	G	—	—	—	—
Buttercups	WA, B&P	F	G	F	F-G	E	E	E	P	P	G	G	G	—	E	—	G	E	G	G	—	E	G	—	G	G	—	—	G
Buttonweed, Virginia	P	—	F	P-F	F	F	E-F	E-F	P	P	F	F	—	—	G	F	F-P	G	G	G	—	G	G	G	G	—	F	—	—
Carpetweed	SA	E	G	F	E	E	E	E	—	—	—	—	—	—	P	—	G	—	G	G	—	E	G	—	G	G	G	G	G
Carrot, wild	A, B	—	G	F	E	G	P-F	E	—	—	G	—	—	—	E	G	F	G	G	G	—	E	G	—	G	—	—	—	—
Chamberbitter	SA, P	G-E	P	—	—	—	—	—	P	—	—	—	P	—	E	—	E	—	—	—	—	—	—	—	—	—	—	—	—
Chickweed, common	WA	E	P	G	G	E	E	E	F-G	P	G	—	G	—	E	—	E	E	E	G	—	E	G	G	G	G	F	G	G
Chickweed, mouse-ear	WA, P	F-G	G	G	G	E	E	E	P	P	G	P	G	—	E	P-F	E-F	E	E	G	—	G	G	G	G	G	F	G	G
Chicory	P	—	G	E	G	E	E	E	—	—	—	—	—	—	E	G	G	—	G	G	—	G	G	—	G	—	—	—	—
Cinquefoil, common	P	—	E-F	E-F	E-F	E-F	E-F	E-F	—	—	F	—	—	—	—	—	—	—	—	G	—	G	G	—	G	G	—	—	—
Clover, crimson	SA	—	G	G	G	E	E	E	—	—	G	G	—	—	—	—	—	E	E	G	E	E	G	—	G	—	F	—	—
Clover, hop	WA	E	F-G	G	G	E	E	E	—	F	G	G	—	—	F	—	E	E	E	G	E	E	G	—	G	—	F	—	—
Clover, white	P	E	F-G	G	G	E	E	E	—	—	G	G	G	—	E	F-G	E-F	E	E	G	E	G	G	G	G	G	F	G	G
Cudweed	WA	G-E	G-E	—	E	G-E	G-E	E	—	G	—	—	G	—	E	—	G-E	G-E	G	—	—	E	G	—	G	G	—	—	—
Daisy, English	P	—	P	F	G	G	F	E	P	P	—	F	—	—	—	—	—	G	G	G	F	F-G	F-G	—	G	—	—	—	—
Daisy, oxeye	P, B	—	F	F	F	F	F	E-F	—	—	—	—	F	—	—	—	—	—	—	—	—	G	G	—	G	—	—	—	—
Dandelion	P	E-F	G	G	G	E	E	E	P	P	G	F-G	P-F	—	E	G	F-E	G	G	G	F-G	G	G	F-G	G	G	F	G	G
Dandelion, catsear	P	—	E-F	F	E	E	E	E	—	—	—	—	—	—	—	—	G	E	E	—	—	G	—	—	G	—	—	—	—
Dayflower, spreading	SA	G-E	F	F	F	F-G	F-G	F-G	G	—	—	—	G	—	G	—	F-G	—	—	G	P	E	G	—	—	—	—	—	—

(continued)

Table 6.10

Expected Control of Broadleaf Weeds with Turf Herbicides (McCarty, 2008) *(Continued)*

Weed	Life cycle	Atrazine/Simazine	2,4-D	MCPP	Dicamba	2,4-D + MCPP	2,4-D + 2,4-DP	2,4-D + MCPP + Dicamba	Bentazon	Bromoxynil	Chlorsulfuron	Clopyralid	Imazaquin	Imazapic	Metsulfuron	Triclopyr	2,4-D + Triclopyr	Triclopyr + Clopyralid	MCPA + Triclopyr + Clopyralid	Carfentrazone + 2,4-D1 MCPP + MCPA &/or Dicamba	Quinclorac	Quinclorac + 2,4-D1 Sulfentrazone + Dicamba	Sulfentrazone + 2,4-D + MCPP + Dicamba	Fluroxypyr	Fluroxypyr + 2,4-D + Dicamba	Sulfentrazone	Carfentrazone	Pyraflufen-Ethyl	Mesotrione
Spurry, corn	P	—	F	—	F-G	F	F	G	—	F-G	—	—	—	—	—	—	F	F	—	—	—	—	—	—	—	—	—	—	—
Spurweed (lawn burweed)	WA	F-G	F	E-F	E	E-F	F-G	E	E	F-G	—	—	—	—	G-E	F-G	E	E	G	—	—	E	—	—	G	G	F	—	G
Strawberry, Indian mock	P	—	P	F	E-F	F	P	E-F	—	—	—	—	—	—	—	—	—	—	—	—	—	G	G	—	G	—	—	—	—
Thistles	B, P	P	G	G	G	E-F	E-F	E	G	G	F	G	G	F	P-F	G	—	G	G	G	—	E	G	—	G	—	F	—	G
Vetch, common	WA, SA	E	G	G	G	G	F	G	—	—	—	G	G	—	E	G	G	E	G	—	G	G	—	—	G	—	—	—	—
Violet, Johnny-jump-up	WA	—	F-P	F-P	E-F	F-P	F	F-P	P	P	—	—	P-F	—	E	F	—	F-G	F-G	—	—	G	—	—	—	—	—	—	G
Violet, wild	P	—	F-P	F-P	E-F	F-P	F	F-P	P	P	F	—	—	—	—	F	F	F-G	F-G	G	—	F-G	F-G	—	—	G	—	—	G
Woodsorrel, creeping	P	F	P	P	G	P-F	P-F	P-F	P	P	—	—	—	—	F-G	F-G	F-G	F	—	G	—	G	—	—	G	G	—	—	—
Woodsorrel, yellow	P	F-G	P	P	G	F-P	F-P	F-P	P	P-F	—	P	—	G	E-F	F-G	—	E-F	—	G	—	G	—	—	G	G	—	—	G
Yarrow	P	—	F	F	E	G	G	E-F	P	P	G	—	—	—	F-G	F-G	G	—	G	G	—	G	G	—	G	—	—	—	—

[1]A = annual; B = biennial; P = perennial; SA = summer annual; WA = winter annual.

[2]E = excellent (>89%) control; F = Fair to good (70 to 89%); G = Good control, sometimes with high rates; however, a repeat treatment one to three weeks later, at the standard or reduced rate, usually is more effective, especially on perennial weeds; P = poor control (<70% in most cases). Not all weeds have been tested for susceptibility to each herbicide listed.

application. When these herbicides are applied alone, average to moderate control will occur. However, when they are mixed with other broadleaf herbicides, a synergistic response occurs, significantly improving weed control.

Postemergence Grass Weed Control

Postemergence grass weed control traditionally has been through single and repeat applications of the organic arsenicals (e.g., MSMA, DSMA, CMA). Two to four applications spaced 7 to 14 days apart are required for complete weed control (McCarty and Yelverton, 2005). Repeat applications are needed for control of annual weeds and for most perennial weeds. The rate and number of applications necessary for control also increases as weeds mature. Organic arsenicals can be phytotoxic on cool-season turfgrasses, especially when used during high temperatures (>90°F; 32°C). Control also is reduced if rainfall occurs within 24 hours of treatment (Bingham, 1985). Recently, new herbicide releases have provided alternatives to the organic arsenicals for postemergence grass weed control (Tables 6.11 and 6.12). Decreased phytotoxicity, as well as reduced number of applications, often are associated with these herbicides. The following subsections discuss herbicides available for various turfgrass species.

Warm-Season Turfgrasses

Bermudagrass and Zoysiagrass. Postemergence control of crabgrass species and goosegrass species usually has been with organic arsenicals. As previously mentioned, repeat applications at short intervals are required, especially as weeds mature. Phytotoxicity usually increases when bermudagrass and zoysiagrass are treated with repeat applications. Zoysia spp. tend to be more sensitive to organic arsenical applications. Phytotoxicity, or leaf yellowing, normally follows for 10 days to 2 weeks after application. Seven to 10 days of phytotoxicity normally accompanies treatments on bermudagrass.

In order to increase herbicidal activity on goosegrass, various combinations of arsenicals with other herbicides have been tested. High rates (e.g., 0.19 to 0.38 lbs ai/a, 0.2 to 0.42 kg ai/ha) of metribuzin 75 DF (an asymmetrical triazine) provide excellent control of goosegrass when the triazine is used alone, but it has marginal bermudagrass tolerance. Lower rates (e.g., 0.094 lb ai/a, 0.1 kg ai/ha) of metribuzin with arsenical herbicides have shown good to excellent goosegrass control (Bingham, 1985; Johnson, 1980). This combination can be safely used only on well-established bermudagrass that is maintained at mowing heights greater than 0.5 inch (1.25 cm) and that is actively growing and not under drought stress. The use of metribuzin increased herbicidal activity on goosegrass, but a certain degree of phytotoxicity and a number of escaped weeds still existed. Metribuzin also has inhibited photosynthesis of bermudagrass for a certain period. Bermudagrass discoloration typically lasts for one to two weeks, but can last for five to eight weeks following an application of metribuzin with arsenical herbicides (Wiecko, 2000).

Diclofop-methyl 3EC, a member of the aryl-oxy-phenoxy herbicide family, is superior in goosegrass control compared with both the individual forms and combinations of the organic arsenicals and metribuzin. Little damage to bermudagrass has resulted, and repeat applications usually are not necessary (McCarty et al., 1991; Murdoch and Nishimoto, 1982). This herbicide seems to be more active on goosegrass that is maintained at lower mowing heights and thus is less mature. The addition of MSMA is possibly antagonistic for goosegrass control, while the addition of metribuzin to diclofop unfavorably injures the turf. Injury observed from metribuzin with diclofop would be less than or equivalent to metribuzin with arsenical herbicides, with injury persisting for at least three weeks after application (Nishimoto and Murdoch, 1999). Goosegrass control is relatively slow and often requires two to three weeks to become effective. Weed control spectrums also appear to be limited, with

Table 6.11

Established Turfgrass Tolerance to Postemergence Grass Herbicides (*Refer to Herbicide Label for Specific Species Listing*)

These are relative rankings and depend on factors such as environmental conditions, turfgrass vigor or health, application timing, etc., and are intended only as a guide.

Herbicides (Trade Names)	Bahiagrass	Bentgrass Fairways	Bentgrass Greens	Bermudagrass	Buffalograss	Carpetgrass	Centipedegrass	Fine Fescue	Kentucky Bluegrass	Kikuyu-Grass	Overseeded Ryegrass/Blends	Ryegrass	Seashore Paspalum	St. Augustinegrass	Tall Fescue	Zoysiagrass
asulam (Asulox)	NR	NR	NR[1]	S-I[2]	NR-I	NR	NR	NR	NR	NR	NR	NR	NR	S-I	NR	NR-I
bispyribac-sodium (Velocity)[3]	NR	NR	NR	S[3]	NR	NR	NR	NR	NR	NR	S[4]	S	NR	NR	NR	NR
clethodim (Envoy)	NR	NR	NR	NR	NR	NR	S	NR	NR	NR	NR	NR	NR	NR	NR	NR
diclofop (Illoxan)	NR	NR	NR	S	NR-S	NR	NR	NR	NR	NR	NR	NR	NR	NR	NR	NR
DSMA, MSMA, CMA	NR	I	NR-I	S-I	I	NR	NR	I	I	NR	NR	S-I	NR-P	NR	I	S-I
ethofumesate (Prograss)[4]	NR	I	NR-I	D	NR	NR	NR	I	S	NR	I	S	NR-S	NR	S	NR
fenoxaprop (Acclaim Extra)	NR-I	I	NR-I	NR-I	NR	NR	NR	S	S	NR	I	S	NR	NR	S	I
fluazifop (Fusilade II)	NR	NR	NR	NR	NR	NR	NR	NR	NR	NR	NR	NR	NR-P	NR	S-I	I
foramsulfuron (Revolver)	NR	NR	NR	S	NR	NR	NR	NR	NR	NR	NR	NR	NR	I	NR	S
mesotrione (Tenacity)	NR	NR	NR	NR	NR	NR	S	S-I	S	NR	NR	S-I	NR	S-I	S-I	NR
metribuzin (Sencor Turf)	NR	NR	NR	S-I	NR	NR	NR	NR	NR	NR	NR	NR	NR-I	NR	NR	NR
pronamide (Kerb)	S	NR	NR	S	NR	NR	S	NR	NR	NR	NR	NR	NR-S	S	NR	S
rimsulfuron (TranXit)	NR	NR	NR	S	NR	NR	NR	NR	NR	NR	NR	NR	NR	NR	NR	NR
sethoxydim (Vantage)	NR	NR	NR	NR	NR	NR	S	S	NR	NR	NR	NR	NR-P	NR	NR	NR
sulfosulfuron (Certainty)	I	NR	NR	S	S	NR	S-I	NR	NR	S	NR	NR	NR	S-I	NR	S
trifloxysulfuron (Monument)	NR	NR	NR	S	NR	NR	NR	NR	NR	NR	NR	NR	NR-P	NR	NR	S
quinclorac (Drive)	NR	I	NR	S-I	S	NR	NR	I	S	NR	S	S	NR-S	NR	S	S

[1]S = Safe at labeled rates; I = Intermediate safety, use at reduced rates; NR = Not Registered for use on and/or damages this turfgrass; D = Dormant turf only.
[2]Asulam is labeled for 'Tifway' (419) Bermudagrass and St. Augustinegrass.
[3]Used on dormant bermudagrass overseeded with perennial ryegrass.

Table 6.12

Guide to Grass Weed Control with Postemergence Turfgrass Herbicides (*Refer to Herbicide Label for Specific Turf Species Use Listing*)

Herbicide[1]	Crabgrass	Goosegrass	Annual Bluegrass	Sandspur	Dallisgrass	Thin Paspalum	Ryegrass	Smutgrass	Bahiagrass	Carpetgrass	Tall Fescue	Bermudagrass	Quackgrass
atrazine (Aatrex)	P-F[2]	P	G-E	F	P	P	G-E	F-G	F	P	F	P-F	F
asulam (Asulox)	G	F	P	F	P	P-F	—	F	P	G	P	P	—
bispyribac-sodium (Velocity)	—	—	G	—	—	—	P	—	—	—	—	P	—
chlorsulfuron (Corsair, TFC)	P	P	P	P	P	P	G	F	P	P	G	P	—
clethodim (Envoy)	E	G-E	G	G	—	—	G-E	—	—	—	P	G	G
diclofop (Illoxan)	P	G-E	P	P	P	P	G	P	P	P	P	P	—
DSMA, MSMA	G	F	P	G	F	F-G	P	P	F	G	P	P	—
ethofumesate (Prograss)	P	P	F-G*	P	P	P	P	P	P	—	P	P-G	—
fenoxaprop (Acclaim)	G-E	G-E	P	G	P	P	P	P	G	—	P	F-G	—
fluazifop (Fusilade II)	G-E	G	F	G	P	P	G-E	P	G	—	P	G	G
foramsulfuron (Revolver)	P	G	E	—	F	—	E	—	—	—	E	P	—
imazapic (Plateau)	G	G	P	P	F	—	F	—	F	—	G	P	P
metribuzin (Sencor)	F-G	G-E	G	—	F	P	F	P	P	—	F	P	—
metsulfuron (Manor)	P	P	P	P	P	P	G	P	G	P	F	P	—
pronamide (Kerb)	P	P	G-E	P	P	P	G-E	P	P	—	G	P	F-G
rimsulfuron (TranXit)	P	P	G	P	P	P	G	P	P	P	P	P	P
sethoxydim (Vantage)	G-E	G	P	G	P-F	P	P	P	G	P	P	F-G	F-G
simazine (Princep T&O)	P-F	P	G-E	P-F	P	P	G-E	F	F	P	F	P-F	F
sulfosulfuron (Certainty)	P	P	G	—	P	P	P	—	P	P	G	P	G
trifloxysulfuron (Monument)	P	P	E	—	F	—	E	—	F	—	E	P	—
quinclorac (Drive)	E	P	P	—	F	P	P	P	P	P	P	P	—

[1]Repeat applications usually 5 to 14 days apart are needed for most herbicides and weeds. This is especially true as weeds mature, producing flowers and seedheads.

[2]E = excellent (>90%) control with one application; G = good (80 to 90%) control with one application; F = Fair to good (70 to 89%), good control sometimes with high rates, however a repeat treatment 1 to 3 weeks later each at the standard or reduced rate is usually more effective; P = poor (<70%) control in most cases.

— = Control unknown as all weeds have not been tested for susceptibility to each herbicide listed.

*Ethofumesate provides good to excellent control of most true annual biotypes of annual bluegrass but only fair control of perennial biotypes.

goosegrass being the most susceptible annual grass weed. Treated areas should not be overseeded with perennial ryegrass for at least six weeks after application (McCarty and Murphy, 1993). Zoysiagrass tolerance to diclofop-methyl has not been reported.

Quinclorac 75 DF (Drive) is a member of the quinolinecarboxylic acid herbicide family and has shown potential for postemergence crabgrass control in bermudagrass and in certain cool-season grasses. Bahiagrass, centipedegrass, and St. Augustinegrass are very sensitive, and extensive injury occurs if they are treated with quinclorac. Crabgrass control is greater when quinclorac is applied at 0.75 lb ai/a (0.84 kg ai/ha) in the spring at the two-leaf to five-tiller plants than when applied in late summer on mature crabgrass, with greater than five tillers. Quinclorac also has provided selective torpedograss suppression in bermudagrass, an option previously not available (McCarty, 1992). A series of two or three applications is normally required for suppression. Selective broadleaf-weed control also is available with quinclorac, but goosegrass control is lacking. Among those broadleaf weeds controlled are black medic, white clover, common dandelion, dollarweed, and speedwell (Yelverton, 2002). Quinclorac also controls barnyardgrass, foxtail, and broadleaf signalgrass. Various combinations of quinclorac with other herbicides have been introduced to provide a wider spectrum of control. For example, combinations of quinclorac plus 2,4-D, dicamba, and sulfentrazone are used to control certain grasses, broadleaf weeds, and nutsedges.

Fenoxaprop-ethyl 0.57 EC, another member of the aryl-oxy-phenoxy herbicide family, has been shown to control annual grass weeds, especially crabgrass. Zoysiagrass has good tolerance to fenoxaprop, while bermudagrass lacks tolerance to it (Higgins et al., 1987). Fenoxaprop shows excellent activity on small crabgrass, but is less effective on tillered crabgrass, while barnyardgrass, goosegrass, foxtail, and sandbur are controlled by fenoxaprop (Yelverton, 2002). Fenoxaprop and fluazifop suppress bermudagrass in cool-season turfgrasses, especially tall fescue. Applications should be made in the spring soon after bermudagrass greens up and should be repeated monthly until control is achieved. Control is increased by tank-mixing triclopyr ester (Turflon Ester 4L) at 1 qt/acre (1 lb ai/a, 1.12 kg ai/ha) (Willis and McCarty, 2006). Phytotoxicity will occur, especially with turf under drought and heat stress.

Foramsulfuron 0.19 L, a sulfonylurea herbicide, has recently been approved for use on both zoysiagrass and bermudagrass and is highly toxic to cool-season grasses. Use rates range from 0.006 to 0.025 lb ai/a (0.007 to 0.028 kg ai/ha). Foramsulfuron provides selective control of perennial ryegrass in bermudagrass. The transition can be very rapid, as quick as two to four weeks (Yelverton, 2003). Therefore, applications should be made as a late transition aid when the underlying bermudagrass is actively growing. Goosegrass control with foramsulfuron in bermudagrass turf occurs within two to four weeks after sequential applications spaced two weeks apart (Higingbottom et al., 2005). With the addition of metribuzin, enhanced goosegrass control was shown with acceptable turf injury, compared with an MSMA plus metribuzin tank mix (Higgingbottom et al., 2005). Occurrence of turf injury is minimal on both zoysiagrass and bermudagrass. Control on other cool-season grasses, such as tall fescue, creeping bentgrass, and *Poa* spp., as well as henbit and centipedegrass, is provided by foramsulfuron (Yelverton, 2003). Foramsulfuron is ineffective in controlling tropical signalgrass and crabgrass (Busey, 2004).

Rimsulfuron 25 DG, another recently introduced sulfonylurea herbicide, is similar to foramsulfuron, with low use rates ranging from 0.031to 0.062 lb ai/a (0.035 to 0.069 kg ai/ha), and is effective in postemergence control of annual bluegrass and perennial ryegrass. Rimsulfuron also selectively transitions both perennial ryegrass and *Poa trivialis* out of overseeded bermudagrass, with a mid-spring application at a rate of 0.032 lb ai/a [0.036 kg ai/ha]. Spring applications of rimsulfuron have little to no effect on bermudagrass green-up (Walker et al., 2003). Synergistic response occurs with rimsulfuron at 0.016 lb ai/a (0.018 kg ai/ha) when mixed with a 28 percent urea ammonium nitrate (3 gals/acre [28 L/ha]). Also, enhanced bermudagrass green-up occurs in the spring, even at rates as high as 0.0625 lb ai/a (0.07 kg ai/ha) (Walker et al., 2003).

Sulfosulfuron 75 WDG (Certainty), like rimsulfuron and foramsulfuron, is a recently introduced sulfonylurea herbicide. It controls nutsedge species, kyllinga species, and a few broadleaf weeds. Sulfosulfuron provides postemergence selective control for annual bluegrass in warm-season turfgrasses. It can be safely applied to common or hybrid bermudagrass, zoysiagrass, centipedegrass, St. Augustinegrass, and kikuyugrass at 0.75 to 2.0 ounces product per acre (0.05 to 0.14 kg/ha). Sulfosulfuron selectively removes tall fescue in bermudagrass, with one application at 1.0 ounce product per acre (0.07 kg/ha) (Murphy and Nutt, 2003). Among those broadleaf weeds labeled for control are catweed bedstraw, buttercup, common chickweed, white clover, henbit, mustard, penny cress, and shepherd's purse.

Centipedegrass

The triazine herbicides traditionally have been used for postemergence grass weed control in centipedegrass. However, repeat applications are necessary for control of mature weeds and may increase the potential of turfgrass injury. Triazine herbicides provide variable postemergence grass weed control, but lose their effectiveness as weeds mature. The organic arsenicals, commonly used in other turfgrass species, are highly injurious to centipedegrass.

Sethoxydim, a member of the aryl-oxy-phenoxy herbicide family, provides good–to-excellent control of many annual grass weeds and also suppresses several perennial grass weeds, including bahiagrass and bermudagrass. Centipedegrass tolerance to sethoxydim is excellent (McCarty et al., 1986). The discovery of sethoxydim's tolerance to centipedegrass has been a major breakthrough in postemergence annual grass control in centipedegrass.

Clethodim, likewise an aryl-oxy-phenoxy herbicide, also suppresses bermudagrass in centipedegrass sod production (Waltz et al., 2001). Annual and perennial weeds such as Johnsongrass, quackgrass, tall fescue, annual bluegrass, and crabgrass can be controlled by clethodim. Clethodim is also widely used in landscape beds to control perennial grasses Daylilies, iris, hosta, and liriope are tolerant of clethodim, as are most broadleaf ornamentals. Control, however, is lacking with clethodim on broadleaf weeds, sedges, and rushes.

St. Augustinegrass

One major problem with growing St. Augustinegrass is its relatively poor tolerance to most postemergence herbicides. Triazine herbicides (e.g., atrazine, simazine) provide fair control of some annual grass weed species, but only in the juvenile growth stage. Repeat applications often are necessary for satisfactory control, usually resulting in some turf phytotoxicity.

Asulam, a member of the carbamate herbicide family, provides control of some annual grass weeds, without serious damage to St. Augustinegrass. Repeat applications 7 to 14 days apart may be required for control of older weeds. This increases the likelihood of damage to St. Augustinegrass. As with most other postemergence herbicides, asulam should be applied when temperatures are below 85°F (29.4°C), good soil moisture is present, and the turf is actively growing.

Recently, mesotrione (Tenacity 4L) has become available for early postemergence grass control in St. Augustinegrass. Two applications two to three weeks apart will be necessary as weeds mature. Mesotrione also controls selective broadleaf weeds.

Bahiagrass

Bahiagrass, like St. Augustinegrass, is somewhat sensitive to most postemergence herbicides. This sensitivity limits the number of herbicides available for use on it. Although labeled as being appropriate for use on bahiagrass, most postemergence broadleaf (e.g., 2,4-D, dicamba, and/or mecoprop) herbicides will cause turf yellowing, especially if applied when temperatures are hot or the turf is under stressful conditions. Normally, the phytotoxicity is not lethal, and turf recovery can be expected within one to two weeks.

Postemergence grass weed control in bahiagrass is possible with hexazinone (Velpar) (Brecke, 1981). Only low-maintenance well-established bahiagrass areas on roadsides, railroads, and utility right-of-ways should be treated. Various broadleaf and grass weeds are controlled if there is adequate soil moisture to activate the herbicide. Early spring applications are recommended. Temporary turf phytotoxicity can be expected following application. On high-maintenance bahiagrass, no selective postemergence herbicide is available to control grass weeds. Spot spraying and selective placement such as rope-wicking with a nonselective herbicide such as glyphosate are the only means of chemical control of grass weeds under these conditions.

Cool-Season Turf Grasses Postemergence grass weed control in cool-season turfgrasses traditionally has been limited to various members of the organic arsenicals (McCarty, 2002). Specific formulations and specific rates are necessary for its use on most cool-season turfgrasses, or else injury may result. Plant damage can be minimized through proper timing, such as during the young weed-growth stage, during mild environmental conditions, and while turfgrass is actively growing. All of these conditions should be considered before any of these herbicides are used.

More recently, fenoxaprop has been found to control tillered crabgrass when adequate soil moisture is present. Fenoxaprop is also tolerated by some cool-season turfgrasses. Seedling and mature perennial ryegrass and tall fescue have shown little injury from fenoxaprop (McCarty et al., 1989). Kentucky bluegrass also is tolerant, but sometimes can be injured when the herbicide is applied in the spring. Young monostands of Kentucky bluegrass have been shown to be injured by fenoxaprop. However, when it is in competition with a severe infestation of crabgrass, the loss through injury may be outweighed by the improved turf cover the following fall (Dernoeden, 1989b). Creeping bentgrass also is injured by fenoxaprop, but recovers within several weeks of application (Higgins et al., 1987. Safeners may reduce fenoxaprop damage on bentgrass, as may the addition of iron with and without methylene urea (Dernoeden, 1989b). A low rate of fenoxaprop 0.57 EC (0.04 lb ai/a, 0.05 kg ai/ha) reduces the degree of bentgrass discoloration (Carroll et al., 1992), but may result in inconsistent crabgrass control.

Quinclorac is the most recent postemergence herbicide that controls certain annual grassy weeds in cool-season grasses. Crabgrass, foxtail, broadleaf signalgrass, and barnyardgrass are grassy weeds controlled with quinclorac, as are certain broadleaf weeds. As with all postemergence herbicides, quinclorac works best on small, juvenile weeds and when good turf-growing conditions exist. Turf tolerance by cool-season turfgrasses is good.

Mesotrione has been widely used in grain crops, but is now available in the turfgrass market. It controls both broadleaf and grassy weeds in selective turf species, mainly cool-season turfgrasses. Mesotrione's mode of action is to block the biosynthesis of two essential pigments. This disruption of pigment synthesis causes bleaching symptoms in susceptible plants that leads to necrosis and plant death. In mixed stands of Kentucky bluegrass and creeping bentgrass, mesotrione at 0.28 kg ai/ha (0.25 lb ai/a) can be used in selectively removing creeping bentgrass, with minimal injury to the bluegrass (Bohwmik and Riego, 2003; Askew et al., 2003). The use of multiple applications at lower rates (0.125 to 0.187 lb ai/a [0.14 to 0.21kg ai/ha]) has also provided excellent control of creeping bentgrass in Kentucky bluegrass, with minimal injury to the bluegrass.

Cool-season grasses like bluegrass, tall fescue, and perennial ryegrass all have shown excellent tolerance to mesotrione, with minimal injury lasting no longer that 14 days (Beam et al., 2006). The new hybrid bluegrass (*Poa pratensis* × *Poa arachnifera*) has also shown excellent tolerance to mesotrione even when treated in the seedling stage (McElroy et al., 2004). It provides excellent control of quackgrass when applied at 0.28 kg ai/ha twice weekly at two- to four-week intervals, but will not acceptably control bermudagrass or mature goosegrass (Askew et al., 2002a; Askew et al., 2003). Nimblewill can be suppressed with the use of mesotrione at rates ranging from 0.03 to 0.28 kg ai/ha, with lower rates requiring multiple applications (Askew, et al., 2002b). At 0.45 kg ai/ha (0.4 lb ai/a), mesotrione provides good control (85 percent) of large crabgrass, 50 days after treatment (Askew et al., 2003). Multiple applications of mesotrione at either 0.125 or 0.25 kg/ha (0.11 or 0.22 lb/a) controlled crabgrass greater than 90 percent (McElroy et al., 2004). Mesotrione is ineffective in controlling smooth crabgrass when applied as a single or sequential preemergence treatment (Dernoeden et al., 2003). Mesotrione has been shown to suppress Virginia buttonweed and ground ivy, but will be ineffective in controlling white clover (Askew et al., 2003). Multiple applications will be needed for mesotrione to be effective and to achieve complete control. Reduced rates at multiple timings have been shown to provide superior control compared with single applications at higher rates.

ANNUAL BLUEGRASS CONTROL

Biology

Annual bluegrass is the most troublesome weed in golf greens. Its low growth habit and its ability to thrive in moist conditions and compacted areas make it difficult to control. Annual bluegrass has a lighter green color than most grass species used to overseed golf greens. It also produces numerous seedheads that reduce the quality of the putting surface. Because of its low tolerance to heat, annual bluegrass dies quickly in warm weather, leaving many playing areas bare until the bermudagrass has time to fill in (Engel and Ilnicki, 1969). Chemical control of annual bluegrass is difficult to achieve due to three factors: (1) The majority of preemergence herbicides cannot selectively prevent annual bluegrass germination while allowing the overseeded grass to establish itself; (2) most effective postemergence herbicides for annual bluegrass also injure bentgrass or the overseeded grass species; and (3) many different biotypes exist, with some being true annuals and others being perennials.

The annual bluegrass biotype (*Poa annua* ssp. *annua* [L.]) has a non-stoloniferous bunch-type growth habit and, generally, a light-green color. It is a true annual and begins to germinate when daytime temperatures consistently drop into the mid-70s Fahrenheit (~24°C) and nighttime temperatures are in the mid-50s Fahrenheit (~13°C) for several consecutive days in late summer and early fall (Table 6.13). Maximum seed germination occurs when full sunlight is present; therefore, thin, weak turf stands often have the earliest annual bluegrass germination. Areas that remain cooler, such as shaded or continuously wet areas, also have earlier annual bluegrass

Table 6.13
GROWTH RESPONSES OF ANNUAL BLUEGRASS AND BENTGRASS TO VARYING ENVIRONMENTAL PARAMETERS

		Factors Favoring *Poa* or Bentgrass Growth		
***Poa* Growth Response**	**Temperature (°F)**	**Factor**	***Poa annua***	**Bentgrass**
Germination	mid-70s, daytime	Moisture	wet	drier
	mid-50s, nighttime	Compaction	high	low
Maximum growth	60 to 70	N rates	high	low
Maximum root growth	65 to 70 (soil temperatures)	Soil Ph levels	high	low
Seedhead development	70 to 80	Clippings	leave	remove
Maximum heat tolerance	85 to 95	Soil pH	6 to 7	5.5 to 6.5
Minimum growth	50 (soil temperature)			
Lethal cold temperature	.5			

seed germination. Another flush of germination typically occurs in early winter, generally from mid-December through mid-January when daytime temperatures are warm and nighttime temperatures are cold. This alternating warm–cold temperature scarifies additional seed, thus encouraging another flush of germination. Many herbicides applied in late summer for the initial fall flush of germination will not satisfactorily control this second germination; hence, repeat applications may be necessary for satisfactory season-long control.

After germination, annual bluegrass grows and then tillers (mostly unnoticeably) throughout the fall and early winter months. Once late winter arrives, annual bluegrass begins to shift its growth from vegetative toward reproductive by forming numerous seedheads that can literally turn a turf stand snow-white in color. The annual biotype reduces its growth in late spring and dies following flowering or when daytime temperatures reach the lower 90s Fahrenheit (~32°C) for several consecutive days. Seed can remain viable in the soil for more than six years, thereby ensuring a continued supply.

Several cultural factors influence annual bluegrass growth and occurrence. The first step, therefore, in a total annual bluegrass management program would be to shift the following to those cultural factors favoring turfgrass growth:

a. *Continuous wet and compacted soils.* Annual bluegrass thrives under wet and/or compacted-soil conditions. Due to its shallow root system, annual bluegrass can tolerate lower soil oxygen levels; turfgrass stands, by contrast, will begin to thin. Appropriate soil mixtures should be used to reduce soil compaction, soils should be frequently aerified, and greens should be mowed with walk mowers. Greens also should be spiked frequently to reduce surface compaction and to sever turfgrass stolons in order to encourage a thicker turf stand. Also, do not overwater.

b. *Excessive nitrogen rates.* A high available nitrogen supply will encourage profuse annual bluegrass occurrence, growth, and tillering.

c. *Excessive soil phosphorus levels.* Annual bluegrass prefers excessive P levels and will out-compete turfgrasses under these conditions. Insufficient P levels are believed to reduce the number and vigor of viable seed. Supply sufficient P as indicated by soil test for the turf, but do not use excessive rates.

d. *Leaving clippings.* Due to the tolerance of annual bluegrass to low mowing height and abundant seed production, leaving clippings only helps spread annual bluegrass seed and thus slowly increases its occurrence.

e. *Overseeding.* In tropical and subtropical regions, overseeding bermudagrass provides winter and early spring turf color for golfers. However, overseeding greatly encourages annual bluegrass encroachment, since prepping bermudagrass via aerifying and verticutting results in poor stand density. This then opens the turf to more sunlight, which encourages annual bluegrass to germinate. The increased watering need for overseeding establishment also encourages annual bluegrass. Overseeding also weakens the bermudagrass in spring, extending the time needed for recovery, allowing annual bluegrass to encroach slowly. Overseeding also restricts the number of herbicide options available, as controlling one cool-season turfgrass growing in another is very difficult, especially if the overseeded grass is *Poa trivialis*, a close relative of *Poa annua.*

In addition, a buffer zone of overseeding around the approach helps intercept annual bluegrass seed from golfers and equipment. Courses with epidemic annual bluegrass levels may opt to skip overseeding fairways until the population is brought under control. Others may overseed fairways, but skip approaches as more control options are available for non-overseeded areas.

Herbicide-Resistant *Poa annua*

Besides those listed management practices that encourage *Poa annua* stands, a recent occurrence of herbicide-resistant annual bluegrass biotypes has further reduced the number of effective control options. Resistant biotypes to dinitroaniline herbicides (DNAH), simazine, atrazine (Princep and Aatrex), and ethofumesate (Prograss) have been noted (Isgrigg et al., 2002; Kelly et al., 1999; Lowe et al., 2001).

Resistance to triazine herbicides is a widespread problem involving nearly 60 weeds species. Two modes of resistance in triazine resistant species exist: target site modification and accelerated metabolism. Target site modification is responsible for *Poa annua* resistance to triazine herbicides. The resistance is caused by amino acid residues changing on the Q_b-binding site on the D1 protein of photosystem II (PSII). This modification reduces the herbicide's ability to compete for the exchangeable plastoquinone Q_b, causing the herbicide to no longer bind and thus creating an immunity to the herbicide (Kelly et al., 1999). The *psbA* gene that codes the D1 protein was sequenced, and it was found that triazine-resistant and triazine-susceptible plants were identical except for one amino acid residue at position 264 (Bettini et al., 1987; Goloubinoff et al., 1984). The substitution of glycine for serine caused a reduction in affinity for the triazine herbicide at the Q_b-binding site, thus creating triazine-resistant *Poa annua* biotypes (Kelly et al., 1999). Similar target-site modification has been shown to cause resistance in the dinitroaniline herbicides (or DNAH). Although not totally identified as the lone source of resistance, a mutation to E-tubulin at the amino acid codon 241, located in DNAH-resistant biotypes of *Poa annua*, has been found converting arginine to lysine (Lowe et al., 2001). This substitution may reduce the binding of the DNAH to the E-tubulin, thus reducing the affinity to the herbicides.

Resistant problems generally begin to show up when a class of compounds is continuously used over a 7- to-10-year period. These herbicides selectively control those susceptible biotypes, gradually allowing the resistant biotypes to spread and increase over time. Just as fungicide groups should be rotated by turf managers to prevent disease resistance, different herbicide families should also be rotated to prevent annual bluegrass resistance.

BEST MANAGEMENT PRACTICES FOR CONTROLLING *POA ANNUA*

The steps necessary to control *Poa annua* are numerous and essential. Preventing weed-seed introduction, providing optimum competitive growing conditions, and, possibly, using herbicides or PGRs are necessary components of a complete control plan (McCarty, 2005).

1. Fumigate all soil mix before planting.
2. Begin with and retain good drainage to prevent soil compaction and excessive soil moisture, which favor the annual bluegrass.
3. Use certified seed, sprigs, or sod that are free of annual bluegrass when planting.
4. Control annual bluegrass in surrounding areas to minimize seed tracking into noninfested sites.
5. Obtain and maintain good turf density to reduce annual bluegrass invasion.
6. Aerify consistently to relieve soil compaction.
7. Use fumigated sand and/or soil when topdressing.
8. Use preemergence herbicides in spring and fall.
9. Use PGRs in spring and fall to reduce annual bluegrass competition and seedhead development.
10. Hand pick or wick nonselective herbicide (e.g., glyphosate) near small (e.g., 1-inch, 2.5 cm, diameter) annual bluegrass plants.
11. Plug larger spots with annual bluegrass-free turf.

The next part discusses various annual bluegrass control options, including the use of selective and nonselective herbicides and plant-growth retardants. Preemergence control of annual bluegrass in turf was first noted in the 1930s when the insecticide lead arsenate was discovered (Sprague and Burton, 1937). Since then, preemergence annual bluegrass control has been reported with numerous herbicides.

Bermudagrass Golf Greens

Preemergence control in bermudagrass golf greens currently is available with several herbicides. Each has its own use precautions. If these are not followed, unsatisfactory results may occur. Bingham et al. (1969) reported that bensulide provides preemergence annual bluegrass control and that acceptable stands of ryegrasses could be obtained when seeding was delayed until one month after its application. This could, however, be influenced by environmental and management practices resulting in a narrow tolerance range. Bensulide application should occur no sooner than 120 days prior to overseeding. This waiting period allows enough bensulide to be in the soil to control the germinating annual bluegrass, but to be low enough not to interfere with germination of the grasses used for overseeding. Control in heavy population areas, however, is often unacceptable with benefin.

Benefin, like bensulide, can be applied before annual bluegrass and overseeding if proper timing is followed. At the low application rate, a minimum of 6 weeks is necessary after treatment before overseeding should be attempted. A period of 12 to 14 weeks is necessary between application of the high rate and overseeding.

Pronamide also provides preemergence control of annual bluegrass in bermudagrass. Like bensulide, pronamide must be applied in advance of annual bluegrass germination and seeding cool-season grasses. The minimum recommended period between application and overseeding is 90 days. It also is recommended that application not be made where drainage flows onto areas planted with cool-season grasses.

Activated charcoal has been used successfully to prevent injury to the desirable overseeded grass when pronamide was applied closer than 90 days to overseeding (Meyers et al., 1973). High rates (e.g., 2.5 to 4 lb/1,000 sq. ft., or 122 to 195 kg/ha) of charcoal are necessary. A disadvantage of this method is the inability to reestablish the ryegrass in the event the charcoal treatment fails. Current formulations of activated charcoal also are inconvenient to handle and apply.

Ethofumesate also provides preemergence and early postemergence annual bluegrass control in bermudagrass fairways overseeded to ryegrass (Dickens, 1979). To prevent turfgrass injury, application rate, timing, and frequency are important. In general, an initial 1 lb ai/a (1.1 kg/ha) is applied 30 to 45 days after overseeding. A sequential application at similar rates will be required 30 days after the initial application for season-long control. If ethofumesate is applied before bermudagrass dormancy in the fall, the bermudagrass stops growing (Dickens, 1979). A delay in spring transition from ryegrass to bermudagrass also occurs when ethofumesate is applied in late winter.

Fenarimol, a systemic fungicide used to control several turfgrass diseases, gradually reduces annual bluegrass populations without adverse effects to overseeded grasses or bermudagrass. Its application should occur prior to overseeding and prior to the germination of annual bluegrass. A treatment scheme has been suggested consisting of one, two, or three applications, with the single or final application taking place two weeks prior to overseeding with perennial ryegrass. If *Poa trivialis* or bentgrass is used for overseeding, the last fenarimol application should be no closer than 30 days prior to seeding. Fenarimol does not appear to deleteriously affect the overseeded grass, but the necessity of properly timed repeat applications can be a drawback for managers who have limited budgets and labor. When bermudagrass growth has slowed due to cooler temperatures or extended cloudy weather, applications of fenarimol can stunt or slow its growth (McCullough et. al., 2005c). Inconsistent annual bluegrass control following fenarimol treatments has been noted (Gaul and Christians, 1988; Johnson, 1988a) and control is rarely 100 percent successful.

Foramsulfuron, a member of the sulfonylurea family, has recently been approved for use on both zoysiagrass and bermudagrass for annual bluegrass control on golf greens. It controls annual bluegrass preemergently, and overseeded bermudagrass and zoysiagrass early postemergently, with either perennial ryegrass or *Poa trivialis.* Use rates range from 0.006 to 0.025 lb ai/a (0.007 to 0.028 kg ai/ha). Applications need to be no later than 14 days prior to overseeding, due to its short half-life in soil (McCarty, 2005). Rimsulfuron, another recently introduced sulfonylurea, is similar to foramsulfuron in its low-use rates ranging from 0.031 to 0.062 lb ai/a (1 to 2 oz/a, 0.035 to 0.069 kg ai/ha) and is effective in controlling annual bluegrass and perennial ryegrass. Annual bluegrass control has been shown in overseeded bermudagrass with perennial ryegrass when applications of rimsulfuron are made 10 to14 days prior to overseeding, as closer applications will result in a reduced stand establishment (McCarty, 2005). Both rimsulfuron and foramsulfuron have extremely short half-lives (Vencill, 2002) and should be applied as close to overseeding as possible. Also, best annual bluegrass control occurs when the overseeding date is delayed as long as possible in fall. This allows the majority of annual bluegrass seed to germinate, making the herbicides more efficacious. Caution should be taken when applying these products on areas that surround bentgrass, overseeded greens, fairways, or tees, since sulfonylurea herbicides can move laterally and cause injury or death to the bentgrass or overseeded grasses. Movement potential increases with high soil pH, increasing slopes (grades), and soils that are saturated at or soon after application.

Bentgrass Golf Greens

Bentgrass is sensitive to most postemergence grass herbicides that are effective against annual bluegrass. Therefore, preemergence herbicides and the use of PGRs are the most common means of controlling this weed. Erratic control of annual bluegrass, however, often results (Callahan and McDonald, 1992). It has been suggested that the presence of perennial biotypes of the species may contribute to this erratic control (Gibeault and Goetze, 1972). Low-growing creeping perennial types become dominant over the annual biotype under frequent close mowing. Moist soil conditions, regular fungicide use, and high soil nutrient levels, which are conditions normally maintained with creeping bentgrass golf greens, also contribute to creeping perennial biotype dominance (Callahan and McDonald, 1992). For this reason, preemergence herbicides may control the annual biotype, allowing the perennial biotype to further dominate the green.

Limited research on preemergence control of the perennial biotype suggests poor short-term control (Callahan and McDonald, 1992). Repeat applications over multiple years (minimum of four years) are necessary for significant reduction of perennial biotypes of annual bluegrass in bentgrass golf greens. Multiple-year treatments during late winter are considered superior for control, compared with late summer treatments.

Ethofumesate has been used safely to control annual bluegrass in bentgrass golf course fairways, but is not recommended or labeled for bentgrass golf greens (Lewis and DiPaola, 1989). Best control with minimum turf injury is achieved with application rates at 0.5 lb ai/a (0.56 kg ai/ha) per application for a total of five applications, or at 0.75 lb ai/a (0.84 kg/ha) per application for three applications. Applications should be 30 days apart. Control is more consistent when treatments begin in early fall. Applications in late fall or winter increase bentgrass injury. Bentgrass, with a shallow root system, grown in compacted, poorly drained soils, or soils that stay cool due to shade, is more prone to injury from the ethofumesate treatments.

For bentgrass fairways, bispyribac-sodium (Velocity 17.6 SC) can be used at low rates and frequent applications. Rates range from 0.71 to 2.1 lb ai/acre (0.8 to 2.4 kg ai/ha) applied 3 (low rate) to 14 (high rate) apart when the bentgrass is actively growing. The more Poa present, lower rates should be used to avoid development of bare areas. Interseed after the last summer application to thicken the bentgrass stand.

For *Poa trivialis* control in bentgrass fairways, bispyribac-sodium rates should range from 1.4 to 3.2 lb ai/acre (1.6 to 3.6 kg ai/ha), on 10- to 14-day intervals. Sulfosulfuron (Certainty 75 WDG) applied three times spaced two weeks apart, with each application at 0.042 lb ai/acre (19 kg ai/ha), also helps control *Poa trivialis.*

With the realization that the elimination of annual bluegrass in golf greens is not always achievable with current herbicide technology, research recently has focused on suppressing the plant's growth and seedhead production. The turf growth regulators (TGRs) paclobutrazol, flurprimidol, and ethephon currently are available for annual bluegrass suppression in bentgrass golf greens. Other materials (e.g., mefluidide, maleic hydrazide) also are available, but only for higher mowed turf. Generally, paclobutrazol or flurprimidol is applied monthly to actively growing bentgrass, starting in fall and continuing through spring. Differential response in species susceptibility to the herbicide is attributed to a greater uptake of the TGR by the shallower-rooted annual bluegrass when compared with that by the deeper-rooted bentgrass (Kageyama et al., 1989). Paclobutrazol is root absorbed (xylem-mobile) and works by reducing the competitive ability of the annual bluegrass for three to eight weeks after application. This allows the creeping bentgrass to out-compete the weed. **These materials should be applied only during periods of active bentgrass root growth.** Plant growth regulators like paclobutrazol encourage a gradual transition in favor of the bentgrass by suppressing the growth and seedhead production of the annual bluegrass by inhibiting the growth hormone gibberellic acid (McCarty, 2005). Paclobutrazol at 0.25 to 0.375 lb ai/a (0.28 to 0.42 kg ai/ha) applied twice in the fall at four-week intervals, followed by an additional applications in early spring, has provided good annual bluegrass suppression in bentgrass turf (McCarty, 2005). At a rate of 0.375 lb ai/a (0.42 kg ai/ha) and applied two to three times in fall and twice in spring, paclobutrazol provides suppression of annual bluegrass populations and seedheads on creeping bentgrass greens. If paclobutrazol is applied two weeks, or less, prior to or after the seeding of bentgrass, the bentgrass growth will be inhibited (Yelverton, et al., 1999). The use of paclobutrazol on bentgrass and annual bluegrass greens can result in a shift of competitiveness of both grasses in favor of bentgrass by inhibiting the photosynthesis rate of annual bluegrass more than that of the bentgrass (Isgrigg et. al., 2002). Paclobutrazol inhibited bentgrass growth when applied two weeks, or less, prior to or after seeding or interseeding bentgrass into existing greens (Yelverton et al., 1999). Prevention of annual bluegrass seedhead formation may be inconsistent with these materials, but they do prevent seedhead stalk elongation (Kaufman, 1989), which may result in improved turf uniformity and appearance. Characteristically, treated bentgrass forms a blue-green to grayish color and often forms a coarser leaf texture that can disrupt normal putting conditions.

Overseeded Perennial Ryegrass or Creeping Bentgrass Fairways Dithiopyr can be used to control annual bluegrass in overseeded bermudagrass fairways with perennial ryegrass. Application needs to be made 6 to 8 weeks prior to overseeding. Any closer application to overseeding will delay or prevent perennial ryegrass establishment. Maximum use rate is 0.5 lb ai/a (0.56 kg ai/ha). Like dithiopyr, prodiamine can be used to control annual bluegrass in overseeded fairways with perennial ryegrass. Initial use rates range from 0.38 to 0.75 lb ai/a (0.43 to 0.84 kg ai/ha). Prodiamine has a very long soil half-life, and applications need to be made 8 to 10 weeks prior to overseeding and must be repeated at 0.38 lb ai/acre (0.43 kg ai/ha) in midwinter. Caution needs to be taken not to overlap, as perennial ryegrass establishment will be adversely effected. A minimum of 250 lb/a (280 kg/ha) of perennial ryegrass seed needs to be used to offset any adverse effects caused by these preemergence herbicides (McCarty, 2005).

Sulfosulfuron, foramsulfuron, rimsulfuron and trifloxysulfuron, all members of the sulfonylurea herbicide family, can be applied 7 to 14 days prior to overseeding bermudagrass fairways with perennial ryegrass for early postemergence control of annual bluegrass. These perform best when overseeding timing is delayed to allow maximum *Poa annua* germination.

Bispyribac-sodium (Velocity) is a postemergence herbicide recently introduced to selectively control annual bluegrass in cool-season grasses such as creeping bentgrass and overseeded perennial ryegrass. It inhibits the acetolactate synthase (ALS) enzyme, which plants require to produce three branch-chain amino acids. This mode of action is very similar to sulfonylurea herbicides and requires up to three weeks for control, especially in cooler weather below 65°F (18.3°C) (McCarty and Estes, 2005, McCullough and Hart, 2005). Rates range from 0.05 to 0.17 lb ai/a (0.06 to 0.18 kg ai/ha). Different strategies have been developed, depending on severity of *Poa annua* in bentgrass fairways. If annual bluegrass populations are less than 30 percent, two applications at 0.17 lb ai/a (0.18 kg ai/ha) spaced two weeks apart should be considered, starting in late spring or early summer (McCarty and Estes, 2005). However, if populations are above 30 percent, a less aggressive approach would be favorable to prevent large bare areas. The less aggressive approach requires three applications at 0.125 lb ai/a (0.14 kg ai/ha) every 7 to10 days or six applications at 0.05 lb ai/a (0.06 kg ai/ha) applied every 3 to7 days (McCarty and Estes, 2005). Bispyribac-sodium also suppresses *Poa annua* seedheads and controls untillered plants, with two applications at 0.17 lb ai/a (0.18 kg ai/ha) in late winter and early spring (Estes and McCarty, 2005). Bispyribac-sodium also has activity on rough bluegrass (*Poa trivialis*) and numerous broadleaf weeds. Among those broadleaf weeds controlled are chickweed (common and sticky), dandelion, henbit, parsley piert, lawn burweed, plantain, yellow woodsorrel, white clover, and hop clover (McCarty and Estes, 2005). Reseeding of perennial ryegrass, Kentucky bluegrass, and/or bentgrass must be delayed at least 2 weeks after treatment of bispyribac-sodium at rates of 0.13 to 0.27 lb ai/a (0.15 to 0.3 kg ai/ha), due to significant stand reductions (Lycan and Hart, 2006.) Currently, bispyribac-sodium is not labeled for use on golf greens.

In bentgrass fairways, paclobutrazol 2 SC at 0.5 lb ai/a (0.56 kg ai/ha) controls annual bluegrass when applied twice at four-week intervals starting in October and reapplied as the bentgrass starts to grow in the spring (Yelverton et. al., 1999). In overseeded perennial ryegrass fairways, single treatments of paclobutrazol at rates of 0.75 lb ai/a (0.84 kg ai/ha) and 1.0 lb ai/a (1.12 kg ai/ha) applied in mid-fall provide good annual bluegrass control (Mahady et. al., 2001). Sequential applications of the tank mixture of paclobutrazol at 0.2 lb ai/a (0.23 kg ai/ha) and trinexapac-ethyl 1 EC at 0.68 lb ai/a (0.77 kg ai/ha) initiated in early winter and applied monthly through winter provide exceptional annual bluegrass control (Mahady et al., 2001). Continued research is needed to further identify the critical threshold rates and timing of plant-growth regulators in their use to control annual bluegrass in cool-season greens and fairways.

Flurprimidol also is used in bentgrass fairways similarly to paclobutrazol. Flurprimidol is applied at 0.25 to 0.5 lbs ai/a (0.28 to 0.56 kg ai/a) and repeated every 30 days during periods of active bentgrass growth (under no stress) for continued suppression.

NUTSEDGE CONTROL

The predominant nutsedge species in turfgrasses are yellow and purple nutsedge. Other, more local members of the *Cyperus* genus include annual or water sedge, perennial and annual kyllinga, globe sedge, Texas sedge, flathead sedge, and cylindrical sedge. Path, or slender rush, a member of the rush (*Juncus*) family, also can occur in some turf situations, especially in highly compacted soils.

Nutsedges, in general, are yellow green to dark green, with triangular stems filled with pith, and bear three-ranked leaves (Figure 6.13), unlike the two-ranked

Figure 6.13
Nutsedges and Kyllinga species have a three-rank leaf arrangement. In the inset, notice that these mostly have triangularly shaped stems filled with pith.

leaves of the grass family. The root systems are fibrous, with deep-rooted tubers or nutlets. Yellow and purple nutsedge are low-growing perennials that resemble grasses. Seedhead color is often used to distinguish between the two major nutsedges. Leaf-tip shape is another distinguishing characteristic. Leaf tips of purple nutsedge are generally wider and more rounded; conversely, yellow nutsedge leaf tips are narrow, forming a needle-like tip (Figure 6.14). Yellow and purple nutsedge are not believed to produce viable seed, but due to their underground tubers and rhizomes, these species have enormous capacity to reproduce and spread. Propagation is primarily by tubers, which generally sprout in the spring. Tubers can remain dormant in the soil for up to three years, and yellow nutsedge tubers can remain dormant for as long as 10 years (Neil, 1995; Stoller and Sweet, 1987). These tubers can sprout repeatedly, and control strategies must include a commitment to prevent new tuber formation (Blum et al., 2000).

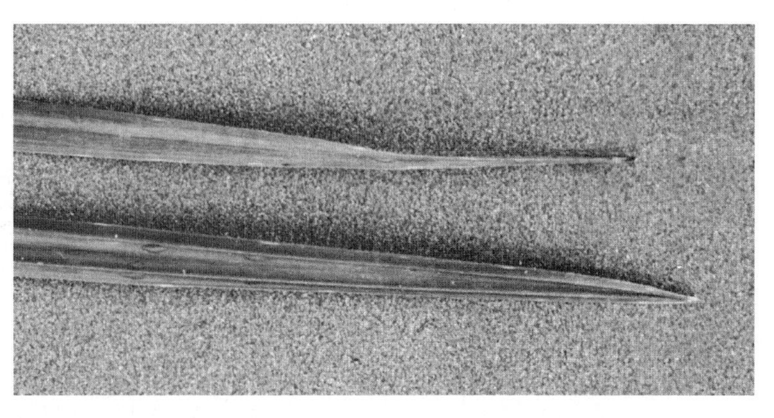

Figure 6.14
Yellow nutsedge (top) leaf tips tend to be narrower and needle-like compared with thicker purple nutsedge leaf tips (bottom).

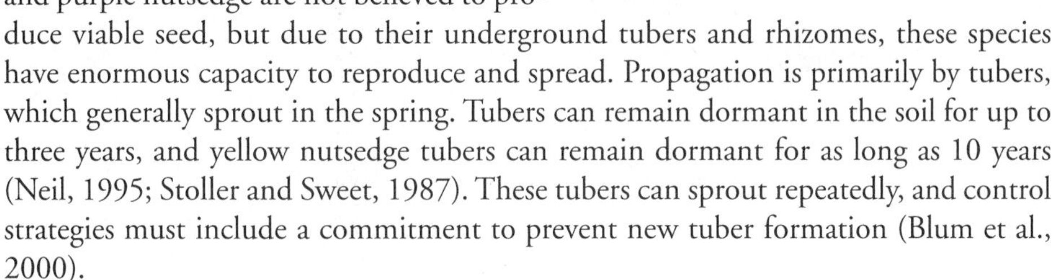

These weeds generally thrive in soils that remain wet for extended periods. The first control step is to correct the cause of continuously wet soils. Do not over-irrigate an area, and if necessary, provide surface and subsurface drainage.

Preemergence control of sedges is available from several materials (Table 6.14). Triazine herbicides (e.g., atrazine, simazine) provide fair preemergence control of select annual sedges, but generally are ineffective on perennial species. Metolachlor (Pennant) provides preemergence control of most annual sedges and selective control of perennial species, especially yellow nutsedge. Purple nutsedge control with metolachlor, however, is not as consistent. Preemergence control of purple nutsedge is currently unavailable. Metolachlor often is combined with triazine herbicides (Prompt) for nutsedge, broadleaf, and certain annual-grass control.

Historically, postemergence chemical control of most sedges was attempted with repeat applications of 2,4-D, the organic arsenicals, or a combination of the two. Although effective, treatments were slow to kill and repeat applications generally were necessary. Extensive damage resulted with certain turf species, such as centipedegrass and St. Augustinegrass.

Postemergence control of yellow nutsedge and annual sedges is possible with bentazon (Basagran), with minimal turf damage. However, bentazon is a contact-type herbicide, controls only the shoot region, and has minimal effect on rhizomes and tubers of yellow nutsedge (Stoller and Sweet, 1987).

Purple nutsedge can be suppressed with imazaquin. Control often increases when it is tank-mixed with MSMA. Certain broadleaf and monocot weeds, such as wild garlic, also are controlled with imazaquin (Ferguson et al., 1992); however, spring green-up of bermudagrass may be delayed following fall or winter treatments.

Repeat applications with either herbicide normally are required to completely control an established nutsedge stand. Treatments over multiple years also may be required to eradicate nutsedges. For example, when applied annually, three years of imazaquin plus MSMA treatments are required for complete purple nutsedge control.

Immature annual sedges can be suppressed with triazine herbicides. Atrazine commonly is used in centipedegrass and St. Augustinegrass sod production shortly after harvesting. It suppresses many annual grasses, broadleaf weeds, and sedges. Timing is important. As the weeds mature, control with atrazine decreases dramatically. Repeat applications of atrazine may be necessary for season-long control.

Table 6.14
RELATIVE SEDGE CONTROL AND TURF TOLERANCE TO VARIOUS HERBICIDES (*REFER TO HERBICIDE LABEL FOR SPECIFIC TURF SPECIES USE LISTING*)

These are relative rankings and depend on many factors such as environmental conditions, turfgrass vigor or health, application timing, etc., and are intended only as a guide.

	Sedge Control					Turf Tolerance (Excluding Greens)								
Herbicide (Trade Names)[1]	Annual Sedge	Purple Nutsedge	Yellow Nutsedge	Annual Kyllinga spp.	Perennial Kyllinga spp.	Bermudagrass	Bentgrass	Bluegrass, Fescue, Ryegrass	Centipedegrass	St. Augustinegrass	Bahiagrass	Zoysiagrass	Kikuyurass	Tall Fescue
Preemergence Control														
Metolachlor (Pennant)	G[2]	P	G	F-G	P	S[3]	NR	NR	NR	NR	NR	S	NR	NR
Oxadiazon (Ronstar 2G)	G	P	P	F	P	S	NR	S	NR	NR	NR	S	NR	S
Postemergence Control														
Bentazon (Basagran T&O)	G	P	G	F-G	F-G	S	S-I	S	S	S	S	S	NR	S
Imazaquin (Image)	G	G	F	G	G	I-S	NR	NR	I	I	NR	S	NR	NR
Imazapic (Plateau)	G	G	G	—	—	I-S	NR	NR	S	NR	NR	NR	NR	NR
Halosulfuron (Sedgehammer)	G	G-E	G-E	G	F-G	S	S	S	S	S	S	S	S	S
MSMA/DSMA/CMA	G	P-F	F	G	G	S-I	I	NR	NR	NR	NR	S-I	NR	I-S
Image + MSMA/DSMA	G	G	G	G	G	S-I	NR	NR	NR	NR	NR	S-I	NR	NR
Sufentrazone (Dismiss)	G	P-F	F	F	F	S	S	I-S	S	NR	S	S	S	S
Sulfosulfuron (Certainty)	G	G	G	G	G	S	NR	NR	S	S	S	S	S	NR
Trifloxysulfuron (Monument)	G	G	G	G	G	S	NR	NR	NR	NR	NR	S	NR	NR

[1]Repeat applications are necessary for complete control from all herbicides. This interval is from 5 days for MSMA/DSMA and 3 to 5 weeks for Certainty, Sedgehammer, Monument, or Image.
[2] E = excellent (>89%) control; F = Fair to Good (70 to 89%), good control sometimes with high rates, however a repeat treatment 1 to 3 weeks later each at the standard or reduced rate is usually more effective; P = poor (<70%) control in most cases.
[3]S = Safe at labeled rates; I = Intermediate safety, use at reduced rates; NR = Not Registered for use on and/or damages this turfgrass; D = Dormant turf only.

A newer material, halosulfuron (Sedgehammer), belongs to the sulfonylurea herbicide family and provides good postemergence control of most major sedges. Halosulfuron 75 WP has low use rates (e.g., 0.035 to 0.07 kg ai/ha, or 0.03 to 0.06 lb ai/acre) and should be repeated in four to six weeks for increased control. Repeat applications are necessary for controlling perennial sedges. Turf tolerance to halosulfuron is generally very good (Table 6.14).

Trifloxysulfuron 75 DP (Monument) is another newer compound that also belongs to the sulfonylurea family and has shown excellent postemergence control of various major sedges, including purple nutsedge and annual and perennial kyllinga species (Table 6.14). Trifloxysulfuron 75 DP use rates range from 0.015 to 0.026 lb ai/a (0.017 to 0.029 kg ai/ha). Two applications at four-week intervals at the rate of 0.026 lb ai/a (0.029 kg ai/ha) have shown to provide excellent purple nutsedge control (Yelverton, 2004). Multiple applications of trifloxysulfuron 75 DP at 0.044 lb ai/a (0.05 kg ai/ha) at six-week intervals also have provided torpedograss control (Teuton et al., 2001). Trifloxysulfuron, however, is very toxic to cool-season grasses and is very effective in removing perennial ryegrass from overseeded bermudagrass fairways. Like other sulfonylureas, control is temperature dependent, requiring three to six weeks when temperatures are less than 60°F (15.6°C) and shorter periods as temperatures increase (Yelverton, 2004).

Another sulfonylurea herbicide, sulfosulfuron 75 DF (Certainty), also provides good to excellent control of nutsedge species, kyllinga species, and selective broadleaf weeds. It can be safely applied to common or hybrid bermudagrass, zoysiagrass, centipedegrass, St. Augustinegrass, and kikuyugrass, at rates that range from 0.035 to 0.17 lb ai/a (0.04 to 0.19 kg ai/ha). Like trifloxysulfuron, sulfosulfuron 75 DF damages or eradicates most cool-season turfgrasses; for example, it has been shown to selectively remove tall fescue in bermudagrass with one application at a rate of 0.08 lb ai/a (0.09 kg ai/ha) (Murphy and Nutt, 2003). Among those broadleaf weeds it controls are catweed bedstraw, buttercup, common chickweed, white clover, henbit, mustard, pennycress, and shepherd's purse.

Reducing Turf Injury from Postemergence Herbicides

Postemergence herbicides often slightly to moderately discolor (injure) turf species, even though these herbicides are labeled for turf use. Injury generally is temporary (one to two weeks), but increases with moisture, heat, or pest stress. It is desirable to minimize injury to turf, especially where high-quality turf is expected to thrive. Iron and nitrogen have been used in conjunction with postemergence herbicides to minimize turf injury (Johnson et al., 1990). Although these additives have shown some short-term prevention of postemergence herbicide injury, complete elimination of turf injury for all grasses or for all herbicides may not occur.

Selectively Controlling Undesirable Turf in Mixed Stands

Because of incompatible leaf texture and color, it normally is undesirable to have a mixture of different grass species in the same area. Selectively controlling one species of grass appearing within a stand of another species generally is more desirable than using a nonselective herbicide and then replanting the turf. Table 6.15 summarizes current knowledge on selectively controlling a particular turf species in a mixed stand. As with most postemergence herbicides, multiple applications normally are required for complete eradication of a certain species, which may result in some injury to the desired turf. Selective control of bermudagrass in seashore paspalum, St. Augustinegrass, bentgrass, and bahiagrass still is difficult to achieve with herbicides alone.

An increasing problem for many golf course superintendents is the encroachment of bermudagrass from the collar region onto bentgrass greens. Siduron and ethofumesate have suppressed bermudagrass (Dickens, 1979), but varying levels of

Table 6.15
POSTEMERGENCE HERBICIDES WHICH SELECTIVELY REMOVE UNDESIRABLE TURF SPECIES FROM A MIXED STAND

Herbicide	Tolerant Turf	Susceptible Turf	Reference
Arsenicals (MSMA/DSMA)	bermudagrass zoysiagrass	centipedegrass St. Augustinegrass	McCarty, 2005
Asulam	St. Augustinegrass bermudagrass (Tifway 419)	centipedegrass	McCarty, 2005
Atrazine, simazine pronamide	bermudagrass centipedegrass St. Augustinegrass	bentgrass bluegrass fine fescue ryegrass	McCarty, 2005
chlorsulfuron	bahiagrass bermudagrass bentgrass fairways fine fescue Kentucky bluegrass	tall fescue	Dernoeden, 1990
clethodim	centipedegrass	bermudagrass Kentucky bluegrass ryegrass *Poa trivialis* tall fescue	McCarty, 2005
ethofumesate + atrazine	St. Augustinegrass	common bermudagrass	McCarty, 2005
fluazifop-P-butyl	tall fescue zoysiagrass	bermudagrass centipedegrass	Johnson, 1987b
fenoxaprop-ethyl	bentgrass* fine fescue Kentucky bluegrass ryegrass zoysiagrass (Meyer) tall fescue	bermudagrass centipedegrass St. Augustinegrass	Dernoeden, 1987; 1989a; 1989b Higgins et al., 1987 Johnson, 1987b McCarty, 2005
foramsulfuron	bermudagrass zoysiagrass (Meyer)	bentgrass centipedegrass Kentucky bluegrass *Poa trivialis* ryegrass tall fescue	McCarty, 2005
imazapic	bahiagrass bermudagrass Kentucky bluegrass K-31 tall fescue	bahiagrass ryegrass tall fescue	McCarty, 2005
mesotrione	centipedegrass fine fescue Kentucky bluegrass St. Augustinegrass tall fescue	bentgrass	McCarty, 2008
metsulfuron	bermudagrass centipedegrass fine fescue Kentucky bluegrass St. Augustinegrass zoysiagrass (Meyer, Emerald)	bahiagrass tall fescue ryegrass	Dernoeden, 1990 McCarty, 2005

Table 6.15 (CONTINUED)

Herbicide	Tolerant Turf	Susceptible Turf	Reference
rimsulfuron	bermudagrass	ryegrass *Poa trivialis* bentgrass	McCarty, 2005
sethoxydim	centipedegrass fine fescue	bermudagrass	McCarty et al., 1986
sulfosulfuron	bermudagrass centipedegrass St. Augustinegrass zoysiagrass	tall fescue ryegrass bluegrass	McCarty, 2005
triclopyr	bahiagrass Kentucky bluegrass ryegrass tall fescue zoysiagrass	bermudagrass centipedegrass St. Augustinegrass	McCarty, 2005
trifloxysulfuron	bermudagrass zoysiagrass	bahiagrass ryegrass tall fescue bluegrass bentgrass	McCarty, 2005

*Herbicide rate and/or mowing height dependent. Varying susceptibility often occurs with cultivar differences.

bentgrass injury may occur. However, three to four weeks of bentgrass phytotoxicity often accompanies ethofumesate treatment. Best control with siduron 50 WP has been with treatment at 12 lbs ai/a (13.4 kg ai/ha), repeated every 30 days. Start treatments in early spring as bermudagrass initiates green-up. This treatment tends to suppress the growth of the bermudagrass or keep it in a semi-dormant growth stage. Continue treatments at 30-day intervals until summer. Treatments during summer tend not to significantly affect the bermudagrass. However, resume treatments when temperatures begin to cool in late summer and repeat monthly until the bermudagrass goes dormant.

Temporary (up to three months) bermudagrass suppression has been achieved with combinations of siduron with flurprimidol, as well as ethofumesate plus flurprimidol (Johnson and Carrow, 1989). This suppression was superior to that achieved by the standard practice of using siduron alone. Spring treatments were less injurious to bentgrass and provided a level of bermudagrass retardation similar to that produced by a late-summer application followed by a spring application.

MOSS BIOLOGY AND CONTROL

Mosses are categorized as *Bryophytes,* a class of primitive plants noted for their lack of roots and vascular (xylem and phloem) systems. This is the first clue indicating why most control products are inconsistent, since these plants lack functional roots and vascular systems to absorb and translocate control materials throughout the plant. Mosses are also 400 million years old, predating the dinosaurs and indicating that their simple structure is very adaptable to changing climatic conditions. Of the over 9,500 moss variations that occur, the major one found on golf greens is *silvery thread moss* (*Bryum argenteum* [Hedw.]). *Bryum,* unlike most mosses, grows well in both wet, shady environments and in hot, dry sites in full sun. These are nonparasitic to plants and spread mainly asexually via plant fragments, and less so as airborne spores. *Bryum* species are able to photosynthesize and fix nitrogen; this is another clue as to why these plants have become problematic in recent years. When mowing heights are lower and nitrogen rates on greens are reduced, sunlight is better able to penetrate to the soil surface where moss can grow. Moss can become very competitive in lower-nitrogen sites, since it can fix its own nitrogen, which explains why little moss is seen on higher mowed and fertilized bentgrass areas such as collars or fairways. Typical occurrence of silvery thread moss on greens is initially on weak turf areas such as ridges or mounds where the grass is thinned from scalping and/or drought. This moss also establishes in summer following periods of rainy, overcast, warm days and is favored by poorly drained, heavily shaded turf sites with acidic, infertile soils.

Moss infestations are also noted to be "a sign of soil poverty." The literature suggests that mosses are more problematic on calcium-rich and potassium-poor soils (Richardson, 1981; Watson, 1964). This nutritional trend has been observed in turf over the last 10 years. If calcium levels are high, treating with Epsom salt (magnesium sulfate) may help, while potassium sulfate treatments often improve potassium-deficient soils. Weekly applications of liquid ammonium sulfate at 0.10 to 0.125 lb N per 1,000 sq. ft. (4.9 to 61 g/m^2) often promotes a more competitive turf, helps desiccate the moss, and also acidifies the soil to discourage moss and patch diseases.

The elimination of fungicides containing heavy metals such as mercury has also encouraged moss, as the heavy metals in these products displace the central magnesium atom in chlorophyll molecules. By reducing or eliminating chlorophyll in moss, they eventually cause weakness and death. Metals with higher atomic weights, such as mercury, cadmium, and silver, are generally more effective than those with lighter atomic weights, such as zinc, copper, and iron.

Control

A holistic approach to control is necessary, as healthy turf is the only means to prevent and cure moss occurrence. As expected, increasing mowing heights and fertility are necessary to promote turf health and thus discourage moss encroachment. Air circulation and light penetration on greens can be improved by installing fans and providing surface and subsurface drainage. Maintaining sufficient potassium and moderate calcium levels should also be considered.

A number of products have been evaluated for moss control, with most providing inconsistent results and/or excessive turf damage. These products include Junction (mancozeb + copper hydroxide); Fore (mancozeb); Kocide 2000 (copper hydroxide); No-Mas and DeMoss, both fatty-acid soaps; dishwashing soap such as Ultra Dawn; iron-containing products such as ferrous ammonium sulfate and iron sulfate; ground limestone; baking soda (sodium hydroxide); Zerotol (hydrogen dioxide); copper sulfate; copper soaps; and zinc sulfate. Most must be used in cooler weather and only at limited rates in order to avoid turf burn or nutrient deficiencies from the buildup of heavy metals. Of the products listed, copper hydroxide, fatty-acid soaps, and iron products have been reported to work best with multiple (five to seven) applications in cooler climates. Recently, TerraCyte (sodium carbonate peroxyhydrate) and silver nitrate have been explored for effectiveness in moss control. TerraCyte trials at 12 lbs per 1,000 sq. ft. (586 kg/ha), with six applications spaced two weeks apart, have provided up to 97 percent control. Junction alone, and combined with the fungicides Fore and Spotrete applied at 6 oz per 1,000 sq. ft. (18.3 kg/ha), also has provided ~80% control, while Ultra Dawn at 4 to 6 fl oz per 1,000 sq. ft. (12.7 to 19.1 L/ha) in 2 gallons (7.57 L) of water, applied every two weeks, has provided excellent control in some tests, but not in others. Inconsistent control, potential for turf injury, and legalities are matters that must be considered before these products are put to use.

Work in the mountains of North Carolina identified chlorothalonil (Daconil Weather Stik or Daconil Zn) as providing acceptable moss control with no turf damage (Yelverton, 2005). Various rates of each product were used, with 4 oz per 1,000 sq. ft. (12.2 kg/ha) Daconil Weather Stik and 6 oz per 1,000 sq. ft. (18.3 kg/ha) of Daconil Zn providing similar results. Because the product is a desiccant, control is best with sequential applications every seven days for at least three consecutive weeks and when air temperatures at the time of application are at least 80°F (26.7°C) and, preferably, 85°F (28.3°C). This minimum temperature requirement explains why cooler regions have not had control results with chlorothalonil as consistent as those reached in warmer regions.

The latest material to show promise as a moss management aid is the herbicide carfentrazone, or Quicksilver T&O 1.9L (McCarty et al., 2008). Carfentrazone acts as a membrane disrupter in plants and thus is another desiccant. With carfentrazone, however, temperatures do not appear to be as important for successful control as with chlorothalonil. Currently, best results have been with 0.031 to 0.099 lb ai/a (0.034 to 0.11 kg ai/ha), with a repeat application in 14 days. Because mosses are nonvascular plants, 2 to 5 gallons of spray solution should be used per 1,000 sq. ft. (8.15 to 20.4 L/m^2) for adequate coverage. As moss control commences, the infested areas should be spiked or sliced, raked, verticut, or heavily groomed and topdressed to remove the impervious dying moss layer and smooth the putting surface. Note, however, that unless cultural practices are changed to shift the competitive edge back to turf, moss reoccurrence is inevitable and repeat applications will be necessary.

CONTROL OF COMMON WEEDS

Annual Blue-Eyed Grass (*Sisyrinchium rosulatum*)

Apply products containing atrazine or simazine twice, 30 days apart. Prompt (a premix of atrazine and bentazon) also works well. Sencor provides excellent control in tolerant turfgrasses. Products containing two- or three-way broadleaf herbicide mixtures applied at least twice, seven days apart, also work.

Annual Bluegrass (*Poa annua* [L.])

Control options/strategies change constantly. Check with your local state turfgrass specialist for the latest recommendations. Preemergence control includes members of the dinitroaniline herbicide family (benefin, oryzalin, pendimethalin, and prodiamine) and other preemergence products (e.g., bensulide, dithiopyr, fenarimol, napropamide, and oxadiazon). Apply in late summer when air temperatures reach 75°F (24°C) for several consecutive days. Early postemergence control includes atrazine, simazine, or metribuzin; clethodim (Envoy); diquat (Reward); foramsulfuron (Revolver); trifloxysulfuron (Monument); sulfosulfuron (Certainty); ethofumesate (Prograss); bispyribac-sodium (Velocity); and pronamide (Kerb). Selective suppression is provided by plant-growth regulators such as paclobutrazol (Trimmit/TGR), flurprimidol (Cutless), and mefluidide (Embark).

Bahiagrass (*Paspalum notatum*)

Manor or Blade (metsulfuron) works in bermudagrass, St. Augustinegrass, centipedegrass, and zoysiagrass. Of these grasses, bermudagrass and St. Augustinegrass tolerate Manor best. In bermudagrass, apply Manor twice, three to four weeks apart, at 1 oz/acre. In centipedegrass, two applications of Sethoxydim G-Pro or Vantage, three weeks apart, suppress bahiagrass growth and seedhead development. In zoysiagrass or tall fescue, repeat applications of Fusilade II or Acclaim Extra may be used. Monument (trifloxysulfuron) will provide bahiagrass suppression in warm-season grasses.

Bermudagrass (*Cynodon dactylon*)

Preplant: Fumigate with methyl bromide (Dowfume, Brom-o-gas, Profume, Terr-o-gas), dazomet (Basamid), or metam-sodium (Vapam). If the site is not fumigated, use three repeat glyphosate 4L treatments every three weeks at 1 gal/a (9.4 L/ha) each. Postemergence control: In zoysiagrass or tall fescue, repeat fenoxaprop-ethyl (Acclaim Extra 0.57L at 1½ pts/A, 1.8 L/ha) or fluazifop-P-butyl (Fusilade T&O 2L at 5 to 6 oz/a, 367 to 439 ml/ha) on 30-day intervals. Fusilade T&O can be used only on commercial turf (golf courses, sports fields, parks, etc.); it is restricted for use on home lawns. Use only on fescue more than four weeks old, starting in spring. Maintain good soil moisture and discontinue treatment during summer stress. In centipedegrass, apply sethoxydim (Vantage 1L at 2 pts/a, 2.3 L/ha) and repeat in three weeks. In St. Augustinegrass, use ethofumesate (Prograss 1.5EC at 2 gal/A, 18.7 L/ha) plus atrazine (Aatrex 4L 2 qts/a, 4.7 L/ha). Begin in spring, two weeks after green-up, and repeat in 30 days.

Bermudagrass Encroachment into Bentgrass Golf Greens

Siduron (Tupersan) and ethofumesate (Prograss) suppress bermudagrass; however, varying levels of bentgrass injury normally occur. Control is generally best in spring and fall when the bentgrass is actively growing and the bermudagrass is not. Temporary (up to three months) bermudagrass suppression has been achieved with combinations of siduron with flurprimidol (Cutless), as well as ethofumesate plus flurprimidol. This suppression has been superior to that achieved by the standard practice of using siduron alone. April treatments are less injurious to bentgrass and provide a level of bermudagrass suppression similar to that produced by a September application followed by an April application. Siduron (Tupersan 50WP) is used at 18 to 24 oz/1,000 sq. ft. (55 to 73 kg/ha) in spring and fall. Water in and repeat in 30 days, with at least three applications in spring and three in fall. With ethofumesate (Prograss 1.5EC) plus flurprimidol (Cutless 50W), apply initial application in March/April, at 3 oz plus 0.6 oz/1000 sq. ft. (9.5 L/ha plus 1.8 kg/ha). Repeat in six weeks at 1.7 plus 0.14 oz/1000 sq. ft. (5.4 L/ha plus 0.43 kg/ha). Repeat again in 30 days. Temporary turf discoloration (~30%) will follow treatments.

Birdseye Pearlwort (*Sagina procumbens*)

Postemergence control in tolerant turfgrasses consists of repeat applications of mixtures of 2,4-D, carfentrazone, dicamba, MCPP, MCPA, pyraflufen-ethyl, and/or sulfentrazone. Other postemergence options include atrazine or simazine, atrazine plus bentazon, chlorsulfuron, imazaquin, metribuzin, metsulfuron, triclopyr alone or combined with Clopyralid, fluroxypyr, or 2,4-D. Postemergence control in bentgrass greens is with fenarimol.

Chamberbitter, Niruri or Gripeweed (*Phyllanthus urinaria*)

Apply products containing atrazine or simazine, twice, 30 days apart. Prompt (a premix of atrazine and bentazon) also works well. Products containing two- or three-way broadleaf herbicide mixtures applied at least twice seven days apart also work in tolerant turfgrasses. Begin treatments in spring when weeds are small.

Common Lespedeza, Annual Lespedeza, or Japanese-Clover (*Kummerowia striata* [*Lespedeza striata*])

Use repeat applications of two- or three-way mixtures of 2,4-D, dicamba, MCPP, or MCPA. 2,4-D alone does not satisfactorily control lespedeza. Other options include atrazine or simazine, metribuzin, triclopyr alone or combined with clopyralid or 2,4-D, atrazine plus bentazon, imazaquin, and metsulfuron. Spotlight, Confront, and Manor/Blade are effective.

Dallisgrass (*Paspalum dilatatum*)

For postemergence control in bermudagrass, repeat MSMA/DSMA applications at 1 to 2 lbs ai/a (1.12 to 2.24 kg ai/ha) every five to seven days, starting in spring. Must stay on schedule. Adding metribuzin (Sencor 75DF at 0.19 to 0.25 lb/a, 13.3 to 17.5 g/ha) to MSMA or DSMA increases control, but also increases turf injury. Tank mixing foramsulfuron (Revolver 0.19L at 26 oz/a, 1.9 L/ha) with MSMA, applied twice three weeks apart, or using an alternating application pattern of MSMA followed by foramsulfuron (Revolver 0.19L) followed by MSMA, two weeks apart, also increases control without increasing turf injury. Water if turf is drought stressed. In other grasses, spot treat or rope wick with glyphosate (4S), using 2 fl oz/gal water (15.6 mL/L). Begin in spring, and repeat in two to three weeks. Avoid desirable plants.

Dollarweed, Pennywort (*Hydrocotyle* spp.)

Repeat applications of two- or three-way mixtures of 2,4-D, dicamba, MCPP, or MCPA. Other suggested options include atrazine or simazine, metribuzin, triclopyr alone or combined with clopyralid or 2,4-D, atrazine plus bentazon, imazaquin, metsulfuron, and quinclorac. Best results are obtained with fall or spring treatments.

Doveweed (*Murdannia nudiflora*) and Spreading Dayflower (*Commelina diffusa*)

Products containing atrazine or simazine, foramsulfuron (Revolver), or metsulfuron (Manor or Blade) should be applied twice, 30 days apart. Prompt (a premix of atrazine and Basagran) also works well. Tank mixes of MSMA or DSMA with Sencor or multiple applications of two- or three-way broadleaf herbicide mixtures also provide good control, but can cause phytotoxicity to certain turfgrass species. Tank mixing pyraflufen–ethyl (Octane), carfentrazone (Quicksilver), or sulfentrazone (Dismiss) with these products increases and hastens their activity. Repeat applications of all herbicides or combinations will be needed for complete control.

Florida Betony or Rattlesnake Weed, Florida Hedgenettle (*Stachys floridana*)

Fumigate contaminated soil. Repeat applications of two- or three-way mixtures of 2,4-D, dicamba, MCPP, or MCPA. Other suggested options include atrazine or simazine, metribuzin, triclopyr alone or combined with clopyralid or 2,4-D, and atrazine plus bentazon.

Goosegrass (*Eleusine indica*)

Preemergence control is best with Ronstar at 3 lbs ai/acre. Mix with pendimethalin or Barricade (Regalstar II) for additional control of crabgrass. First application is in early spring when soil temperatures at 4 inches reach 63°F for 24 consecutive hours. POST control is with repeat applications of Illoxan 3EC at 1 to 1.4 qts/a, Sencor 75DF (0.19 lbs/a) + MSMA (1 lb ai/a), Fusilade 2EC (4 to 6 oz/a), Acclaim Extra (13 to 39 oz/a), Revolver 0.19L at 26oz/a, or Dismiss 4L at 8 oz/acre. Illoxan and Revolver may be used on bermudagrass greens. Avoid treating drought- and heat-stressed turf.

Ground Ivy (*Glechoma hederacea*)

Reduce shade source and grow shade-tolerant turfgrasses. Herbicides include three-way combinations of 2,4-D + MCPP + dicamba. Other herbicides include those containing 2,4-DP, fluroxypyr, or triclopyr; dicamba alone also works well. Mid- to late-fall applications are best, followed by another application in spring.

Knotweed, Prostrate (*Polygonum aviculare* [L.])

Repeat applications of dicamba or two- or three-way mixtures of 2,4-D, dicamba, MCPP, or MCPA. Other suggested options include atrazine or simazine, metribuzin, and triclopyr alone or combined with clopyralid or 2,4-D. Oxadiazon may provide good preemergence control if applied at or before the time for crabgrass control. In cool-season turfgrasses, fall preemergence applications provide good control.

Kyllinga spp.

Annual kyllinga species can be controlled with Basagran, Image, Manage, Certainty, Monument, or repeat applications of MSMA or DSMA. Perennial species require repeat applications of Image, Image + MSMA, Certainty, Monument, or Manage.

Lawn Burweed, or Spurweed (*Soliva pterosperma*)

Preemergence or postemergence applications of simazine or atrazine in mid-fall provide excellent control. Prompt and Sencor also work well in tolerant turfgrasses. Repeat applications of two- or three-way broadleaf herbicide mixtures Velocity or Monument also provide control. The key to control is timing: Apply in fall when weeds are small.

Mat Lippia, or Matchweed (*Phyla nodiflora*)

Use products containing atrazine or simazine applied twice, 30 days apart. Prompt (a premix of atrazine and bentazon) also works well. Products containing two- or three-way broadleaf herbicide mixtures applied at least twice, seven days apart, also work in tolerant turfgrasses.

Poa trivialis in Bentgrass

Use bispyribac-sodium (Velocity 80WP at 2.6 oz/acre, 182 g/ha) applied twice, three weeks apart, or sulfosulfuron (Certainty 75 WDG at 0.25 oz/a, 17.5 g/ha) applied three weeks apart. Expect short-term phytotoxicity. *Poa trivialis* is suppressed in perennial ryegrass with fenoxaprop-ethyl (Acclaim Extra) every two to three weeks from April to September, or with ethofumesate (Prograss) in October and November. Lower rates must be used in bentgrass; thus, poorer control often results. Spot treat with glyphosate (Roundup Pro, others) in late summer just prior to overseeding.

Purple Nutsedge (*Cyperus rotundus*)

Monument at 0.56 oz/acre or Certainty at 1.25 oz/acre in bermudagrass and zoysiagrass is recommended. Make a repeat application six to eight weeks after the first. Two applications of Sedgehammer 75 DF at 1.33 oz/acre rate four to six weeks apart provide suppression for most of the growing season. The spot treatment rate for Sedgehammer is 0.9 gram per gallon. Sedgehammer is safe on all turfgrasses but is not labeled for golf greens. Image 70 DG (imazaquin) is an effective herbicide for suppressing sedges in turfgrass. It is approved for use on bermudagrass, zoysiagrass, St. Augustinegrass, and centipedegrass. The recommended rate is 0.26 oz of product per 1,000 sq. ft. This translates to 11.4 oz/acre. The active-ingredient rate is 0.5 lb/acre. A repeat application will be needed for full-season suppression. Make the repeat application in four to six weeks or when regrowth appears. A tank mix of 0.5 lb/acre of Image + 2.0 lb/acre MSMA has given better suppression than either product alone. Do not use the MSMA tank mix on St. Augustinegrass or centipedegrass. Image + MSMA will cause significantly more injury to zoysiagrass than to bermudagrass. The coarse-leaved japonica-type zoysiagrasses tend to be more herbicide tolerant than the fine-leaved matrella types.

Sandbur (Sandspur) (*Cenchrus* sp.)

Perform preemergence control in early spring, with split applications 60 days apart of preemergence herbicides such as prodiamine (Barricade), dithiopyr (Dimension), pendimethalin (Pendulum), oxadiazon (Ronstar), or oryzalin (Surflan). Postemergence control in bermudagrass and zoysiagrass includes MSMA (1 lb ai/a, 1.12 kg ai/ha). Repeat in 10 days. In centipedegrass, apply sethoxydim (Vantage 1L at 2 pts/a, 2.3 L/ha). Repeat in 21 days. In fescue and zoysiagrass, repeat fenoxaprop-ethyl (Acclaim Extra 0.57L at 1½ pts/a, 1.8 L/ha) or fluazifop (Fusilade T&O 2L at 5 to 6 oz/a, 365 to 439 ml/ha) in 30-day intervals. Use only on fescue greater than four weeks old, starting in spring. Maintain good soil moisture and discontinue during summer stress.

Smutgrass (*Sporobolus indicus*)

Selective control has been very elusive. Best control strategies in warm-season grasses are repeated summer atrazine or simazine applications 10 days apart. Add a crop-oil concentrate. However, expect temporary turfgrass damage with this. Non-selective control is achieved by spot spraying or rope wicking glyphosate (Roundup Pro). If rope wicking, treat in two directions.

Spurges (*Chamaesyce* sp.)

Summer annuals species include spotted, prostrate, garden, and round-leaf spurges. These often act as indicator plants for high-nematode-containing soils. Metsulfuron (Manor or Blade 60DF at 0.25 oz/a, 17.5 g/ha) provides best control. Two and three-way mixes of 2,4-D, dicamba, and MCPP also work. Repeat applications of the mixes may be necessary as plants mature.

Tall Fescue Clumps (*Festuca arundinacea*)

Postemergence control in Kentucky bluegrass, fine fescue, zoysiagrass, or bermudagrass includes chlorsulfuron (Corsair 75DF) at 4 to 5 oz/a, 280 to 350 g/ha, or as a spot treatment at 0.33 g/L water. In bermudagrass and zoysiagrass, foramsulfuron (Revolver 0.19L at 26 oz/a, 1.9 L/ha) or trifloxysulfuron (Monument 75DF at 0.32 oz/a, 22.4 g/ha) can be used. In dormant bermudagrass, spot treat with glyphosate (Roundup Pro 4L at 2 oz/gal water, 15.6 ml/L), avoiding desirable green plants. Repeat in 60 days.

Thin or Bull Paspalum (*Paspalum setaceum*)

Repeat applications of MSMA or DSMA at 1 to 2 lb ai/a (1.12 to 2.24 kg ai/ha) are required every seven days until complete control is achieved.

Torpedograss (*Panicum repens*)

Nonselective control is achieved with at least three applications of glyphosate (Roundup Pro), each spaced three weeks apart. Another nonselective control option involves fumigating with methyl bromide and replanting. Selective control (or suppression) is available with quinclorac (Drive) and trifloxysulfuron (Monument). These should be applied two or three times, spaced three to four weeks apart. Expect some minor temporary turfgrass discoloration.

Violets (*Viola* spp.)

Use products containing triclopyr and clopyralid, or triclopyr plus 2,4-D. Multiple applications seven days apart are often required. Partial control is obtained with 2,4-D plus 2,4-DP. Mid- to late-fall applications are best, followed by mid-spring to early summer.

Virginia Buttonweed (*Diodia virginiana*)

Postemergence suppression is achieved with two-way or three-way herbicides with 2,4-D, dicamba, and/or MCPP. 2,4-D is most effective; therefore, use combination products with a high concentration of it. Repeat in four to five weeks. A combination of chlorsulfuron (Corsair 75DF at 3 oz/a, 210 g/ha) plus 2,4-D at 0.5 lb ai/a (0.56 kg ai/ha); 2,4-D plus clopyralid plus dicamba (Millennium Ultra 3.75L at 2.5 pts/a, 2.9 L/ha); trifloxysulfuron (Monument 75DF at 0.56 oz/a, 39 g/ha) plus clopyralid (Confront 3L at 1 pt/a, 1.17 L/ha); or trifloxysulfuron plus 2,4-D has worked. Repeat in 60 days.

Wild Garlic and Wild Onion (*Allium* sp.)

Imazaquin (Image 1.5L at 2 pts/a, 2.3 L/ha) should be applied in December. Repeat with 1 pt/a (1.2 L/ha) in early March. Add 0.25 percent nonionic surfactant. Also, use 2,4-D LV ester alone or in two- or three-way combination products. Treat in November, March, and again the following November. Trifloxysulfuron (Monument) and metsulfuron (Manor/Blade) also work well. In dormant turf, apply glyphosate (Roundup Pro 4L) at 1 pt/a (1.17 L/ha) and repeat in three to four weeks.

Yellow Nutsedge (*Cyperus esculentus*)

Apply Monument 75 DF (trifloxysulfuron) at 0.56 oz/acre or Certainty 75 WDG (Sulfosulfuron) at 1.25 oz/acre in bermudagrass and zoysiagrass. Make a repeat application if regrowth appears. Add 0.25 percent nonionic surfactant. Basagran T/O (bentazon) may be used at 0.75 to 1.5 fl oz/1,000 sq. ft or 2 to 5 pt/acre at the four- to six-leaf stage of nutsedge growth. Apply when the temperature is at least 75°F. Add crop oil or surfactant to Basagran. Complete coverage is essential. Repeat the application in 7 to 10 days. Do not apply more than 6 pt/acre of Basagran in one season. Sedgehammer 75 DF (halosulfuron) is effective on purple nutsedge, yellow nutsedge, and kyllinga and should be applied to nutsedge in the three- to eight-leaf stage at 1.33 oz/acre. Make a second application in four to six weeks or when regrowth appears. Add 0.25 percent nonionic surfactant. Do not mow for two days before and two days after application. For spot treatment, mix 0.9 gram of Sedgehammer in one gallon of water with 1/3 fl oz nonionic surfactant. Image 70 DG (imazaquin) is an effective herbicide for suppressing purple or yellow nutsedge in turfgrass. It is approved for use on bermudagrass, zoysiagrass, St. Augustinegrass, and centipedegrass. The recommended rate is 0.26 oz of product per 1,000 sq. ft. (11.4 oz/acre). A repeat application will probably be needed for full-season suppression. Make the repeat application in four to six weeks or when regrowth appears. A tank mix of Image + 2.0 lb/acre MSMA has given better suppression than either product alone. Do not use the MSMA tank mix on St. Augustinegrass or centipedegrass. Image + MSMA will cause longer-term injury to zoysiagrass than to bermudagrass. Meyer and El Toro zoysiagrasses are more tolerant of MSMA than Emerald or Matrella zoysigrasses are.

GENETICALLY MODIFIED TURFGRASSES

In recent years, innovations in agriculture have led to the development of transgenic crops, also known as genetically modified organisms (GMOs). Genes have been isolated from other plants, animals, and even microbes and placed into crops to develop a plant more tolerant to herbicides and insects. This technology was first applied to row crops, using *Bacillus thurgenensis* (Bt). This microorganism was inserted into the plants to make them more tolerant to leaf-feeding insects in order to reduce the use of insecticides. Recently, this technology has been used to create herbicide-tolerant plants. The first commercial plants to be genetically modified were corn, soybeans, and cotton, making these plants tolerant to the popular herbicide glyphosate (Roundup). These Roundup-ready crops have revolutionized weed control in most agriculture markets.

Glyphosate inhibits the synthesis of aromatic amino acids, thus stopping protein production. Specifically, glyphosate inhibits enolpyruvylshikimate phoshate synthase *(EPSPS)*, an enzyme, located in the chloroplast, which is involved in the shikimic acid pathway of plants and microorganisms (della-Cioppa et al., 1986). Roundup tolerance in most tolerant plants is conferred by a gene (*CP4 EPSPS*) from *Agrobacterium* sp. strain *CP4* that produces a version of the *EPSPS* with reduced affinity for glyphosate (Padgette et al., 1996). Agrobacterium infection and particle bombardment are ways of introducing the desired gene into the target plant species (Fei and Nelson, 2004). Three steps are involved in creating a transgenic plant through particle bombardment. First, embryogenic callus by tissue culture must be induced, followed by particle bombardment of the embryogenic callus; finally, on tissue culture medium with a selection agent, subsequent selections are made to recover transgenic plants (Fei and Nelson, 2004). A drawback of this method is that potential genetic variations can be generated, creating phenotypic changes. Furthermore, the selection agent may exert additional stresses that favor somaclonal variation (Bregitzer et al., 1998).

Recently, this technology has been transferred into the turfgrass industry. In the late 1990s, a cooperative effort between Monsanto Company and O. M. Scotts Company produced the first Roundup Ready turfgrass by successfully inserting the *CP4 EPSPS* gene into creeping bentgrass. The reasoning for this development was to aid in controlling *Poa annua* [L.] and other grassy weeds in creeping bentgrass turf. The development of such a modified turfgrass would offer great promise for the golf industry because it would provide a simplified and more effective control for aggressive grassy weeds including annual bluegrass, roughstalk bluegrass, bermudagrass, and other noxious grassy and broadleaf weeds.

Roundup Ready bentgrass (RRB), however, presents concerns. Scientists are wary of the environmental impact this perennial can have on natural ecosystems. Since creeping bentgrass will be the first perennial plant to share the resistant gene, concerns exist that this wind-pollinated species can cross-pollinate with its wild relatives and form natural resistant hybrids, or "superweeds." The EPA found long-distance pollen movement from multiple planting sites of this genetically modified creeping bentgrass. Most of the gene flow was found within 1 to 2 miles (1.6 to 3.2 km) in the direction of prevailing winds, but some were found as far as 13 miles (21 km) away (Watrud et al., 2004). Also, the cross pollination potential can occur much farther away. Research has documented pollen movement and gene flow from GMO bentgrass to resident *Agrostis* plants; however, none of the sampled plants suspected of being cross pollinated has proven positive for possessing the *CP4 EPSPS* marker (Watrud et al., 2004). Male sterility genes are now available that make the pollen sterile, thus removing much of the concern about gene flow.

There are, however, many potential benefits to the use of RRB. The first is the obvious advantage of controlling annual bluegrass and other noxious weeds by using the nonselective herbicide glyphosate. RRB can reduce the overall use of fungicides, insecticides, fumigants, and plant-growth regulators, with the elimination of annual bluegrass in bentgrass turf reducing the need for these pesticides. A potential reduction in labor would also occur, since fewer overall pesticide applications would be required. Managers would be able to successfully control these noxious weeds in their bentgrass turf, thus reducing the time and attention required to culturally maintain them. All of these benefits would lead to an improved playing surface and consistent performance due to the increased uniformity, compared with those sites infested by noxious weeds.

Roundup Ready bentgrass is not yet approved for sale or distribution. It is currently in the review process by the Animal and Plant Health Inspection Service (APHIS) and the U.S. Department of Agriculture (USDA) to determine whether it poses any risks to the environment or to agriculture.

The ability to develop dwarfed plants is another area that is being explored by genetic engineers. Geneticists are able to characterize the gibberellin biosynthesis and metabolism genes *GA20ox*, *GA3ox* and *GA2ox* (Taiz and Zeiger, 2002). This has enabled them to modify the transcription of these genes regarding the gibberellin production in plants and has resulted in smaller or shorter plants (Hedden and Phillips, 2000). Increased dwarfness is important in cereal crops like wheat because, in dense communities, wheat often grows too tall and is prone to lodging. Inhibition in the rise of gibberellin can also prevent bolting in rosette plants. This is significant especially in sugar beet production, since seeds can be sowed earlier to form a larger storage root without the concern of bolting, which prevents the formation of the root. This approach has also been applied to turfgrasses in order to keep the grass short and seedhead free, thus reducing the need to mow.

Plants are transformed with antisense constructs of the *GA20ox* or *GA3ox* genes, which encode the enzymes leading to the synthesis of GA_1, thus reducing the levels of GA_1. Another way to reduce GA_1 levels is by overexpressing the gene responsible for GA_1 metabolism: *GA2ox* (Taiz and Zeiger, 2002). Varying degrees of overexpression of the foreign gene can provide varying degrees of dwarfness.

IMPENDING LOSS OF THE ORGANIC ARSENICAL HERBICIDES

The United States Environmental Protection Agency (USEPA) has announced the intent to cancel all agricultural uses of members of the organic arsenical herbicide family. This includes all formulations of MSMA, DSMA, CMA, and CAA. Weed-control strategies will change significantly when this ban occurs, especially when turf managers are dealing with postemergence control of perennial grasses such as dallisgrass, knotgrass, thin (or bull) paspalum, tropical signalgrass, and others (McCarty and Yelverton, 2005). Preemergence herbicides will become more important in controlling annual grasses, while selective placement of glyphosate will become more important in managing perennial grasses. Some other herbicide will be available to replace the organic arsenical herbicides, but costs will drastically increase and control will largely decrease.

7

Turfgrass Plant Growth Regulators

Plant growth regulators (PGRs), or plant growth inhibitors, are increasingly being used on golf courses and other turfgrass sites to suppress seedhead(s) and vegetative growth of desirable turfgrasses, enhance turfgrass quality, and manage annual bluegrass (*Poa annua* [L.]) growth and development. Depending upon the turfgrass and situation, PGRs may reduce mowing costs, prevent scalping, increase turf density, and decrease the need to mow steep slopes (Figure 7.1). Traditionally, PGRs were used in the United States to suppress bahiagrass, Kentucky bluegrass, and tall fescue seedhead production in low-maintenance areas such as highway roadsides, airports, and golf course roughs (Figure 7.2); however, products are now registered for use in most high-maintenance turfgrasses.

PGRs are recommended for use only on certain turfgrass species (Table 7.1). Additionally, the use of a PGR is often determined by the type of turfgrass area and level of maintenance. For example, imazapic (Plateau) is recommended for use on low-maintenance sites such as roadsides and airports. However, paclobutrazol (Trimmit), flurprimidol (Cutless), and trinexapac-ethyl (Primo) may be used on putting green turfgrasses. The product label should always be consulted for information concerning turfgrass species and application sites before use.

Prior to the development of PGRs for fine turfgrasses, several undesirable characteristics were associated with PGRs used on low-maintenance, or rough, turfgrass sites. These included (a) phytotoxicity (burn) of treated leaves for four to six weeks following applications, (b) reduced recuperative potential when the PGR-treated turfgrass was physically damaged, and (c) increased weed pressure due to reduced competition from treated turfgrasses. However, because most PGRs were historically used in low-maintenance areas, these undesirable characteristics did not pose a problem to most managers. Fine turfgrass PGRs suppress vertical top growth, but have less of an effect on the lateral or horizontal spread of stolons. The most noticeable effect is usually a reduction in the amount of clippings and a reduction in mowing frequency. On tee boxes and fairway landing areas, turf recovery from golf club divots and other injuries occur while vertical top growth remains suppressed. Depending upon the product, fine turf PGRs also enable superintendents to reduce mowing frequency on fairways, suppress annual bluegrass in creeping bentgrass greens, improve ball lie or playability, and suppress the growth of bermudagrass during overseeding with a cool-season turfgrass.

Figure 7.1
Plant growth regulators (PGRs) help regulate (often, retard) shoot and seedhead growth. This is especially useful in hard-to-mow areas.

Figure 7.2
PGRs were initially used to maintain low-maintenance areas such as roadsides.

PGR CLASSIFICATION

Similar to herbicides, PGRs are placed into groups on the basis of mode of action, or the way they inhibit growth of turfgrasses. Classification schemes can vary; however, three distinct groups of PGRs exist (Table 7.2).

Cell-Division Inhibitors (Also Called Type I PGRs)

Cell-division inhibitors are primarily foliage absorbed and inhibit cell division and differentiation in meristematic regions. They inhibit both vegetative growth and seedhead development. Growth inhibition is rapid, occurring within 4 to 10 days and lasting three

Table 7.1
Plant Growth Regulators for Various Turfgrass Species

Plant Growth Regulator		Root or Foliar Absorbed	Suppression Characteristics		Maintenance Level of Turfgrass Site[1]	Turfgrass Species
Common Name	Trade Names		Foliage	Seedhead		
Chlorsulfuron	Telar	Both	Yes	Yes	Low	bahiagrass, bermudagrass, fescues, Kentucky bluegrass
Ethephon	Proxy	Foliar	Yes	Yes	Low, Medium, High	bentgrass, fescues, Kentucky bluegrass, ryegrass, bermudagrass
Flurprimidol	Cutless	Root	Yes	Partial	Low, Medium, High	bentgrass, bermudagrass, Kentucky bluegrass, ryegrass, St. Augustinegrass, zoysiagrass
Flurprimidol + trinexapac-ethyl	Legacy	Root & foliar	Yes	Partial	Medium, High	bentgrass, bermudagrass, Kentucky bluegrass, ryegrass, St. Augustinegrass, zoysiagrass
Glyphosate	Roundup Pro, others	Foliar	Yes	Yes	Low	bahiagrass, bermudagrass, Kentucky blue-grass, fescues
Imazapic	Plateau	Both	Yes	Yes	Low	bahiagrass, bermudagrass, centipedegrass
Imazapic + Glyphosate	Journey	Both	Yes	Yes	Low	bahiagrass, bermudagrass
Imazethapyr+ Imazapyr	Event	Both	Yes	Yes	Low	bahiagrass
Maleic hydrazied	Slo-Gro	Foliar	Yes	Yes	Low	bahiagrass, fescues, Kentucky bluegrass, ryegrass
Mefluidide	Embark, Embark T&O 0.2S	Foliar	Yes	Yes	Low, Medium Low, Medium, High	bermudagrass, centipedegrass, fescues, Kentucky bluegrass, kikuyugrass, ryegrass, St. Augustinegrass, zoysiagrass, creeping bentgrass fairways (Embark T&O only)
Metsulfuron	Escort, Manor, Blade	Both	Yes	Yes	Low	bermudagrass, fescues, Kentucky bluegrass
Paclobutrazol	Trimmit	Root	Yes	Partial	Low, Medium, High	bentgrass, bermudagrass, fescues, Kentucky bluegrass, ryegrass, St. Augustinegrass
Sethoxydim	Poast, Vantage	Foliar	Yes	Yes	Low, Medium, High	centipedegrass, fine fescues (roadside tall fescue)
Sulfometuron	Oust	Both	Yes	Yes	Low	bahiagrass, bermudagrass
Trinexapac-ethyl	Primo	Foliar	Yes	No	Low, Medium, High	bahiagrass, bentgrass, bermudagrass, centipe-degrass, fescues, Kentucky bluegrass, kikuyu-grass, ryegrass, St. Augustinegrass, zoysiagrass

[1] Low-maintenance turfgrass sites: roadsides, airports, storage sites, hard-to-mow areas, etc.
Medium-maintenance turfgrass sites: industrial grounds, parks, cemeteries, golf course roughs, home lawns.
High-maintenance turfgrass sites: putting greens, tees, fairways, athletic fields, high-quality home lawns, and commercial properties.

Table 7.2
Characteristics of Various Plant Growth Regulators Used in Turfgrass Management

Mode of Action	PGR Common Name	PGR Trade Name	Absorption Site	Comments
Inhibit cell division (or mitosis)	Maleic hydrazide	Retard, Royal Slo-Gro, Liquid Growth Retardant	Foliar	Effective seedhead suppressors. Growth inhibition is rapid, within 4 to 10 days, and lasting 3 to 4 weeks.
	Mefluidide	Embark, Embark 0.2S		
Inhibit fatty acid biosynthesis	Sethoxydim	Poast, Vantage, Sethoxydim-Pro	Foliar	Usually low in cost, but turfgrass tolerance is low and rate dependent.
Inhibit amino acid biosynthesis	Glyphosate	Roundup Pro, others	Foliar	
Inhibit amino acid biosynthesis	Imazapic	Plateau	Foliar and root	Use is restricted to low-maintenance turfgrasses.
	Imazapic + Glyphosate	Journey	Foliar and root	
	Imazethapyr + Imazapyr	Event	Foliar and root	
	Sulfometuron	Oust	Foliar and root	
	Metsulfuron	Escort, Manor, Blade	Foliar and root	
	Chlorsulfuron	Telar	Foliar and root	
Promotes ethylene, which reduces cell elongation	Ethephon	Proxy	Foliar	Effects will not be seen until 7 to 10 days after application. Duration of activity ranges from 4 to 7 weeks. Rainfall or high spray volumes are not needed for activation.
Interferes with gibberellin biosynthesis	Flurprimidol	Cutless	Root	Initial growth response is slower compared with cell division inhibitors, but duration of activity is usually longer, 3 to 7 weeks. Rainfall, irrigation, or high application volumes are required for activating root-absorbed PGRs.
	Flurprimidol + trinexapac-ethyl	Legacy	Root and foliar	
	Paclobutrazol	Trimmit	Root	
	Trinexapac-ethyl	Primo	Foliar	
Chemicals for Growth and Color Promotion				
Promotes gibberellin biosynthesis	Gibberellic acid	RyzUp 4% active solution, ProGibb	Foliar	Used to promote growth and prevent discoloration (e.g., purpling) during periods of cold stress and light frosts on bermudagrass such as Tifdwarf or Tifgreen. Also reverses overdosing of GA-inhibiting PGRs. Do not apply when night temperatures exceed 65°F (18.3°C).

to four weeks, depending on application rate. Mefluidide (Embark, Embark T&O 0.2S) and maleic hydrazide (Royal Slo-Gro, others) are examples of cell-division inhibitors. These products are used primarily on low- and medium-maintenance turfgrass areas, as phytotoxicity (yellowing) can be a problem. On golf courses, cell-division inhibitors may be useful for reducing mowing on steep slopes, ditches, and other difficult-to-mow areas and for reducing annual bluegrass seedhead development on putting greens.

Herbicides Various herbicides are used at low rates to suppress growth or seedhead development of turfgrasses. Depending upon the chemical, herbicides inhibit turfgrass growth and development through interruption of amino acid synthesis (glyphosate, sulfometuron, chlorsulfuron, metsulfuron, imazapic, imazethapyr + imazapyr) or fatty acid biosynthesis (sethoxydim). Turfgrass tolerance can be marginal and is highly rate dependent. Herbicides are used primarily on low-maintenance turfgrasses, to reduce mowing and control weeds.

Gibberellin Biosynthesis (or GA) Inhibitors (Also Called Type II PGRs)

Gibberellin is a plant-produced hormone that is needed for cell elongation and for normal growth and development (Figure 7.3). There are numerous gibberellins needed for

Figure 7.3
Most modern PGRs used in fine turf block the production of gibberellic acid in plants, a hormone that regulates cell elongation. Shown on the left is PGR-treated turf, compared with untreated turf on the right.
(Courtesy Jan Weinbrecht.)

normal plant growth and development. When gibberellin production is inhibited, plant cells do not elongate, internodes become shortened, and overall plant growth is reduced. Two types of gibberellin biosynthesis inhibitors are available for use on golf courses. Trinexapac-ethyl (Primo), a Class A gibberellin biosynthesis inhibitor, is foliar absorbed and inhibits the synthesis of gibberellin late in its biosynthetic pathway. Paclobutrazol (Trimmit) and flurprimidol (Cutless) are root absorbed Class B gibberellin biosynthesis inhibitors that inhibit gibberellin biosynthesis early stages of its pathway. This early blockage prevents the synthesis of numerous subsequent gibberellins. Inhibition during the early stages of gibberellin biosynthesis can lead to increased injury when environmentally stressed turfgrasses are treated with Class B gibberellin biosynthesis inhibitors. Additionally, turfgrasses may exhibit various morphological responses such as the widening of creeping bentgrass leaf blades. Studies have shown that inhibiting gibberellin biosynthesis late in the pathway, as with trinexapac-ethyl, is less physiologically disruptive and injurious to turfgrasses, although as with all PGRs, plant phytotoxicity can still occur at higher rates. Although less injury may result from Class A gibberellin biosynthesis inhibitors, they also tend to provide less seedhead and seed stalk suppression or control. Most recently, tank-mix combinations of PGRs are becoming available. The first is a combination of flurprimidol and trinexapac-ethyl (Legacy 1.52MEC).

Ethephon (Proxy), a PGR recently introduced into the turfgrass market, is hydrolyzed by treated plants into the gas ethylene. Ethylene is a growth hormone noted as a fruit ripener and retardant in floriculture crops. Ethephon appears to evoke more response on C3 grasses than on C4 grasses. In turfgrass, ethephon at 3.4 lbs ai/a (3.74 kg ai/ha) provides good Kentucky bluegrass, perennial ryegrass, and hard fescue growth suppression for approximately four weeks. Reduction of mowing frequency and clipping yield by 50 percent for up to seven weeks will result, as ethephon will slow the overall growth of the turf. Kentucky bluegrass and creeping red fescue exhibited shorter leaf blades when treated with ethephon for up to 12 weeks (Christians, 1985). Reduction ranged from 20 to 72 percent on Kentucky bluegrass and 26 to 55 percent on creeping red fescue, compared with similar untreated turf (Howieson and Christians, 2005). Tall fescue and perennial ryegrass both exhibited significant reduction in leaf blade width at 33 to 42 percent and 28 percent, respectively, compared with untreated turf, bermudagrass, manilagrass, and zoysiagrass, all of which developed shortened leaf-blade width when treated with ethephon (Howieson and Christians, 2005). Due to its action in the shortening of the leaf blades, ethephon causes a determinant to grow to canopy height, thus decreasing vertical grass growth and mowing frequency (Diesburg and Christians, 1989). Clipping weights also are reduced on Kentucky bluegrass and creeping red fescue when treated by ethephon. Enhanced elongation of internodes is observed on creeping red fescue, Kentucky bluegrass, perennial ryegrass, and tall fescue when treated with ethephon (Howieson and Christians, 2005). This increased elongation causes stolon-like stems to become prostrate in nature (Christians, 1985). This morphological change may allow these grasses to have an increased recuperative rate, which makes ethephon ideal for use on golf courses, sports turf, and heavy-traffic areas. Although most grasses experience an enhancement in internode length, a reduction in internode length on bermudagrass was provided by ethephon two to three weeks after an application. Creeping bentgrass, manilagrass, and zoysiagrass internode spacing was not affected by ethephon (Howieson and Christians, 2005). However, van Andel and Verkerke (1978) reported that ethephon-treated bentgrass had shorter internodes, with as many 52 percent fewer cells than nontreated plants had. This reduction is likely a result of the reduced cell division in internode tissue. Treated turf often has a decrease in chlorophyll content, resulting in a lighter green color and a possible decrease in root length. Rhizome development also may be somewhat inhibited. No affect on tillering has been observed.

Limited work in bermudagrass suggests that reduced plant growth occurs at rates above 6 lbs ai/a (6.6 kg ai/ha). On centipedegrass, at 4.5 lbs ai/a (5.04 kg ai/ha), ethephon suppresses growth for four to six weeks and enhances turf quality by providing a more uniform turf canopy, with fewer seedhead(s) and a desirable blue-green turf color.

SITE OF ABSORPTION

PGRs are absorbed, or enter the turfgrass plant, by roots or foliage (or shoots), or with some products, both roots and foliage (Table 7.2). Root-absorbed PGRs, such as paclobutrazol and flurprimidol, require irrigation or rainfall after application to move the material into the turfgrass root zone. In contrast, trinexapac-ethyl is rapidly absorbed by turfgrass foliage, and irrigation after application is not necessary. Compared with cell-division inhibitors, less likelihood of leaf burn exists due to improper spray pattern overlaps with the gibberellin biosynthesis inhibitor PGRs. Most foliar absorbed materials (e.g., mefluidide, maleic hydrazide, and herbicides) require uniform, even coverage to prevent phytotoxicity and must be absorbed by turfgrass leaves before irrigation or rainfall occurs.

GROWTH SUPPRESSION

Cell-division-inhibitor PGRs quickly (within five to seven days) suppress vegetative growth, but usually provide a shorter period of growth suppression than gibberellin biosynthesis inhibitors (three to six weeks). However, unlike gibberellin biosynthesis inhibitors, cell-division inhibitors are highly effective in suppressing seedhead development. The growth suppression activity of gibberellin biosynthesis inhibitors often is not immediately evident. Compared with cell-division inhibitors, paclobutrazol and flurprimidol are slower (10 to 14 days) in suppressing turfgrass growth, but their duration of activity is usually longer, lasting from four to eight weeks, depending on application rate. Trinexapac-ethyl typically reduces common and hybrid bermudagrass clipping weights by 50 percent at seven days after application. Depending upon application rates and schedules, trinexapac-ethyl also provides long-term (four to eight weeks) growth suppression. Another key difference is that while gibberellin biosynthesis inhibitors decrease seedhead stalk height, they have little effect on the actual formation of seedhead(s) (Figure 7.4)

Figure 7.4
Less obvious Poa annua *seedhead(s) (left) following PGR treatment, compared with the untreated (right).*
(courtesy A.J. Powell and David Williams.)

APPLICATION TIMING

Timing of application with PGRs is critical to achieving desired results. When used for seedhead suppression, the PGR must be applied before seedhead formation and emergence. Applications made after seedhead emergence will not be effective. For bahiagrass, mow the area as seedhead(s) initially emerge (usually from mid- to late spring), in order to provide a uniform, even appearance to the site. For tall fescue or Kentucky bluegrass, mow the area in early spring (late winter to early spring). The PGR treatment should be applied about 7 to 10 days following mowing or just prior to new seedhead appearance. Additional applications six to eight weeks later may be required if new seedhead(s) begin to emerge.

If PGRs are being used on creeping-bentgrass golf greens, applications should be made during periods of active root growth. In most areas of the United States, this would be during the mid-fall and spring months. Applications should not be made during stressful midsummer and midwinter months. On warm-season turfgrasses such as bermudagrass, the appropriate PGR should be applied to actively growing turfgrasses after full spring green-up and several mowings. Applications can be repeated during summer months if additional growth regulation is needed.

WEED CONTROL CONSIDERATIONS

An integrated weed management program must accompany any PGR use, as these herbicides normally do not suppress weed growth, particularly broadleaf weeds. In addition, after the PGR has been applied, annual and perennial weeds can become a problem, as PGR-treated turfgrasses are less competitive with weeds. On high-maintenance turfgrasses, it usually is advisable to continue preemergence herbicide use for control of annual grass weeds. For postemergence control, 2,4-D, dicamba, or various two- and three-way herbicide mixtures normally are used to control broadleaf weeds. Other postemergence herbicides such as MSMA, for annual grass weed control, or nutsedge control herbicides, may also be needed in some situations. Postemergence herbicides often cause temporary phytotoxicity to turfgrasses. Postemergence herbicides can be tank-mixed with PGRs; however, turfgrass injury is often greater than when either type of product is used alone. Therefore, on high-maintenance turfgrasses where color and appearance may be of utmost importance, it is advisable not to tank-mix postemergence herbicides with PGRs. Additionally, if a postemergence herbicide has injured the turfgrass, PGR application should be delayed until the turfgrass has fully recovered. Reference to the PGR label and personal experience are the best guides to determine suitability of tank-mixing PGRs and postemergence herbicides.

CURRENTLY AVAILABLE PLANT-GROWTH RETARDANTS

LOW-MAINTENANCE TURFGRASSES

Sulfometuron-Methyl (Oust 75DG)

Used on bahiagrass at 0.02 lb ai/acre (0.24 kg ai/ha), sulfometuron-methyl is foliar and root absorbed and should be applied to bahiagrass in late spring or early summer, 7 to 10 days after the first mowing and before seedhead(s) appear. Do not use a surfactant. Do not apply to wetlands or where runoff water may drain onto cultivated lands or forests. Do not apply to turf less than three years old. Treated areas may appear less dense and temporarily discolored. Only one application per year should be made, as repeat application within the same year can reduce bahiagrass density. Read and follow all label recommendations before use. Often tank-mixed with glyphosate, chlorsulfuron (Telar), 2,4-D plus glyphosate (Campaign), and/or hexazinone (Velpar). **Do not exceed recommended rate.** Sulfometuron-methyl is not recommended for use on high-maintenance turfgrasses.

Chlorsulfuron (Telar 75DG)

Used in tall fescue, bahiagrass, Kentucky bluegrass, and bermudagrass, at rates ranging from 0.012 (cool-season turfgrasses) to 0.05 (warm-season turfgrasses) lb ai/acre (0.013 to 0.56 kg ai/ha). Chlorsulfuron is foliar and root absorbed and should be applied to well-established tall fescue prior to seedhead formation. Do not apply to turfgrasses less than one year old. A nonionic surfactant should be added to the spray mixture. A tank-mix of chlorsufluron (Telar 75DG) at 0.012 lb ai/a (0.013 kg/ha) plus mefluidide (Embark 2S) at 0.125 lb ai/a (0.14 kg ai/ha) can be used to suppress growth and seedhead emergence in bluegrass and fescue stands. Apply this tank-mix after spring green-up, but before tall fescue seedhead emergence. Grass seed may be planted in areas six months after treatment, but cultivation is recommended. Chlorsulfuron may also be tank-mixed with glyphosate and sulfometuron-methyl. Chlorsulfuron should not be used on turfgrasses that are stressed due to drought, insects, disease, cold temperatures, or poor fertility, as injury may result. Chlorsulfuron is not recommended for use on high-maintenance turfgrasses.

Metsulfuron (Escort 60 DF)

Used in tall fescue, Kentucky bluegrass, and bermudagrass, at rates ranging from 0.009 to 0.075 lb ai/acre (10.5 to 84 g ai/ha). Metsulfuron is foliar and root absorbed. Low rates (0.009 to 0.018 lb ai/a [10.5 to 21 g ai/ha]) may be used for growth and seedhead suppression in well-established tall fescue. Applications to tall fescue should be made in the spring after two to three inches (5 to 7.5 cm) of new growth, but before seedhead emergence. In bermudagrass, metsulfuron is used primarily to control weeds at rates ranging from 0.009 to 0.075 lb ai/a (10.5 to 84 g ai/ha). A tank mix of metsulfuron (Escort 60DF) at 0.009 to 0.012 lb ai/a (10.5 to 14 g/ha) plus mefluidide (Embark 2S) at 0.031 to 0.062 lb ai/a (0.045 to 0.07 kg ai/ha) can be used to suppress growth and seedhead emergence in tall fescue. Do not apply metsulfuron to turfgrasses less than one year old. A nonionic surfactant should be added to the spray mixture. Metsulfuron should not be used on turfgrasses that are stressed due to drought, insects, disease, cold temperatures, or poor fertility, as injury may result. Metsulfuron should not be used on cool-season grasses or bahiagrass. Use lower rates ranging from 0.009 to 0.038 lb ai/a (10.5 to 42 g ai/ha) on centipedegrass for bahiagrass control and certain broadleaf weeds including chickweed, clover, dandelion, purslane, spurge, woodsorrel, and wild garlic/onion.

Sethoxydim (Vantage 1.0 lb/gal.; Poast 1.5 lbs/gal.)

Apply at 0.1875 lbs ai/A (0.21 kg ai/ha) to established low-maintenance tall fescue for seedhead suppression. Sethoxydim is foliar absorbed. Applications should be made in the early spring before tall fescue seedhead emergence. A crop oil concentrate at 2.0 pts/acre (2.34 L/ha) or Dash HC spray adjuvant at 1.0 pt/A (1.17 L/ha) must be added to Poast. It is not necessary to add a spray adjuvant to Vantage. Unlike other herbicides used for growth suppression on low-maintenance grasses, sethoxydim has no herbicidal activity on broadleaf weeds. Appropriate broadleaf-weed-control practices are usually necessary following the use of sethoxydim. Vantage may also be used in high-maintenance centipedegrass and fine fescues (creeping red, chewings, sheep, and hard) for annual grass and bahiagrass control, as well as for suppression of common bermudagrass.

Maleic Hydrazide (Retard 2.25 lbs/gal.; Royal Slo-Gro 1.5 lbs/gal.; Liquid Growth Retardant 0.6 lbs/gal.)

Apply the respective product at 3.0 lbs ai/acre (3.36 kg ai/ha). Maleic hydrazide is foliar absorbed. Apply to bahiagrass in late spring or 7 to 14 days after first mowing, but before seedhead(s) appear. Do not use a surfactant. Do not apply to turf less than three years old and do not reseed within three days after application. Treated areas may appear less dense and temporarily discolored. Do not use on St. Augustinegrass, and do not apply to bahiagrass under drought conditions. Read and follow all label recommendations before use. A 12-hour rainfree period is required for optimum activity. A repeat application may be required six weeks after initial application, to provide season-long suppression. Also use on fescue, bluegrass, and ryegrass. Maleic hydrazide is not recommended for use on high-maintenance turfgrasses

Glyphosate (Roundup Pro 4L, Others)

Apply only to bahiagrass at 0.125 to 0.25 lb ai/acre (0.14 to 0.28 kg ai/ha). Glyphosate is also sold in combination with 2,4-D as Campaign 2.5L and with imazapic as Journey 2.25L. The addition of 2,4-D to glyphosate increases broadleaf-weed control. Glyphosate is foliar absorbed. **Note: Glyphosate is a nonselective herbicide; thus, only low rates of glyphosate should be used in bahiagrass, or severe injury will occur.** Make the initial application at a rate of 0.1875 to 0.25 lb ai/a (0.21 to 0.28 kg ai/ha) in spring, after full green-up of bahiagrass. (Timing will vary according to location.) Vegetative growth and seedhead(s) may be suppressed for approximately 45 days. Treated areas may appear less dense and temporarily discolored. Do not apply to turf less than three years old. Repeat applications of Roundup Pro 4 L at 0.125 to 0.1875 lb ai/a (0.14 to 0.21 kg ai/ha) at six weeks can be used to extend the period of growth and seedhead regulation to approximately 120 days. Glyphosate will suppress annual grasses such as ryegrass, wild barley, and wild oats at 0.125 to 0.156 lb ai/a (0.14 to 0.175 kg ai/ha) in general-use areas. Use as a mowing substitute on Kentucky bluegrass at 0.1875 lb ai/a (0.21 kg ai/ha), and at 0.25 lb ai/a (0.28 kg ai/ha) on fescues. Read and follow label recommendations prior to use. Glyphosate is not recommended for use on high-maintenance turfgrasses.

Imazapic (Plateau 2ASU) or Imazapic Plus Glyphosate (Journey 2.25L)

Apply imazapic to bahiagrass and bermudagrass at 0.031 to 0.062 lb ai/a (0.035 to 0.069 kg ai/ha). Imazapic is foliar and root absorbed. Apply to bahiagrass in spring two to three weeks before seedhead formation or 7 to 10 days after mowing. These herbicides also provide some broadleaf-weed and annual-grass control. Do not apply to wetlands. Treated areas may appear less dense and temporarily discolored. Do not use on St. Augustinegrass, tall fescue, or drought-stressed bahiagrass. Add a surfactant or methylated seed oil according to label recommendations. Read and follow label directions before use. Imazapic is not recommended for use on high-maintenance turfgrasses.

Imazethapyr + Imazapyr (Event 1.46L)

Apply to low-maintenance bahiagrass, tall fescue, perennial ryegrass, and bluegrass at 0.09 to 0.11 lb ai/acre (0.1 to 0.12 kg ai/ha). Apply after turf has completed spring transition, is actively growing, and has at least 2 inches (5 cm) of vertical growth. Add a surfactant at 0.25% v/v. Do not use on stands less than one year old or on highly managed turf. Do not reseed within three months after application. Read and follow label directions before use.

Mefluidide (Embark 2S)

Apply at 0.38 to 0.5 lb. ai/a (0.43 to 0.56 kg ai/ha) approximately two weeks before seedhead appearance. This herbicide is foliar absorbed. Do not use on turf less than four months old and do not reseed within three days after application. Treated turf may appear less dense and temporarily discolored. An eight-hour rain-free period is needed after application. Remove excess grass clippings and fallen leaves before application. The addition of 0.25 to 0.5% v/v nonionic surfactant may increase seedhead control, but also may increase turf discoloration. For low maintenance bermudagrass, use 1.0 lb. ai/a (1.12 kg ai/ha) application rate.

LOW- TO MEDIUM-MAINTENANCE TURF

Mefluidide (Embark 2S, Embark 0.2S)

Mefluidide is recommended for use on turfgrasses (common bermudagrass, tall fescue, Kentucky bluegrass, St. Augustinegrass) under low to medium levels of maintenance. Apply at 0.125 to 1.0 lb ai/A (0.14 to 1.12 kg ai/ha) [Embark 2S—0.5 to 4 pts./A, 0.58 to 4.68 L/ha; Embark 0.2S—5 to 20 pts/A, 5.84 to 23.4 L/ha]. Mefluidide is primarily foliar absorbed. Apply in spring after green-up until approximately two weeks before seedhead appearance. Optimum results may not be obtained if rainfall or irrigation occurs within eight hours following application. Do not apply to turf within four growing months after seeding, and do not reseed within three days after application. Treated turf may appear less dense and temporarily discolored. Adding 1 to 2 qts (0.95 to 1.89 L) of a nonionic surfactant per 100 gal (378.5 L) of spray solution may enhance suppression; however, turfgrass injury (discoloration) may also be increased. Use 0.125 lb ai/a (0.14 kg ai/ha) [0.5 pt/A, 0.58 L/ha] of Embark or 0.05 to 0.125 lb ai/a (0.056 to 0.14 kg ai/ha) [2 to 5 pts/A, 2.34 to 5.84 L/ha] of Embark 0.2S in mid- to late winter to suppress *Poa annua* seedhead(s) in fairways. Iron applications 10 days or less before mefluidide application may lessen discoloration. Read and follow label recommendations before use (Figure 7.5).

Figure 7.5
Poa annua *seedhead control (right) following treatment of a Class B gibberellic acid inhibitory PGR, compared with the untreated (left).*
(courtesy Fred Yelverton.)

Flurprimidol (Cutless 50WP)

Apply at 0.375 to 1.5 lbs ai/acre (0.42 to 1.68 kg ai/ha). Flurprimidol is root absorbed. Apply to Tifway, Tifgreen, and common bermudagrass or zoysiagrass golf course fairways, and hard-to-mow and trimmed areas. Provides four to eight weeks of suppression. Flurprimidol must be uniformly applied and irrigated in with 0.5 inch (1.27 cm) water within 24 hours of application. Flurprimidol does not completely control seedhead(s), only seedhead stalk elongation. Temporary turf discoloration may follow this treatment. St. Augustinegrass, bahiagrass, and common bermudagrass require the higher rate. Repeat applications every four weeks on Tifway bermudagrass; treatment at 0.5 lb ai/A (0.56 kg ai/ha) will minimize turf injury. Apply to bentgrass, bluegrass, and perennial ryegrass in late spring to early summer or late summer to early fall. Application rates range from 0.375 to 1.5 lbs ai/acre (0.42 to 1.68 kg ai/ha), depending on turf species. A second application needs to be timed at least three months before expected winter dormancy. Do not apply to golf greens. Do not exceed 0.75 lb ai/acre (84 kg ai/ha) application on coarse textured soils. Resume mowing three to five days after application.

Trinexapac-Ethyl (Primo MAXX 1MEC, Primo 25WSB)

Apply at rates ranging from 0.023 to 0.0.172 lb ai/acre (0.026 to 0.193 kg ai/ha) [Primo MAXX—3 to 22 fl.oz./a, 0.22 to 1.61 L/ha; Primo 25 WSB—refer to label], depending upon the turfgrass species, mowing height, and length of suppression desired. Trinexapac-ethyl is foliar absorbed. Low rates are for hybrid bermudagrass, centipedegrass, zoysiagrass, and St. Augustinegrass; medium rates are for common bermudagrass and tall fescue, while the high rate is for bahiagrass. Trinexapac-ethyl (Primo MAXX) and 1EC at 0.023 to 0.047 lb ai/a (0.026 to 0.05 kg ai/ha) [3.0 to 6.0 fl oz/a, 0.22 to 0.44 L/ha] and Primo WSB at 0.021 to 0.042 lb ai/a (0.024 to 0.048 kg ai/ha) [1.35 to 2.7 oz/a, 0.09 to 0.19 kg/ha] may also be used on creeping bentgrass and hybrid bermudagrass putting greens. A one-hour rain-free period is needed after application. Mowing one to seven days after application improves appearance. Repeat applications may be applied as needed or approximately three to six weeks apart to maintain growth suppression, but do not exceed 2.375 lb ai/a (2.66 kg ai/ha) [19 pts/a, 22.2 L/ha] of Primo MAXX and 1EC, or 2.71 lb ai/a (3.05 kg ai/ha) [174 oz/a,12.72 kg/ha] of Primo WSB, per year. Trinexapac-ethyl provides only partial seedhead suppression. Temporary turf discoloration may follow treatment. It is not necessary to add a surfactant to trinexapac-ethyl.

Paclobutrazol (Trimmit 2SC)

Apply at 0.5 to 0.75 lb ai/acre (0.56 to 0.84 kg ai/ha) to actively growing hybrid bermudagrass and St. Augustinegrass that have recovered from dormancy. Paclobutrazol is also available on several dry-fertilizer-formulation carriers. Paclobutrazol is root absorbed. Apply with 0.5 to 0.9 lb nitrogen per 1,000 sq. ft. (24.4 to 44.0 kg N/ha) of a nonburning fertilizer. Apply 0.25 in. (0.64 cm) of water within 24 hours after application. Reapplication can be made within the same growing season, but no sooner than eight weeks after the initial application. No more than three applications can be made annually. Do not seed within six weeks prior to or two weeks after application.

This product may also be used on overseeded golf greens and fairways during winter for turf enhancement, for annual bluegrass suppression, and to suppress the growth of perennial biotypes of annual bluegrass in creeping bentgrass greens. Paclobutrazol should be applied in early winter for seedhead suppression of annual bluegrass and should be repeated every 3 weeks until seedhead production subsides.

Do not apply to saturated soils, and treat only dry foliage. On cool-season grasses, apply in spring after green-up, but one month before onset of high temperatures. If applied in late summer/early fall, apply at least one month before the anticipated first killing frost. Do not use on areas containing greater than 70 percent annual bluegrass. Read and follow recommendations before use.

HIGHLY MAINTAINED TURF

Ethephon (Proxy 2L)

Ethephon (Proxy), a PGR recently introduced into the turfgrass market, is hydrolyzed by treated plants into ethylene. In turfgrass, ethephon at 3.4 lbs ai/a (3.74 kg ai/ha) provides good Kentucky bluegrass, perennial ryegrass, and hard fescue with growth suppression for approximately four weeks. Reduction of mowing frequency and clipping yield by 50 percent for up to seven weeks will result, as ethephon will slow the overall growth of the turf (Eggens et al., 1989). Treated turf often has a decrease in chlorophyll content, resulting in a lighter green color and a possible decrease in root length. Rhizome development, however, maybe somewhat inhibited. Tall fescue growth appears less influenced by ethephon. No affect on tillering has been observed; thus, a dense turf in maintained.

Limited work in bermudagrass suggests reduced growth at rates above 6 lbs ai/a (6.6 kg ai/ha). On centipedegrass, 4.5 lbs ai/a (5.04 kg ai/ha) ethephon suppresses growth for four to six weeks and enhances turf quality in terms of both a more uniform turf canopy with fewer seedhead(s) and a desirable blue-green turf color.

Flurprimidol (Cutless 50WP)

Apply at 0.12 to 0.25 lb ai/acre (0.22 to 0.28 kg ai/ha) to actively growing creeping bentgrass in the spring after third or fourth mowing or in the fall months. Repeat, if necessary, at three- to six-week intervals, but do not exceed 1 lb ai/a (1.12 kg ai/ha) per growing season. Delay overseeding for two weeks after application. Make final fall application eight weeks before onset of winter dormancy. Flurprimidol is not as effective as paclobutrazol in suppressing the growth of perennial biotypes of annual bluegrass. Apply to nonputting greens for bentgrass, Kentucky bluegrass, and perennial ryegrass in late spring to early summer and/or late summer to early fall at 0.5 to 0.75 lb ai/a (0.56 to 0.84 kg ai/ha). Make second application at least three months before expected winter dormancy. Discoloration of annual bluegrass will be noticeable in 7 to 10 days after application and will last for three to six weeks. Do not apply to putting greens at these rates. Area treated should receive 0.5 in. (1.27 cm) of irrigation within 24 hours after application. Resume mowing 3 to 5 days after application.

Mefluidide (Embark 2S, Embark 0.2S)

Apply at 0.05 to 0.125 lb ai/a (0.056 to 0.14 kg ai/ha) [(Embark 2S—0.5 pt/A, 0.58 L/ha; Embark 0.2S—2 to 5 pt/A, 2.34 to 5.84 L/ha] to suppress annual bluegrass seedhead development. Mefluidide must be applied after uniform green-up, but before first appearance of seedhead(s). Application timing varies between geographical locations, but is generally during early winter through early spring. (Actual timing of application depends upon location and climatic conditions.) Mefluidide is primarily foliage absorbed. Do not apply to turf within four growing months after seeding, and do not reseed within three days after application. Treated turf may appear less dense and temporarily discolored. Adding 1 to 2 qts (0.95 to 1.89 L) of a nonionic surfactant per

100 gal (378.5 L) of spray solution may enhance suppression; however, discoloration may also be increased. Iron applications may lessen discoloration. Mefluidide formulations are not recommended for use on golf course putting greens. Read and follow label recommendations before use.

Paclobutrazol (Trimmit 2SC)

Apply 0.1 to 0.75 lb ai/acre (0.112 to 0.84 kg ai/ha) in late winter to early spring after growth of desired grasses has resumed and one to two mowings have occurred. Do not apply after late winter, to avoid delaying green-up of bermudagrass. Apply with 0.5 to 0.9 lb nitrogen per 1,000 sq. ft. (24.4 to 44.0 kg N/ha) of a nonburning fertilizer. Paclobutrazol is root absorbed, and 0.25 inch (0.64 cm) rainfall or irrigation water should be applied within 24 hours of application. Fall and spring applications of paclobutazol may also be used over a period of years to suppress the growth of perennial biotypes of annual bluegrass in creeping bentgrass greens. Repeat applications may be made three to four weeks apart. On bentgrass greens, do not apply more than 0.25 lb ai/acre (0.28 kg ai/ha) per application. Do not apply more than three times annually. Do not use if *Poa annua* populations exceed 70 percent.

Trinexapac-Ethyl (Primo MAXX 1MEC, Primo 25WSB)

Trinexapac-ethyl (Primo MAXX) at 0.02 to 0.04 lb ai/a (0.026 to 0.05 kg ai/ha) [3.0 to 6.0 fl oz/a, 0.22 to 0.44 L/ha] may be used on creeping bentgrass/*Poa* and hybrid bermudagrass putting greens. Trinexapac-ethyl is foliar absorbed, and a one-hour rain-free period is needed after application. Repeat applications may be applied as needed or approximately one to three weeks apart to maintain growth suppression. A current rule of thumb is to apply the equivalent of 0.007 lb ai/a (0.007 kg ai/ha) [1 oz/a, 0.073 L/ha] of trinexapac-ethyl per week during periods of active growth. Some prefer to split this into 0.016 lb ai/a (0.0175 kg ai/ha) [2 oz/a, 0.15 L/ha] every two weeks, or even 0.02 lb ai/a (0.026 kg ai/ha) [3 oz/a, 0.22 L/ha] every three weeks. It is advised to use the low rate initially to minimize turf "bronzing." Subsequent applications can be used at higher rates. It is not necessary to add a surfactant to trinexapac-ethyl. Various tank-mixtures of trinexapac-ethyl with paclobutrazol, flurprimidol, or ethephon are commonly employed to reduce *Poa annua* seedhead development (McCullough et al., 2006a).

Trinexapac-ethyl has repeatedly been shown not to reduce root growth and possibly increase it (Fagerness and Yelverton, 2001; McCullough, 2006; Tucker et al., 2006). Regular use of trinexapac-ethyl also increases ball-roll distances (Fagerness et al., 2000; McCullough et al., 2005a and 2005b), improves turf performance in low-light environments (Bunnell et al., 2005; Goss et al., 2002; Quan and Engelke, 1999; Steinke and Stier, 2003), reduces clippings and improves surface uniformity (Lickfeldt et al., 2001; McCullough et al., 2005a), and redistributes nutrient partitioning where trinexapac-ethyl-treated turf has higher Ca and Mg concentrations in leaves and higher N, P, and K concentrations in roots, compared with the untreated (McCullough et al., 2006a and 2006b) (Figure 7.6). Other benefits of regular trinexapac-ethyl use in turfgrass have been noted (Baldwin et al., 2006; Fagerness et al., 2004; Johnson, 1997; Richardson, 2002).

Flurprimidol + Trinexapac-Ethyl (Legacy 1.52 MEC)

A pre-tank-mix combination available for most medium- to high-maintenance turfgrass areas, Flurprimidol provides extended growth suppression due to its slower root uptake, while improved turfgrass quality is provided with trinexapac-ethyl.

Figure 7.6
Improving shade tolerance of bermudagrass with the use of a gibberellic acid-inhibiting PGR.

PLANT-GROWTH PROMOTERS

Biostimulants are plant-hormone-containing substances that can stimulate growth when exogenously applied (Schmidt, 1992). Biostimulant products have great potential if they actually enhance root and shoot growth under field conditions. The addition of exogenous plant hormones can increase antioxidant activity in cool-season grasses, as well as improve drought and heat tolerance (Liu et al., 2002; Zhang and Schmidt, 1999; Zhang et al., 2002). Long-term (2 yr.) use on ultradwarf bermudagrass maintained at low nitrogen levels increases turf rooting (Tucker et al., 2006).

An available plant growth promoter is RyzUp, or ProGibb, from Abbott Laboratories. RyzUp is gibberellic acid, which encourages cell division and elongation. When used, RyzUp helps initiate or maintain growth and prevent color changes (e.g., purpling) during periods of cold stress and light frosts on bermudagrass such as Tifdwarf and Tifgreen. Do not apply when night temperatures exceed 65°F (18.3°C). Oftentimes, fall golf tournaments may experience an early light frost before the overseeding has become established. RyzUp helps the turf recover from this discoloration. RyzUp and ProGibb also help reverse overdosing of anti-gibberellic acid PGRs, such as trinexapac-ethyl, flurprimidol, and paclobutrazole.

8

Weed Control in Landscapes and Nonturf Areas

Many turfgrass managers are familiar with the yearly task of controlling such weeds as crabgrass and dandelion in their turf. However, weeds also can be a major problem in areas like landscape beds, sidewalks, driveways, and other mulched areas. These weeds can cause reductions in plant growth or quality due to their competition for water, nutrients, and light, and they may harbor insects and diseases that can cause severe damage to landscape ornamentals. However, the primary reason for weed control in landscape plantings is aesthetics. When weeds are growing in landscape beds, they can detract from the beauty of the site (Figure 8.1). Landscaping around many businesses is there to attract new business, and when these landscapes are neglected and become overrun with weeds, a poor image for the business results.

Weed control in ornamental plantings involves developing a management program that includes site assessment, determining the type of plantings to place within the landscape, weed control options for those ornamental species selected, site preparation, and implementation of a weed management program.

SITE ASSESSMENT

Cultural Factors

The first step is to evaluate the landscape and assess the cultural factors that would enable weeds to habitat a landscape bed. Landscapers need to note areas of heavy traffic and drainage patterns, take soil samples measuring pH, and perform nutrient analysis to determine cultural issues that would affect plant health. Weeds present need to be charted, especially perennial ones that are generally harder to control. If weeds like bermudagrass, spurge, and ground ivy are present in the surrounding turf area, those weeds will eventually encroach upon the landscape beds. Weeds, especially perennial weeds like nutsedge, quackgrass, and bermudagrass, are usually easier to control prior to the site being planted.

The use of fertilizer in ornamentals is important to the growth and development of the plant, but used in excess, fertilizer can cause burning of roots and unnecessary stress. High soil phosphorus can aid in the germination of many annual weeds. Certain soil levels of pH also can promote certain types of weeds. Many mulches used, such as pine bark and straw, will gradually decrease the soil's pH immediately below the mulch layer.

Drainage is important to an ornamental bed's health and development. Poorly drained soils can result in root rot and a shortened root system, leading to a decline in plant health. Certain weeds are prone to water-logged

Figure 8.1
Weeds distract from flower and ornamental beds.

areas. The installation of drain tiles or french drains to drain or reshape the water away from the beds can mitigate this problem.

Spreading mulch is one of the best cultural practices to use in landscape beds. Mulch not only helps the soil retain important soil moisture, but also acts as a barrier for weed emergence. More details regarding specific mulches will be discussed shortly. (See "Weed Management Options.")

Proper Plant Selection

To determine the chemical weed-control options available to the landscaper, plants on-site are typically divided into four distinct groups: woody trees and shrubs, woody groundcovers, herbaceous perennials and bulbs, and annual flowers.

Woody Trees and Shrubs Woody trees and shrubs are the easiest to work around, since many are open and evenly spaced throughout the bed, creating an easy avenue for manually removing the weeds. This group consists of all the trees and shrubs found in foundation plantings, such as azaleas, camellias, dogwoods, oaks, rhododendrons, crepe myrtles, etc. They usually are easy to spot-treat around with a nonselective herbicide like glyphosate without injuring the plant. Woody tree and shrub ornamentals have a wide range of preemergence and postemergence herbicides available for use. Care must be exercised around younger trees, since they will have tender, green bark that can be damaged or penetrated by contact or noncontact herbicides.

Woody Ground Covers Unlike woody tree and shrub beds, weeds in ground covers can be more difficult to control due to their low and sprawling growth habit. This group consists of plants like ivy, junipers, liriope, and periwinkle. Their low and sprawling nature makes it hard to spot-treat existing weeds in these types of beds and makes the ornamental plants more prone to herbicide injury. Eventually, these ornamentals will spread and shade out the bed, preventing establishment of annual weeds. However, this will take years, and weed control will be necessary until then. These ornamental plantings are tolerant to a wide range of preemergence herbicides, and, depending on the species, various postemergence herbicides can be used. Perennial grasses may be selectively controlled after planting with either fluazifop or clethodim.

The use of nonselective herbicides is not recommended after planting due to the difficulty of application without injuring desirable plants.

Herbaceous Perennials and Bulbs Planting of herbaceous perennials and bulbs can pose a difficult task for landscapers. These plants, unlike woody ornamentals, have stems that are soft or succulent and green, as opposed to brown and woody. Generally, soft, green, herbaceous growth will die back to the ground in the winter months. The roots remain viable through the cold months, and new growth will emerge in spring. Herbaceous perennials and bulbs consist of plants like butterfly bush, lantana, daffodils, hosta, tulips, ornamental grasses, and many more ornamentals too numerous to list. Many of these plants are sensitive to spot treatment with nonselective herbicides because they are generally planted close together. A limited number of preemergence herbicides also exist that are effective for these types of plants. Perennial weeds are especially difficult to control in these plantings; however, perennial grasses can be treated with fluazifop. Nonselective herbicides should be avoided after planting, due to the injury potential to the desired plantings.

Annual Flowers In annual flowers, weed control is similar to weed control in herbaceous plantings. Annuals are seasonal and will be switched after each season. This group consists of petunias, violas, pansies, and other seasonal annual color selections. Nonselective herbicides are not safe to use, because they can severely injure, and most likely kill, the annual bedding plants. Control of perennial weeds in annual flower beds is very difficult if not provided prior to planting. Annuals can be easily controlled with mulches, preemergence herbicides, and/or hand weeding. Annual and perennial grasses can be controlled with selective products like sethoxydim or fluazifop.

WEED MANAGEMENT OPTIONS

Mulches

Organic Mulches One of the best methods in a successful weed-control program in landscape plants is the spreading of mulch. Many types of mulches are used, and these are divided into two groups (Table 8.1). The first group, organic mulch, is naturally occurring and includes leaves, grass clippings, pine needles, and wood bark. These materials are easily found and relatively inexpensive. Organic mulches are used in almost all landscape plantings, with or without solid black plastic or geotextile weed barriers under them. They can be applied to weed-free soil soon after planting, or to areas recently cleared of weeds, to effectively control weeds growing from seeds that germinate at or near the soil surface. Mulch should be applied to a depth of 2 to 4 inches (5 to 10 cm) for woody and shrub bedding plants, while a 1-inch (2.5 cm) depth is ideal for herbaceous and annual color beds. Mulch should not be applied too heavily, nor piled around the base of the plantings. Piling mulch will cause moisture accumulation and heat buildup, resulting in trunk damage and root rot. Also, mulch should be clean and free of weed seeds when it is selected. For example, hay straw generally contains weed seeds and will create problems in the future.

Perennial weeds growing from established roots or rhizomes are capable of growing through organic mulches. They must be controlled with postemergence, translocated herbicides, or be pulled repeatedly. Using a preemergence herbicide in combination with organic mulches improves weed control.

Organic mulches have become extremely big business in recent years and have served as an excellent way to use waste products no longer dumped into most landfills. Fresh and composted bark products and wood residue are available from lumber and paper companies. Arboricultural operations and land-clearing companies

Table 8.1
COMPARISON OF VARIOUS MULCHES USED IN LANDSCAPES AND ORNAMENTAL BEDS (BRIGGS AND WHITWELL, 2001)

Mulch Type	Appearance	Erosion Resistance	Compaction Resistance	Availability	Comments
Organic					
Lawn Clippings	Fair	Fair	Poor	Good	Compost first
Leaves	Fair	Fair	Poor	Good	Compost first
Peat Moss	Good	Fair	Good	Good	Expensive
Pine Needles	Good	Fair	Good	Good	Easy to apply
Bark	Good	Good	Good	Good	Used in large amounts
Sawdust	Fair	Fair	Fair	Fair	Should not apply too deep
Inorganic					
Plastic	Poor	Good	N/A	Fair	Has to be anchored
Gravel	Poor	Good	Good	Fair	Must complement landscape
Crushed Rock	Poor	Good	Good	Fair	Must complement landscape
Landscape Fabric	Fair	Good	N/A	Fair	Can be expensive

previously disposed of their wood wastes in landfills. Mountains of wooden pallets were also disposed of in this way. But two things happened that changed that. Many states prohibited the disposal of organic residues into their landfills, and the market for mulch has improved to the point that what once was a waste product is now a marketable product.

The preferred mulches, from an aesthetic perspective, are the well-composted shredded-bark products that are dark brown in color. Many of the chipped-wood products are light in color until aged or composted for six months or more. Some producers now paint the chips to produce mulches that are various shades of brown, black, yellow, red, or even blue.

Many types of organisms grow in organic mulches. One, commonly called dog vomit fungus, produces fruiting bodies that are colorful, but unattractive to some people. Mushrooms of various sizes and shapes are commonly found in organic mulches. The only ones that cause any real problems are called stinkhorns because they emit a foul odor. The birdsnest fungus produces fruiting bodies that look like masses of little birdsnests, containing spore masses that look like little eggs. None of these fungi are harmful to the plants in the beds or the soil. If they are thought to be too objectionable, they can be picked up and discarded in a compost pile or in the garbage.

The one fungus associated with organic mulches that does cause a serious problem is the artillery fungus. It produces a fruiting body that forcibly ejects a black, sticky spore mass. The fruiting body is phototropic, meaning that it orients itself toward light or bright, shiny surfaces that reflect light. These surfaces include windows, white siding, and shiny automobiles. The spore masses typically can be shot up to 8 feet (2.4 m), but with assistance from wind can reach the tops of two-story homes. The glue with which the spore masses attach to surfaces is extremely strong. After they have dried, the spore masses are very difficult to remove, and even when removed, they leave a dark stain behind. Clients have sued landscape managers for the cost of having their siding replaced or automobiles repainted, and currently no good solution exists.

Figure 8.2
Natural and artificial mulches are used to help reduce landscape weeds and to retain soil moisture.

Inorganic or Synthetic Mulches The second group is inorganic mulch, which includes synthetic materials. Examples are plastics, geotextile landscape fabrics, rocks, and ground tires (Figure 8.2). For many years, the primary inorganic mulch used in landscapes to prevent weed growth was simple, inexpensive, black, solid polyethylene film. Only black films should be used. Clear or white films allow weeds to grow under them. When the geotextile weed barriers were introduced, claims were made that solid black plastic is harmful to plants because it does not allow for adequate air and water exchange. The geotextiles are either woven or perforated to allow some water to penetrate and to enable limited air exchange. On the basis of these supposed advantages, many landscape managers changed from using solid black plastic to the more expensive geotextile weed barriers.

However, after several years of trying the newer weed barriers, many landscape managers are returning to using the solid black plastic. No harmful effects were ever documented from the use of solid black plastic. It seems that sufficient water penetrates the plastic at holes around the plants and at seams where the sheets of plastic overlap. The only observed problem related to poor aeration from plastic mulches is a tendency for the roots to grow closer to the soil surface, and irrigation must be placed below the material, since water cannot penetrate. One effective approach is to use a combination of mulches, such as geotextile fabric mulch covered with wood mulch. Pine straw could even be placed on top of the wood mulch to create a layered weed barrier. The barrier mulch will prevent emerging seedlings from reaching the surface and will block much-needed sunlight from penetrating the soil surface, thus preventing weed-seed germination and new seedling development. These materials are not used in annual color beds or in woody ground covers, since such plants are expected to spread. Landscape fabrics will degrade quickly if left exposed to the sun; therefore, the addition of a 1- to 2-inch (2.5 to 5 cm) layer of organic mulch is recommended. Protected with loose mulch geotextile fabric, loose mulch can last about 15 years.

Simultaneously with the discovery that no problems evidently were associated with plain black plastic mulch material, some problems associated with the long-term use of the geotextiles were observed. Some weeds, such as nutsedge, could still grow through the holes in the weed barriers. Also, since they are unattractive when left uncovered, organic mulches are often placed on top of them. When this organic mulch decomposes, weeds can grow on top of the mulch. Roots of these weeds can penetrate the geotextile mulch, and their removal is extremely difficult. Attempts at pulling them often caused lifting of the weed barrier.

Roots of trees and shrubs in the bed can also grow in the decaying mulch on top of the geotextile, and then grow down through it, creating a plastic/root complex that cannot be removed without damaging the bedding plants.

Solid black plastic used as a weed barrier is less expensive than the geotextile barriers and does not allow penetration by weed roots. Weeds growing in the decaying mulch on top of solid black plastic are easily pulled, and when the plastic must be removed, it easily can be pulled out of the bed without seriously disturbing the plants. Current recommendations are to use solid black plastic under organic or stone mulch on relatively flat surfaces. On slopes, where mulches would slide off of solid black plastic, a weed barrier that has some texture to hold the mulch in place is suggested.

Gravel or stones should not be used alone. They do not provide adequate weed control, and they tend to mix with the soil under them during heavy rains or freeze/thaw cycles. Gravel and stones should always be placed on top of a layer of plastic or geotextile for weed control and soil separation. Stones placed on top of geotextile weed barriers or black plastic make an attractive mulch and almost totally eliminate weed problems.

Ground-up rubber tires have been marketed as essentially permanent mulches. Like gravel, they should have a plastic layer or geotextile weed barrier between them and the soil. Another currently available product is ground tire chunks held together with a type of epoxy that allows for water penetration. They are sold in flat sheets about 1 inch (2.54 cm) thick that are cut and placed around trees in turf areas to keep the grass away from the base. Weeds will not penetrate this product, and it lasts indefinitely.

Unfortunately, mulching rarely will provide 100 percent control of weeds, and supplemental hand weeding and chemical control will be necessary.

Hand Weeding

Many weeds, especially annuals, can easily be removed from landscaped beds by hand pulling (Figure 8.3). This method is very laborious and time consuming, but may be the only method available, especially in annual color beds, herbaceous plants, and ground covers where herbicide injury is a possibility. When hand-pulling weeds, make sure to remove the roots. Many plants, especially perennial weeds, can regenerate if all the roots and storage organs like rhizomes, tubers, and bulbs are not totally removed.

Chemical Controls

The rest of this chapter will concentrate primarily on chemical options for removing weeds from landscape areas. A wide variety of preemergence and postemergence herbicides are labeled for use in ornamental beds and landscape situations.

A preemergence herbicide must be applied after planting and before weeds germinate. These products will not control established perennials or annual weeds already emerged. Generally, preemergence herbicides will need to be activated with either rainfall or irrigation in order to move the material into the soil and/or mulch layer, where the germinating weed seeds exist. Granular preemergence products are preferred due to their ease of application, better uniformity around landscape planting, and less risk of ornamental plant injury. If a granular product is used, it should be applied to dry foliage, as

Figure 8.3
Hand-weeding is often necessary in some landscapes.

Figure 8.4
Weeds in mulched areas are often spot-sprayed with nonselective herbicides, being careful not to treat desirable plants.

granules will stick to wet foliage and may cause injury to the ornamental plant. Some preemergence products are more effective when applied under the mulch layer, as they provide longer residual and better control.

Weeds which escape preemergence treatments or emerge through the mulch will need to be controlled by postemergence herbicides. Postemergence herbicides are most effective on young, actively growing weeds and do not provide residual control. These herbicides fall into two classes, contact and systemic. Contact herbicides injure only the part of the plant with which they come in contact (Figure 8.4). Injury symptoms will develop quickly, within 1 or 2 days. Systemic herbicides are translocated throughout the plant after they are absorbed by either the leaves or roots, with injury symptoms developing within 7 to 14 days after an application. Postemergence herbicides can be further classified into selective and nonselective. Selective herbicides target a specific weed and will do little to no damage to others; however, nonselective herbicides will injure or kill almost all plants. Selective herbicides are classified into three groups of control products: grass, broadleaf, and nutsedge. When applying a postemergence herbicide, especially a nonselective herbicide, avoid contact with the foliage of the desired ornamental, as injury is likely to occur.

Proper herbicide selection is the key to a successful chemical program. The selection should be based on the weeds to be controlled, current weed population, specific ornamentals present, and length of time of control desired.

In woody trees and shrubs beds, it is best to control any perennial weeds prior to planting. The use of a thick layer of mulch or a geotextile base with a thin layer of mulch, followed by the use of a preemergence herbicide, will provide excellent control of most annual weeds. Escaped weeds can be manually pulled or spot-treated with either a nonselective herbicide or a labeled selective postemergence herbicide. In woody ground covers, geotextiles are not practical, since the ground cover is expected to root and spread to surrounding areas. A layer of mulch and a preemergence herbicide will be required to control annual weeds. Over time, the ground cover will grow in density and exclude most weeds; but encroachment is likely during establishment. Supplemental hand-weeding will be required, since nonselective herbicides are difficult to use without causing injury to the desirable plants. Selective postemergence control of annual and perennial grasses is available.

In annual flower beds, it is best to use species with weed control compatibility. Perennial weeds should be controlled prior to bed establishment. With the use of mulch, preemergence herbicides, and hand-weeding, most annual weeds are easily controlled. Geotextiles are not recommended, due to the seasonal rotation of plants within these types of beds. Nonselective herbicides need to be avoided due to the severe chance of plant injury. Postemergence control of perennial and annual grassy weeds can be controlled with selective grass herbicides. Weed-control options in herbaceous perennial beds are similar to those in annual flower beds, but it is more important to eradicate perennial weeds, since the beds will not be renovated for several years and only a select few species are included on herbicide labels.

Mixed plantings of woody and herbaceous vegetation can present a complex challenge in achieving a successful weed-management program, due to the diversity of planting. Preplant preparation is needed because postplanting treatments will be severely limited. Groupings of plants with similar herbicide needs can be an option allowing defined-use areas. Another strategy is to use a tight planting grid to shade the entire area, disallowing weeds to establish.

PREEMERGENCE HERBICIDES

Preemergence Grass Herbicides

Bensulide controls several annual grasses and broadleaf weeds in woody trees and shrubs and herbaceous ornamentals, especially ground covers (Tables 8.2 and 8.3). Sprayable and granular formulations are available with rates ranging from 7.5 to 12.5 lb ai/acre (8.4 to 14 kg ai/ha). The major weeds controlled are crabgrass, annual bluegrass, goosegrass, barnyardgrass, and henbit (Figure 8.5). Bensulide has no activity on perennial weeds and should be applied under mulch to optimize control, with a 0.5- to 1.0-in. (1.3 to 2.5 cm) of irrigation following application. Bensulide can be applied over the top on tolerant ornamentals.

Figure 8.5
Granular preemergence herbicides are often used in landscapes to prevent weed establishment. (Courtesy John Boyd.)

Table 8.2

Preemergence Herbicides Used in Ornamentals

Key:
Based on label:
G = good control (80–100%),
F = fair control (50–80%); use higher rates,
P = poor control (0–50%); some suppression

Genus Species	Common Name	Bensulide*	DCPA	Diclobenil*	Dithiopyr	Imazaquin	Isoxaben	Metolachlor	Napropamide	Oryzalin	Oxadiazon	Pendimethalin*	Prodiamine	Pronamide	Simazine*	Trifluralin*	Benefin + Oryzalin	Oxadiazon + Pendimethalin	Oxyfluorfen + Oryzalin	Oxyfluorfen + Oxadiazon	Trifluralin + Isoxaben	Trifluralin + Isoxaben + Oxyfluorfen
Grasses (or Grasslike)																						
Aegilops cylindrica	Goatgrass, jointed													G								
Allium canadense	Onion, wild																					
Allium vineale	Garlic, wild																					
Avena fatua	Oats, wild				G				G	G	G			G	G	G	G	G			G	G
Brachiaria platyphylla	Signalgrass, broadleaf							G		G		G	G		G	G	G				G	G
Bromus catharticus	Rescuegrass				G								G			G					G	G
Bromus mollis	Chess, soft				G				G							G					G	G
Bromus rigidus	Brome, rigput				G				G		G					G		G			G	G
Bromus secalinus	Brome, smooth			G	G											G					G	G
Bromus secalinus	Cheatgrass				G				G							G					G	G
Bromus tectorum	Brome, downy				G				G					G	G	G					G	G
Cenchrus spp.	Sandbur species		F		G			F	G													
Cenchrus incertus	Sandbur, field					G						G				G	G	G			G	G
Cynodon dactylon	Bermudagrass (from seed)																					
Cyperus compressus	Sedge, annual					F					G							G				
Cyperus esculentus	Nutsedge, yellow					F		G														
Cyperus globosus	Sedge, globe					F																
Cyperus rotundifolia	Nutsedge, purple					F																
Dactylis glomerata	Orchardgrass			G										G								
Dactyloctenium aegyptium	Crowfootgrass				G			G		G		G	G				G					

(continued)

Table 8.2
(CONTINUED)

Key:
Based on label:
G = good control (80–100%),
F = fair control (50–80%); use higher rates,
P = poor control (0–50%); some suppression

Genus Species	Common Name	Bensulide*	DCPA	Diclobenil*	Dithiopyr	Imazaquin	Isoxaben	Metolachlor	Napropamide	Oryzalin	Oxadiazon	Pendimethalin*	Prodiamine	Pronamide	Simazine*	Trifluralin*	Benefin + Oryzalin	Oxadiazon + Pendimethalin	Oxyfluorfen + Oryzalin	Oxyfluorfen + Oxadiazon	Trifluralin + Isoxaben	Trifluralin + Isoxaben + Oxyfluorfen
Grasses (or Grasslike)																						
Digitaria ciliaris	Crabgrass, southern			G	G	F		G			G	G			G	G	G				G	G
Digitaria ischaemum	Crabgrass, smooth	G	G	G	G	F		G	G	G	G	G	G		G	G	G	G			G	G
Digitaria sanguinalis	Crabgrass, large	G	G	G	G	F		G	G	G	G	G	G	G	G	G	G	G	G	G	G	G
Echinochloa colunum	Junglerice								G	G	F	G	G		G	G	G				G	G
Echinochloa crus-galli	Barnyardgrass	G	F		G			G	G	G	G	G	G	G	G	G	G	G	G	G	G	G
Eleusine indica	Goosegrass	G	F		G			G	G	G	G	G	G	G	G	G	G	G	G	G	G	G
Elytrigia (Agropyron) repens	Quackgrass (from rhizome)			G										G								
Elytrigia (Agropyron) repens	Quackgrass (from seed)			G										G								
Equisetum spp.	Horsetail (Equisetum)			G																		
Eragrostis spp.	Lovegrass species		G							G		G	G	G		G	G				G	G
Eragrostis cilianensis	Stinkgrass								G							G					G	G
Eriochloa contracta	Cupgrass, prairie							G														
Eriochloa gracilis	Cupgrass, Southwestern							G	G	G						G	G				G	G
Eriochloa villosa	Cupgrass, wooly											G	G									
Festuca arundinaceae	Fescue, tall (from seed)					F						G										
Holcus lanatus	Velvetgrass, common											G										
Hordeum jubatum	Barley, foxtail				G				G					G								
Hordeum leporinum	Barley, wild (hare)				G				G							G					G	G
Hordeum pusillum	Barley, little				G					G							G					
Kyllinga brevifolius	Kyllinga, green					F																

(continued)

Table 8.2
PREEMERGENCE HERBICIDES USED IN ORNAMENTALS (*CONTINUED*)

Key:
Based on label:
G = good control (80–100%),
F = fair control (50–80%); use higher rates,
P = poor control (0–50%); some suppression

Genus Species	Common Name	Bensulide*	DCPA	Diclobenil*	Dithiopyr	Imazaquin	Isoxaben	Metolachlor	Napropamide	Oryzalin	Oxadiazon	Pendimethalin*	Prodiamine	Pronamide	Simazine*	Trifluralin*	Benefin + Oryzalin	Oxadiazon + Pendimethalin	Oxyfluorfen + Oryzalin	Oxyfluorfen + Oxadiazon	Trifluralin + Isoxaben	Trifluralin + Isoxaben + Oxyfluorfen
Grasses (or Grasslike)																						
Leptochloa spp.	Sprangletop species												G									
Leptochloa filiformis	Sprangletop, red								G	G		G					G					
Leptochloa uninervia	Sprangletop, Mexican								G			G				G				G	G	G
Lolium multiflorum	Ryegrass, annual				G				G	G				G	G		G				G	G
Lolium perenne	Ryegrass, perennial				G	G								G								
Melinis repens; Rhynchelytrum roseum	Natalgrass			G																		
Oryza satiza	Red rice							G														
Panicum	Panicum, Fall	G	P					G	G	G	G	G	G	G	G	G	G	G	G		G	G
Panicum capillare	Witchgrass		G	G				G	G	G		G	G		G	G	G				G	G
Panicum fasciculatum	Panicum, browntop		F							G		G	G				G					
Panicum maximum	Guineagrass, seedling								G													
Panicum texanum	Panicum, Texas		P	G					G	G		G	G			F	G				F	F
Paspalum dilatatum	Dallisgrass				G																	
Paspalum fimbriacatum	Paspalum, Panama										F											
Paspalum notatum	Bahiagrass					F																
Paspalum urvillei	Vaseygrass		P																			
Pennisetum clandestinum	Kikuyugrass				G														G			
Phalaris canariensis	Canarygrass								G					G								
Phalaris minor	Canarygrass, littleseed		P																			

(continued)

Table 8.2
(Continued)

Key:
Based on label:
G = good control (80–100%),
F = fair control (50–80%); use higher rates,
P = poor control (0–50%); some suppression

Genus Species	Common Name	Bensulide*	DCPA	Diclobenil*	Dithiopyr	Imazaquin	Isoxaben	Metolachlor	Napropamide	Oryzalin	Oxadiazon	Pendimethalin*	Prodiamine	Pronamide	Simazine*	Trifluralin*	Benefin + Oryzalin	Oxadiazon + Pendimethalin	Oxyfluorfen + Oryzalin	Oxyfluorfen + Oxadiazon	Trifluralin + Isoxaben	Trifluralin + Isoxaben + Oxyfluorfen
Grasses (or Grasslike)																						
Phleum pratense	Timothy			G																		
Poa annua	Bluegrass, annual	G	F	G	G	F		G	G	G	G	G	G	G	G	G	G	G	G	G	G	G
Poa pratensis	Bluegrass, Ky			G										G								
Polypogon monspeliensis	Rabbitfootgrass		P																			
Rottboellia cochinchinensis	Itchgrass											G	G									
Setaria faberi	Foxtail, giant	G	P	G				G	G	G		G	G		G	G	G				G	G
Setaria glauca/viridis	Foxtail (yellow, green)	G	G	G	G			G	G	G	G	G	G	G	G	G	G	G	G		G	G
Setaria verticillata	Foxtail, bristly	G		G					G		G		G		G	G	G	G				
Sorghum bicolor	Shattercane							F								F					F	F
Sorghum halapense	Johnsongrass (rhizome)																					
Sorghum halapense	Johnsongrass (seedling)		F					F	G	G		G	G			F	G				F	F
Sporobolus indicus	Smutgrass				G																	
Vulpia myorus	Fescue, rattail						F								G	G					G	G
Broadleaf Weeds																						
Abrus precatorius	Rosarypea			G																		
Acalypha rhomboidea	Copperleaf, rhombic		F																			
Acalypha virginica	Copperleaf, Va.		F																			
Acanthospermum hispidum	Starbur, bristly																					
Acroptilon repens	Knapweed, Russian			G																		
Ageratum conyzoides	Ageratum																		G			

(continued)

Table 8.2
Preemergence Herbicides Used in Ornamentals (*Continued*)

Key:
Based on label:
G = good control (80–100%),
F = fair control (50–80%); use higher rates,
P = poor control (0–50%); some suppression

Genus Species	Common Name	Bensulide*	DCPA	Diclobenil*	Dithiopyr	Imazaquin	Isoxaben	Metolachlor	Napropamide	Oryzalin	Oxadiazon	Pendimethalin*	Prodiamine	Pronamide	Simazine*	Trifluralin*	Benefin + Oryzalin	Oxadiazon + Pendimethalin	Oxyfluorfen + Oryzalin	Oxyfluorfen + Oxadiazon	Trifluralin + Isoxaben	Trifluralin + Isoxaben + Oxyfluorfen
Broadleaf Weeds																						
Alchemilla arvensis	Parsley-piert				G	G																
Amaranthus spp.	Pigweed spp.	G	F	G	G	G	G	G	G	G	G	G	G		G	G	G	G	G	G	G	G
Amaranthus palmeri	Amaranth, palmer														G							
Amaranth spinosus	Amaranth, spiny										G							G				
Amaranthus rudis	Waterhemp, common																					
Ambrosia acanthicarpa	Bursage, annual						G														G	G
Ambrosia artemisiifolia	Ragweed, common			G			G		F	F					G		F				G	G
Ambrosia trifida	Ragweed, giant			G											G							
Ampelamus albidus	Milkweed, honeyvine						P														F	F
Amsinckia spp.	Fiddleneck			G			G		G	G	G	G			G		G	G			G	G
Anagallis arvensis	Pimpernel, scarlet						F														G	G
Anthemis cotula	Mayweed (chamomile)						G														G	G
Apium leptophyllum	Celery, wild						G														G	G
Arbutilon theophrasti	Velvetleaf									F		G					F				G	G
Artemesia vulgaris	Mugwort (from rhizome)			G																		
Artemesia vulgaris	Mugwort (from seed)			G																		
Aster spp.	Aster			G			G														G	G
Atriplex hortensis	Orach, red																					G
Barbarea vulgaris	Rocket, yellow			G																		
Bassia hyssopifolia	Bassia, five-hook														G							

(continued)

Table 8.2
(Continued)

Key:
Based on label:
G = good control (80–100%),
F = fair control (50–80%); use higher rates,
P = poor control (0–50%); some suppression

Genus Species	Common Name	Bensulide*	DCPA	Diclobenil*	Dithiopyr	Imazaquin	Isoxaben	Metolachlor	Napropamide	Oryzalin	Oxadiazon	Pendimethalin*	Prodiamine	Pronamide	Simazine*	Trifluralin*	Benefin + Oryzalin	Oxadiazon + Pendimethalin	Oxyfluorfen + Oryzalin	Oxyfluorfen + Oxadiazon	Trifluralin + Isoxaben	Trifluralin + Isoxaben + Oxyfluorfen
Broadleaf Weeds																						
Berteroa incana	Alyssum, hoary														G							
Bidens bipinnata	Spanishneedles			G											G							
Borreria laevis	Buttonweed																		G			
Brassica juncea	Mustard, Indian				G		G														G	G
Brassica kaber	Mustard, wild			G	G		G			F				G	G		F				G	G
Brassica nigra	Mustard, black				G		G			F							F				G	G
Bryum spp.	Moss																	G				
Calandrinia ciliata	Redmaids (Dessert rock purslane)						G			G					G		G				G	G
Calystegia sepium	Bindweed, hedge																					
Capsella bursa-pastoris	Shepherdspurse	G	P	G	G		G			G	G	G	G	G	G		G	G	G	G	G	G
Cardamine hirsuta	Bittercress, hairy				G	G	G				G										G	G
Cardamine oligosperma	Bittercress, lesser				G		G			G	G						G				G	G
Cardamine pennsylvanica	Bittercress, Pennsylvania				G						G							G	G	G		
Carduus nutans	Thistle, musk						F													G	G	
Cassia occidentalis	Senna, coffee			G																		
Centaurea maculosa	Knapweed, spotted																					
Cerastium vulgatum	Chickweed, mouseear					G	G					G	G	G							G	G
Chenopodium spp.	Goosefoot		P	G			F							G		G					G	G
Chenopodium album	Lambsquarters	G	G	G			G		G	G	G	G	G	G	G	G	G	G	G		G	G

(continued)

Table 8.2
PREEMERGENCE HERBICIDES USED IN ORNAMENTALS *(CONTINUED)*

Key:
Based on label:
G = good control (80–100%),
F = fair control (50–80%); use higher rates,
P = poor control (0–50%); some suppression

Genus Species	Common Name	Bensulide*	DCPA	Diclobenil*	Dithiopyr	Imazaquin	Isoxaben	Metolachlor	Napropamide	Oryzalin	Oxadiazon	Pendimethalin*	Prodiamine	Pronamide	Simazine*	Trifluralin*	Benefin + Oryzalin	Oxadiazon + Pendimethalin	Oxyfluorfen + Oryzalin	Oxyfluorfen + Oxadiazon	Trifluralin + Isoxaben	Trifluralin + Isoxaben + Oxyfluorfen
Broadleaf Weeds																						
Chorispora tenella	Mustard, blue																					G
Cirsium arvense	Thistle, Canada			G																		
Cirsium vulgare	Thistle, bull			G																		G
Citrullus lanatus	Citron melon			G																		
Commelina communis	Dayflower																					
Convolvulus arvensis	Bindweed, field						P														F	F
Conyza bonariensis	Fleabane, blackleaved						G														G	G
Conyza canadensis	Horseweed (marestail)				G		G			F							F		G	G	G	G
Conyza ramosissima	Fleabane, dwarf						G														G	G
Coronopus spp.	Swinecress						F				G										G	G
Cotula australis	Brassbuttons, Southern						G														G	G
Croton glandulosus	Croton, tropic																					
Cuscuta spp.	Dodder		F											G								
Datura stramonium	Jimsonweed						F														G	G
Daucus carota	Carrot, wild			G			G														G	G
Descurainia sophia	Flixweed																					G
Desmodium tortuosum	Beggarweed, Florida																					
Deuscurainia pinnata	Tansymustard														G							
Deuscurainia pinnata ssp. *brachycarpa*	Tansymustard, green						G														G	G

(continued)

Table 8.2
(Continued)

Key:
Based on label:
G = good control (80–100%),
F = fair control (50–80%); use higher rates,
P = poor control (0–50%); some suppression

Genus Species	Common Name	Bensulide*	DCPA	Diclobenil*	Dithiopyr	Imazaquin	Isoxaben	Metolachlor	Napropamide	Oryzalin	Oxadiazon	Pendimethalin*	Prodiamine	Pronamide	Simazine*	Trifluralin*	Benefin + Oryzalin	Oxadiazon + Pendimethalin	Oxyfluorfen + Oryzalin	Oxyfluorfen + Oxadiazon	Trifluralin + Isoxaben	Trifluralin + Isoxaben + Oxyfluorfen
Broadleaf Weeds																						
Diodia virginiana	Buttonweed, Virginia					F																
Eclipta prostrata	Eclipta					G	G														G	G
Enilia songhifolia	Flora's paintbrush														G							
Epilobium spp.	Willoweed; fireweed						F				G				G			G	G	G	G	G
Erechtites hieracifolia	Fireweed (Am. Burnweed)																		G			G
Eremocarpus setigerus	Mullein, turkey						F														G	G
Erigeron annuus	Fleabane																					
Erigeron strigosus	Fleabane, rough																			G		
Erodium botrys	Filaree, broadleaf											G			G							G
Erodium cicutarium	Filaree, redstem						G		G	G		G			G		G				G	G
Erodium moschatum	Filaree, whitestem						F			G		G			G		G				G	G
Eupatorium capillifolium	Dogfennel			G			F														G	G
Euphorbia spp.	Spurge, annual			G								G										
Euphorbia cordifolia	Spurge, roundleaf			G														G				
Euphorbia esula	Spurge, leafy			G																		
Euphorbia heterophylla	Poinsettia, wild																					
Euphorbia hirta	Spurge, garden			G	G						G								G	G		
Euphorbia humistrata	Spurge, prostrate		F	G	G		F			G	F	G	G				G	G	G	G	G	G
Euphorbia hyssopifolia	Spurge, hyssop			G			G														G	G
Euphorbia maculata	Spurge, spotted		F	G	G		G			F							F		G	G	G	G

(continued)

Table 8.2
Preemergence Herbicides Used in Ornamentals (*Continued*)

Key:
Based on label:
G = good control (80–100%),
F = fair control (50–80%); use higher rates,
P = poor control (0–50%); some suppression

Genus Species	Common Name	Bensulide*	DCPA	Diclobenil*	Dithiopyr	Imazaquin	Isoxaben	Metolachlor	Napropamide	Oryzalin	Oxadiazon	Pendimethalin*	Prodiamine	Pronamide	Simazine*	Trifluralin*	Benefin + Oryzalin	Oxadiazon + Pendimethalin	Oxyfluorfen + Oryzalin	Oxyfluorfen + Oxadiazon	Trifluralin + Isoxaben	Trifluralin + Isoxaben + Oxyfluorfen
Broadleaf Weeds																						
Euphorbia nutans	Spurge, nodding		F	G																		
Euphorbia peplus	Spurge, petty			G			F				G										G	G
Euphorbia vermiculata	Spurge, hairy			G															G			
Fatuoa villosa	Mulberry weed																					
Galinsoga ciliata	Galinsoga, hairy						G	G			G										G	G
Galium aparine	Bedstraw																					G
Geranium carolinianum	Geranium, Carolina				G	G	G														G	G
Geranium molle	Dovetail					G																
Gisekia spp.	Gisekia			G																		
Glechoma hederacea	Ivy, ground																					
Gnaphalium spp.	Cudweed			G		F	G		G			G								G	G	G
Helianthus annuus	Sunflower						F														G	G
Helianthus tuberosus	Artichoke, Jerusalem			G																		
Heterotheca grandiflora	Telegraphplant						G														G	G
Heterotheca subaxillaris	Camphorweed			G																		
Hibiscus trionum	Mallow, Venice						P														F	F
Hippurus vulgaris	Marestail																				G	G
Hydrocotyl spp.	Pennywort (from seed)					F	G														G	G
Hypochoeris radicata	Dandelion, false			G							G							G				
Indigofera hirsuta	Indigo, hairy																					

(continued)

Table 8.2
(Continued)

Key:
Based on label:
G = good control (80–100%),
F = fair control (50–80%); use higher rates,
P = poor control (0–50%); some suppression

Genus Species	Common Name	Bensulide*	DCPA	Diclobenil*	Dithiopyr	Imazaquin	Isoxaben	Metolachlor	Napropamide	Oryzalin	Oxadiazon	Pendimethalin*	Prodiamine	Pronamide	Simazine*	Trifluralin*	Benefin + Oryzalin	Oxadiazon + Pendimethalin	Oxyfluorfen + Oryzalin	Oxyfluorfen + Oxadiazon	Trifluralin + Isoxaben	Trifluralin + Isoxaben + Oxyfluorfen
Broadleaf Weeds																						
Ipomoea coccinea	Morning glory, red (scarlet)									F							F					
Ipomoea hederacea var. integriuscula	Morning glory, entireleaf									F							F					
Ipomoea nil (hederacea)	Morning glory, ivyleaf						G			F							F				G	G
Ipomoea purpurea	Morning glory, annual (tall)						P			F				G	G		F				F	F
Jacquemontia tamnifolia	Morning glory, smallflower																					
Kochia scoparia	Kochia						F					G	G			G					G	G
Lactuca serriola	Lettuce, prickly						G		G	F					G		F				G	G
Lamium amplexicaule	Henbit	G	P	G	G	G	G			G		G	G	G	G		G				G	G
Lamium maculatum	Deadnettle, spotted		F																			
Lamium purpureum	Deadnettle, purple	G	F	G		G																
Lepidium perfoliatum	Pepperweed, yellowflower														G							G
Lepidium virginicum	Pepperweed, Virginia			G			G								G					G	G	G
Lespedeza spp.	Lespedeza				G																	
Lysimachia angusta	Loosestrife																					
Malva neglecta	Mallow, common									F							F					G
Malva parviflora	Cheeseweed; little mallow		F				G		G	G		G		G					G		G	
Malva rotundifolia	Mallow, dwarf						G														G	G
Marchantia spp.	Liverwort										F											
Matricaria matricarioides	Pineappleweed			G	G		G		G						G						G	G

(continued)

Table 8.2
PREEMERGENCE HERBICIDES USED IN ORNAMENTALS (CONTINUED)

Key:
Based on label:
G = good control (80–100%),
F = fair control (50–80%); use higher rates,
P = poor control (0–50%); some suppression

Genus Species	Common Name	Bensulide*	DCPA	Diclobenil*	Dithiopyr	Imazaquin	Isoxaben	Metolachlor	Napropamide	Oryzalin	Oxadiazon	Pendimethalin*	Prodiamine	Pronamide	Simazine*	Trifluralin*	Benefin + Oryzalin	Oxadiazon + Pendimethalin	Oxyfluorfen + Oryzalin	Oxyfluorfen + Oxadiazon	Trifluralin + Isoxaben	Trifluralin + Isoxaben + Oxyfluorfen
Broadleaf Weeds																						
Medicago lupulina	Medic, black				G	G	F														G	G
Medicago polymorpha	Burclover, California						F								G						G	G
Medicago trunculata	Medic																		G			
Melilotus spp.	Clover, sweet						G														G	G
Melochia corchorifolia	Redweed																					
Mimosa predica	Sensitive plant										G											
Mollugo verticillata	Carpetweed		G	G	G	G	P	G	G	G	G	G	G	G	G	F	G	G			F	G
Montia perfoliata	Miners lettuce			G																		G
Morrenia odorata	Milkweed vine			G																		
Murdannia nudiflora	Doveweed																					
Oenethera spp.	Evening primrose			G		G	F				G	G							G		G	G
Oxalis corniculata	Woodsorrel, creeping				G		F														G	G
Oxalis pes-caprae	Oxalis, buttercup				G																	
Oxalis stricta	Woodsorrel, yellow			G	G		G			G	G	G	G				G	G	G	G	G	G
Passiflora incarnata	Maypops			G																		
Phyllanthus tenellus	Phyllanthus, long stalked						G			G											G	
Physalis spp.	Groundcherry						G			F											G	G
Phytolacca americana	Pokeweed, common						G														G	G
Picris echioides	Oxtongue, bristly						F				G							G			G	G
Pilea microphylla	Artillery weed																		G			

(continued)

Table 8.2
(Continued)

Key:
Based on label:
G = good control (80–100%),
F = fair control (50–80%); use higher rates,
P = poor control (0–50%); some suppression

Genus Species	Common Name	Bensulide*	DCPA	Diclobenil*	Dithiopyr	Imazaquin	Isoxaben	Metolachlor	Napropamide	Oryzalin	Oxadiazon	Pendimethalin*	Prodiamine	Pronamide	Simazine*	Trifluralin*	Benefin + Oryzalin	Oxadiazon + Pendimethalin	Oxyfluorfen + Oryzalin	Oxyfluorfen + Oxadiazon	Trifluralin + Isoxaben	Trifluralin + Isoxaben + Oxyfluorfen
Broadleaf Weeds																						
Plantago spp.	Plantain (from seed)			G																		
Plantago aristata	Plantain, bracted						G														G	G
Plantago elongata	Plantain, slender						G														G	G
Plantago lanceolata	Plantain, narrowleaf (buckhorn)			G			G			G											G	
Polygonum lapathifolium	Ladysthumb		P				G			F				G			F				G	G
Plantago major	Plantain, broadleaf			G			G														G	G
Polygonum argyrocoleon	Knotweed, silversheath		P	G			F						G								G	G
Polygonum aviculare	Knotweed, prostrate		F	G	G		G		G	G		G	G	G		G	G	G			G	G
Polygonum convolvulus	Buckwheat, wild		P																			G
Polygonum cuspidatum	Knotweed spp.			G																		
Polygonum pennsylvanicum	Smartweed, Pennsylvania			G			G			F	G	G			G		F				G	G
Portulaca oleracea	Purslane, common		G	G	G		G	F	G	G	G	G	G	G	G	G	G	G	G		G	G
Pteridium aquilinum	Bracken fern																					
Pyllanthus urinaria	Chamberbitter						G				G							G			G	G
Ranunculus spp.	Buttercup					G																
Raphanus raphanistrum	Radish, wild			G			G							G							G	G
Richardia scabra	Pusley, Florida		G	G			P	G		G		G	G		G	F	G				F	F
Ricinus communis	Castor bean																		G			
Rumex acetosella	Sorrel, red					G	G							G							G	G

(continued)

Table 8.2
Preemergence Herbicides Used in Ornamentals (*Continued*)

Key:
Based on label:
G = good control (80–100%),
F = fair control (50–80%); use higher rates,
P = poor control (0–50%); some suppression

Genus Species	Common Name	Bensulide*	DCPA	Diclobenil*	Dithiopyr	Imazaquin	Isoxaben	Metolachlor	Napropamide	Oryzalin	Oxadiazon	Pendimethalin*	Prodiamine	Pronamide	Simazine*	Trifluralin*	Benefin + Oryzalin	Oxadiazon + Pendimethalin	Oxyfluorfen + Oryzalin	Oxyfluorfen + Oxadiazon	Trifluralin + Isoxaben	Trifluralin + Isoxaben + Oxyfluorfen
Broadleaf Weeds																						
Rumex crispus, obtusifolius	Dock, curly or broadleaf		P	G			P														F	F
Sagina procumbens	Pearlwort, birdseye																		G	G		
Salsola iberica	Thistle, Russian			G			G								G	G					G	G
Sarcostemma cyanchoides	Milkweed, climbing									F							F					
Scleranthus annuus	Knawel					G									G							
Senecio aureus	Ragwort, golden										G											
Senecio vulgaris	Groundsel, common			G			F	F	G	G	G				G		G	G	G	G	G	G
Sesbania exaltata	Sesbania, hemp																					
Sibara virginica	Sibara						G														G	G
Sida spinosa	Sida, prickly			G			G			F							F				G	G
Sisymbrium altissimum	Mustard, tumble																					G
Sisymbrium irio	Rocket, London		P		G		G			G		G		G			G				G	G
Solanum nigrum	Nightshade, black		F				G	G		F				G	G		F				G	G
Solanum ptycanthum	Nightshade, Eastern black																			G		
Solanum sarrachoides	Nightshade, hairy							F						G	G							G
Soliva pterosperma	Burweed, lawn					G	G					G										
Sonchus arvensis	Sowthistle, perennial																		G	G		
Sonchus asper	Sowthistle, spiny						F														G	G
Sonchus oleraceus	Sowthistle, annual	P				G		G	F	G							F	G	G		G	G
Spergula arvensis	Spurry, corn																					G
Spergularia rubra	Spurry, red sand																					G

(continued)

Table 8.2
(Continued)

Key:
Based on label:
G = good control (80–100%),
F = fair control (50–80%); use higher rates,
P = poor control (0–50%); some suppression

Genus Species	Common Name	Bensulide*	DCPA	Diclobenil*	Dithiopyr	Imazaquin	Isoxaben	Metolachlor	Napropamide	Oryzalin	Oxadiazon	Pendimethalin*	Prodiamine	Pronamide	Simazine*	Trifluralin*	Benefin + Oryzalin	Oxadiazon + Pendimethalin	Oxyfluorfen + Oryzalin	Oxyfluorfen + Oxadiazon	Trifluralin + Isoxaben	Trifluralin + Isoxaben + Oxyfluorfen
Broadleaf Weeds																						
Stachys floridanum	Betony, Florida																					
Stellaria media	Chickweed, common		G	G	G	G	G		G	G		G	G	G	G	G	G	G	G	G	G	G
Taraxacum officinale	Dandelion (seedling)			G		F	G												G	G	G	G
Thlaspi arvense	Pennycress, field																					
Tribulus terrestris	Puncturevine									G		G				G						
Trifolium aureum (procumbens)	Clover, hop											G										
Trifolium pratense	Clover, red																					G
Trifolium repens	Clover, white					G	G												G		G	G
Utrica dioica (gracilis, procera)	Nettle, stinging															G					G	G
Urtica urens	Nettle, burning		G				F							G							G	G
Veronica spp.	Speedwell spp.										G				G			G				
Veronica arvensis	Speedwell, corn				G							G										
Veronica peregrina	Speedwell, Purslane						G														G	G
Veronica persica	Speedwell, birdseye (Persian)																	G			G	
Veronica serpyllifolia	Speedwell, thymeleaf						G														G	G
Viola spp.	Violet; pansy		G			F																
Xanthium strumarium	Cocklebur, common																					G
Youngia japonica	Hawksbeard, Asiatic																					

*see text for other trade names of these products.

Compiled by J. C. Neal, J. A. Briggs, and T. Whitwell. 2005. North Carolina State University and Clemson University.

Table 8.3
Preemergence Herbicides Used in Woody Ornamental

Key:
F = registered for field (or landscape) use,
C = registered for container use,
FC = registered for field and container use,
Ø = label prohibits use on this species,
F*, C*, FC* = registered for some species (sp.) or cultivars (cvs.); see label,
Ø* = prohibited for some cultivars in a species; see label Genus species.

See footnote for additional symbol keys.

Genus Species	Common Name	Bensulide*	DCPA	Dichlobenil	Dithiopyr	Imazaquin	Isoxaben	Metolachlor	Napropamide	Oryzalin	Oxadiazon G	Pendimethalin*	Prodiamine*	Pronamide*	Simazine	Trifluralin	Benefin + Oryzalin	Isoxaben + Trifluralin	Oxadiazon + Pendimethalin	Oxyfluorfen + Oxadiazon	Oxyfluorfen + Oryzalin	Trifluralin + Isoxaben + Oxyfluorfen
Abelia spp.	abelia species	F	F			Ø		F	FC											FC		
A. × *grandiflora*	abelia, glossy				F*		FC	FC		F		FC	FC*			FC	F	FC			FC	FC
Abeliophyllum distichum	forsythia, white																					
Abies spp.	fir species		F					F	FC	F*			F	F			F*					
A. balsamea	fir, balsam						FC		F	F		FC			F	FC	F	FC				
A. concolor	fir, white						F			F		FC			F		F	F				
A. fraseri	fir, fraser				F					F		FC			F	FC	F					
A. grandis	fir, grand									F							F					
A. koreana	fir, Korean																					
A. procera	fir, procera																					
Acacia spp.	acacia species						FC*			F*	FC					FC*	F*	FC*	FC*	FC		FC*
Acer spp.	maple species		F	F			F*	F	F	F	FC			F			F	F*	FC	FC		F*
A. ginnala	maple, amur						FC									FC		FC				FC
A. griseum	maple, paperbark											FCg										
A. negundo	boxelder			F											F							
A. palmatum	maple, Japanese				F		FC*		F			FC	FC			FC*		FC*			FC*	FC*
Acer pennsylvanicum	maple, striped																					
A. platanoides	maple, Norway				F							FC	FCa			FC						
A. rubrum	maple, red				F		F	FC	F	F		FC				FC		F				F
A. saccharinum	maple, silver				F		FC			F						FC		FC				FC
A. saccharum	maple, sugar				F							FC				FC						

(continued)

Table 8.3
(Continued)

Key:
F = registered for field (or landscape) use,
C = registered for container use,
FC = registered for field and container use,
Ø = label prohibits use on this species,
F*, C*, FC* = registered for some species (sp.) or cultivars (cvs.); see label,
Ø* = prohibited for some cultivars in a species; see label Genus species.

See footnote for additional symbol keys.

Genus Species	Common Name	Bensulide*	DCPA	Dichlobenil	Dithiopyr	Imazaquin	Isoxaben	Metolachlor	Napropamide	Oryzalin	Oxadiazon G	Pendimethalin*	Prodiamine*	Pronamide*	Simazine	Trifluralin	Benefin + Oryzalin	Isoxaben + Trifluralin	Oxadiazon + Pendimethalin	Oxyfluorfen + Oxadiazon	Oxyfluorfen + Oryzalin	Trifluralin + Isoxaben + Oxyfluorfen
Actinidia deliciosa	kiwi						F			F			FC			F	F	F				
Aesculus spp.	buckeye species											FC*		F*								
Akebia quintata	akebia, fiveleaf												FC									
Amelanchier laevis	serviceberry								F			FC										
Arctostaphylos uva-ursi	bearberry (kinnikinick)			F	F*		FC				FC	FC						FC			FC	
Aucuba spp.	golddust plant							F	F													
Aucuba japonica	aucuba, Japanese							FC*				FC	FC									
Berberis spp.	barberry species		F	F	F*			F			FC			F	F				FC	FC		
B. × *gladwynensis*	barberry						FC*					FC	FC			FC		FC*				FC
B. julianae	barberry, wintergreen												FC									
B. × *mentorensis*	barberry, mentor						FC						FC			FC		FC				FC
B. thunbergii	barberry, Japanese				F*		FC*			FC*		FC	FC			FC	FC*	FC*			FC*	
B. verruculosa	barberry, warty												FC									
Betula spp.	birch species		F	F				F	F					F						FC		
B. nigra	birch, river				F		FC	FC		F		FC				FC	F	FC				FC
B. papyrifera	birch, paper						F			F	FC					FC	F	F				F
B. pendula	birch, Eur. white								F	F		FC				FC	F					
Bougainvillea spp.	bougainvillea				F*		FC*		FC	F*	FC*		FC			FC*	F*	FC*			FC*	FC*
Buddleia spp.	butterfly bush									FC*	FC*		FC*						FC*			
Buxus spp.	boxwood species	F	F	F			FC*	FC	FC					F				FC*	FC	FC		FC*
B. harlandii	boxwood, Harland																					

(continued)

Table 8.3
Preemergence Herbicides Used in Woody Ornamental (*Continued*)

Key:
F = registered for field (or landscape) use,
C = registered for container use,
FC = registered for field and container use,
Ø = label prohibits use on this species,
F*, C*, FC* = registered for some species (sp.) or cultivars (cvs.); see label,
Ø* = prohibited for some cultivars in a species; see label Genus species.

See footnote for additional symbol keys.

Genus Species	Common Name	Bensulide*	DCPA	Dichlobenil	Dithiopyr	Imazaquin	Isoxaben	Metolachlor	Napropamide	Oryzalin	Oxadiazon G	Pendimethalin*	Prodiamine*	Pronamide*	Simazine	Trifluralin	Benefin + Oryzalin	Isoxaben + Trifluralin	Oxadiazon + Pendimethalin	Oxyfluorfen + Oxadiazon	Oxyfluorfen + Oryzalin	Trifluralin + Isoxaben + Oxyfluorfen
B. microphylla	boxwood				F*		F*			FC*	FC	FC	FC			FC*	FC*	F*			FC*	FC*
B. sempervirens	boxwood				F*		FC			FC	FC	FC				FC	FC	FC			FC*	FC
Callistemon spp.	bottlebrush				F*		FC*		FC	FC*	FC*	FC*	FC*			FC*	FC*	FC*				FC*
Calluna spp.	heather species			F			FC*		FC		FC*		FC*			FC*		FC*	FC*, Ø*		Ø*	
Camellia spp.	camellia species		F	F				F	FC										FC	FC		
C. japonica	camellia, Japanese				F*		FC				FC	FC				FC		FC			FC*	FC
C. sasanqua	camellia, Sasanqua				F*						FC					FC						
Campsis × *tagliabuana*	trumpetcreeper												FC									
Caragana arborescens	peashrub, Siberian			F											F							
Carissa grandiflora	plum, Natal								F		FC	FC									FC*	
Carya illinoinensis	pecan			F			F		F	FC		FC				F	FC	F				
Cassia spp.	senna						FC*			F*		FCg	FC*			FC*		FC*				
Castanea spp.	chestnut species		F																			
C. mollissima	chestnut, Chinese															FC						
Ceanothus spp.	wild lilac			F			FC			FC	FC*	FC	FC*			FC	FC	FC				
Cedrus spp.	cedar species								FC					F	F*							
C. atlantica	cedar, Atlas									FC	FC								FC			
C. deodara	cedar, deodar									FC												
Celtis occidentalis	hackberry			F							FC											
Cercis canadensis	redbud, eastern		F				FC			FC				F		FC	FC	FC		FC	FC	FC
Chaenomeles spp.	quince, flowering			F	F		FCe			FC	FC	FC*		F				FCe				

(continued)

Table 8.3
(Continued)

Key:
F = registered for field (or landscape) use,
C = registered for container use,
FC = registered for field and container use,
Ø = label prohibits use on this species,
F*, C*, FC* = registered for some species (sp.) or cultivars (cvs.); see label,
Ø* = prohibited for some cultivars in a species; see label Genus species.

See footnote for additional symbol keys.

Genus Species	Common Name	Bensulide*	DCPA	Dichlobenil	Dithiopyr	Imazaquin	Isoxaben	Metolachlor	Napropamide	Oryzalin	Oxadiazon G	Pendimethalin*	Prodiamine*	Pronamide*	Simazine	Trifluralin	Benefin + Oryzalin	Isoxaben + Trifluralin	Oxadiazon + Pendimethalin	Oxyfluorfen + Oxadiazon	Oxyfluorfen + Oryzalin	Trifluralin + Isoxaben + Oxyfluorfen
Chamaecyparis spp.	falsecypress			F*						F*							F*			FC		
C. obtusa	falsecypress, Hinoki				F*		FC*			F*	FC					FC*	F*	FC*			FC*	FC*
C. pisifera	falsecypress, Sawara						FC*			F*	FC	FC	FC			FC	F*	FC*			FC*	FC*
Chamaedorea spp.	palm						FC*			F*						FC*	F*	FC*		FC*		FC*
Cheirodendron spp.	lapalapa										FC											
Cistus spp.	rockrose			F							FC*					FC						
Citrus spp.	citrus, ornamental						F	F	F	FC	FC*		FC			F*	FC*	F				
Clethra spp.	clethra species						FC*			FC						FC*		FC*	FC			
Cornus spp.	dogwood species		F	F			FC*	FC	FC	F*		FC*		F	F	FC*	F*	FC*	FC	FC	FC*	FC*
C. florida	dogwood, flowering				F		FC			F	FC	FC	FC			FC	F	FC			FC	FC
C. kousa	dogwood, Kousa									FC	FC	FC				FC	FC					
C. sericea	dogwood, redosier				F*		F*			F*	FC	FC	FC			FC*	F*	F*				F*
Corylus spp.	filbert (hazelnut)			F			FC*		F	F						F	F	FC*				
Cotinus coggygria	smokebush						FC*									FC		FC*				
Cotoneaster spp.	cotoneaster species		F	F			FC*	FC	FC	FC*	FC(i)			F	F	FC*	FC*	FC*	FC	FC		FC*
C. apiculatus	cotoneaster, cranberry				F		FC			FC		FC	FC			FC	FC	FC				FC
C. buxifolius	cotoneaster, boxleaf									F			FC				F					
C. dammeri	cotoneaster, bearberry						FC			FC		FC	FC			FC	FC	FC			FC*	FC
C. horizontalis	cotoneaster, rock						FC			FC		FC				FC	FC	FC				FC
C. microphyllus	cotoneaster, little-leaf									F			FC				F					
C. salicifolius	cotoneaster, willowleaf																				FC*	

(continued)

Table 8.3
Preemergence Herbicides Used in Woody Ornamental *(Continued)*

Key:
F = registered for field (or landscape) use,
C = registered for container use,
FC = registered for field and container use,
Ø = label prohibits use on this species,
F*, C*, FC* = registered for some species (sp.) or cultivars (cvs.); see label,
Ø* = prohibited for some cultivars in a species; see label Genus species.

See footnote for additional symbol keys.

Genus Species	Common Name	Bensulide*	DCPA	Dichlobenil	Dithiopyr	Imazaquin	Isoxaben	Metolachlor	Napropamide	Oryzalin	Oxadiazon G	Pendimethalin*	Prodiamine*	Pronamide*	Simazine	Trifluralin	Benefin + Oryzalin	Isoxaben + Trifluralin	Oxadiazon + Pendimethalin	Oxyfluorfen + Oxadiazon	Oxyfluorfen + Oryzalin	Trifluralin + Isoxaben + Oxyfluorfen
Crataegus spp.	hawthorn species		F		F*		F*		F			FC	FC	F		FC*		F*				
Cryptomeria japonica	cryptomeria, Japanese						FC			FC		FC					FC	FC				
× *Cupressocyparis leylandii*	cypress, Leyland		F		F		FCe			FC	FC	FC						FCe	FC			
Cupressus spp.	cypress species		F				FC*		FC	FC*	FC							FC*		FC		
C. glabra	cypress, smooth						F			FC						FC	FC	F				F
C. sempervirens	cypress, Italian				F*		FC			FC		FC	FC				FC	FC			FC*	FC
Cytisus spp.	broom species				F*		FC*	F*		F*	FC*					FC*	F*	FC*		FC		FC*
Daphne spp.	daphne	F					FC				FC*					FC*		FC				
Deutzia spp.	deutzia		F	F			FC*			F*,Ø*	FC*	FC*				FC	F*,Ø*	FC*	FC*		FC*	
Elaeagnus spp.	elaeagnus species		F				FC*	F				FC*						FC*				
E. angustifolia	Russian-olive		F	F			FC			FC	FC				F	FC	FC	FC				
E. pungens	elaeagnus, thorny						FC*						FC			FC		FC*			FC*	FC*
Erica spp.	heather species		F*				FC*				FC(i)					FC*		FC*				Ø*
E. cinerea	heather, bell				F*		FC*				FC					FC		FC*				
Eriobotrya japonica	loquat										FC										FC	
Eucalyptus spp.	eucalyptus species				F*		FC*		FC	F*	FC	FC*			F	FC*	F*	FC*				
E. camaldulensis	Eucalyptus						F			F						FC	F	F				
E. cinerea	Argyle apple						F			F						FC	F	F				
Euonymus spp.	euonymus species		F	F			FC*	F	FC	FC*				F				FC*	FC	FC, Ø*		FC*
E. alatus	euonymous, winged									F		FC				FC	F					
E. alatus 'Compactus'	burning bush, dwarf						Ø*											Ø*			FC	Ø

(continued)

Table 8.3
(Continued)

Key:
F = registered for field (or landscape) use,
C = registered for container use,
FC = registered for field and container use,
Ø = label prohibits use on this species,
F*, C*, FC* = registered for some species (sp.) or cultivars (cvs.); see label,
Ø* = prohibited for some cultivars in a species; see label Genus species.

See footnote for additional symbol keys.

Genus Species	Common Name	Bensulide*	DCPA	Dichlobenil	Dithiopyr	Imazaquin	Isoxaben	Metolachlor	Napropamide	Oryzalin	Oxadiazon G	Pendimethalin*	Prodiamine*	Pronamide*	Simazine	Trifluralin	Benefin + Oryzalin	Isoxaben + Trifluralin	Oxadiazon + Pendimethalin	Oxyfluorfen + Oxadiazon	Oxyfluorfen + Oryzalin	Trifluralin + Isoxaben + Oxyfluorfen
E. fortunei	wintercreeper				F*		FC*	FC		FC	FC	FC	FC			FC	FC	FC*			FC*	FC*
E. japonicus	euonymous, Jap.						FC*			FC		FC	FC			FC*	FC*	FC*			Ø*	FC*
E. kiautschovicus	euonymous, spreading						FC	FC		F	FC		FC			FC	F	FC			FC*	FC
Fagus spp.	beech species													F								
F. sylvatica	beech, European						FCe				FC							FCe				
Fatshedera lizei	tree-ivy						FCf			FC							FC	FCf				
Fatsia japonica							FC			FC			FC					FC				
Ficus benjamana	fig									F						FC	F				FC*	FC
Ficus spp.	fig species				F*		F*		F	F*	FC*					FC*	F*	F*			FC*, Ø*	
Forsythia spp.	forsythia species		F	F			FC*	FC	FC					F		FC		FC*		FC		FC*
F. × intermedia	forsythia, border				F*		FC			F	FC	FC	FC				F	FC	FC		FC*	FC
F. suspensa	forsythia, weeping												FC									
F. viridissima	forsythia, greenstem												FC									
Fraxinus spp.	ash species		F	F			FC*	F	F	F				F			F	FC*	FC			
F. americana	ash, white				F							FC				FC						
F. pennsylvanica	ash, green				F						FC	FC										
Gardenia spp.	gardenia species			F					FC		FC			FC						FC		
G. jasminoides	cape jasmine			F	F*	F	FC	FC		FC		FC	FC			FC	FC	FC	FC		FC*	
Gelsemium sempervirens	Carolina jessamine				F		FC	F				FC				FC		FC			FC	FC
Ginkgo biloba	ginkgo						F	F		FC	FC	FC		F		FC	FC	F	FC			

(continued)

Table 8.3
Preemergence Herbicides Used in Woody Ornamental *(Continued)*

Key:
F = registered for field (or landscape) use,
C = registered for container use,
FC = registered for field and container use,
Ø = label prohibits use on this species,
F*, C*, FC* = registered for some species (sp.) or cultivars (cvs.); see label,
Ø* = prohibited for some cultivars in a species; see label Genus species.

See footnote for additional symbol keys.

Genus Species	Common Name	Bensulide*	DCPA	Dichlobenil	Dithiopyr	Imazaquin	Isoxaben	Metolachlor	Napropamide	Oryzalin	Oxadiazon G	Pendimethalin*	Prodiamine*	Pronamide*	Simazine	Trifluralin	Benefin + Oryzalin	Isoxaben + Trifluralin	Oxadiazon + Pendimethalin	Oxyfluorfen + Oxadiazon	Oxyfluorfen + Oryzalin	Trifluralin + Isoxaben + Oxyfluorfen
Gleditsia triacanthos	honeylocust, thornless						F	F	F	F		FC		F	F	FC	F	F				
Hamamelis virginiana	witchhazel						FCe				FC							FCe				
Hedera helix	ivy, English	F	F	F	F*		FC	FC	FC	F	FC	FC	FC			FC	F	FC				
Hemigraphis alternata	red-ivy																					
Hibiscus syriacus	rose-of-sharon						FC*	F	FC	F*	FC* (i)	FC	F			FC*	F*	FC*	FC			
Hydrangea spp.	hydrangea		F				Ø	F		FC*		FC*g	FC*					Ø	FC	FC	Ø*	Ø
Hypericum spp.	St.-John's-wort species	F					FC	F	F	FC	FC*	FC*				FC	F	FC	FC	FC	FC	FC
Ilex spp.	holly species	F	F	F*, Ø*	F*		FC*	F	FC	FC*	FC			F	F	FC	FC*	FC*	FC	FC, Ø*	FC*	
I. × *attenuata 'Fosteri'*	holly, Foster's hybrid				F*		FC*	FC				FC						FC*				FC*
I. aquifolium	holly, English						F*			F							F	F*				F*
I. aquifolium × *cornuta*	holly, English x Chinese																					
I. cassine	dahoon																					
I. cornuta	holly, Chinese				F	F*	FC	FC		FC		FC	FCb				FC	FC			FC*	FC*
I. crenata	holly, Japanese			Ø	F	F*	FC*	FC		FC		FC	FC				FC	FC*			FC*, Ø*	FC*
I. crenata 'Green Lustre'	holly, 'Green Luster'			Ø			FC											FC				FC
I. glabra	inkberry						FC*			F							F	FC*				FC*
I. latifolia	holly, lusterleaf																					
I. meserveae	holly, Meserve hybrids				F*		FC*			F							F	FC*				FC*
I. opaca	holly, American											FC	FC								FC	

(continued)

Table 8.3
(Continued)

Key:
F = registered for field (or landscape) use,
C = registered for container use,
FC = registered for field and container use,
Ø = label prohibits use on this species,
F*, C*, FC* = registered for some species (sp.) or cultivars (cvs.); see label,
Ø* = prohibited for some cultivars in a species; see label Genus species.

See footnote for additional symbol keys.

Genus Species	Common Name	Bensulide*	DCPA	Dichlobenil	Dithiopyr	Imazaquin	Isoxaben	Metolachlor	Napropamide	Oryzalin	Oxadiazon G	Pendimethalin*	Prodiamine*	Pronamide*	Simazine	Trifluralin	Benefin + Oryzalin	Isoxaben + Trifluralin	Oxadiazon + Pendimethalin	Oxyfluorfen + Oxadiazon	Oxyfluorfen + Oryzalin	Trifluralin + Isoxaben + Oxyfluorfen
I. pernyi	holly, Perny												FC									
I. verticillata	holly, winterberry																					
I. vomitoria	holly, yaupon			Ø	F	F*	FC*			FC		FC	FC				FC	FC*			FC*	FC
Illicium spp.	anise-tree						FC*	F				FC*				FC*		FC*				
I. floridanum	anise-tree, Florida						FC									FC		FC				
Itea virginica	sweetspire, VA																					
Jasminum spp.	jasmine species						FC*	F			FC*	FC*	FC*			FC*		FC*	FC*	FC	FC*	FC*
Juglans spp.	walnut species		F	F*			F*		F	F*			FC	F		F*	F*	F*				
J. nigra	walnut, black						F		F	F		FC				F	F	F				
Juniperus spp.	juniper species	F	F	F	F*		FC*	F	FC	FC	FC	FC		F	F	FC	FC	FC*	FC	FC		FC*
J. chinensis	juniper, Chinese				F*	F*	FC*	FC		FC*		FC*	FC					FC*			FC*	FC*
J. conferta	juniper, shore					F*	FC			FC*		FC	FC					FC			FC	FC
J. davurica	juniper, Dahurian												FC									
J. excelsa	juniper, Greek																					
J. horizontalis	juniper, creeping				F*	F*	FC*, Ø*	FC		FC*		FC	FC					FC*, Ø*			FC*	FC*, Ø*
J. procumbens	juniper, Japanese garden						FC*			FC*								FC*				FC*
J. sabina	juniper, Savin				F*		FC*	FC		FC*								FC*			FC*	FC
J. squamata	juniper, singleseed					F*	FC*			FC*								FC*			FC*	FC*
J. virginiana	redcedar, eastern			F	F		FC	FC		F	FC	FC			F	FC	F	FC	FC			FC
Justicia brandegeana	shrimp plant									FC	FC		FC				FC					

(continued)

Table 8.3

PREEMERGENCE HERBICIDES USED IN WOODY ORNAMENTAL *(CONTINUED)*

Key:
F = registered for field (or landscape) use,
C = registered for container use,
FC = registered for field and container use,
Ø = label prohibits use on this species,
F*, C*, FC* = registered for some species (sp.) or cultivars (cvs.); see label,
Ø* = prohibited for some cultivars in a species; see label Genus species.

See footnote for additional symbol keys.

Genus Species	Common Name	Bensulide*	DCPA	Dichlobenil	Dithiopyr	Imazaquin	Isoxaben	Metolachlor	Napropamide	Oryzalin	Oxadiazon G	Pendimethalin*	Prodiamine*	Pronamide*	Simazine	Trifluralin	Benefin + Oryzalin	Isoxaben + Trifluralin	Oxadiazon + Pendimethalin	Oxyfluorfen + Oxadiazon	Oxyfluorfen + Oryzalin	Trifluralin + Isoxaben + Oxyfluorfen
Kalmia latifolia	mountainlaurel		F	F	F		FC	FC		F		FC		F		FC	F	FC	FC			
Koelreuteria paniculata	golden raintree			F					F	F	FC						F					
Kolkwitzia amabilis	beautybush			F																		
Laburnum spp.	golden-chain tree								F													
Lagerstroemia indica	crapemyrtle				F*	F*	FC	F	FC	FC	FC(i)	FC	FC			FC	FC	FC	FC			
Lantana spp.	lantana		F		F*		FC				FC*		FC*			FC*		FC			FC*, Ø*	
Larix spp.	larch species								F*		FC*					FC*						
Leucothoe spp.	leucothoe species			F	F*		FC*	FC*	FC	F*	FC*	FC*				FC*	F*	FC*	FC			
L. axillaris	leucothoe, coast						FC			F	FC	FC				FC	F	FC				Ø
Ligustrum spp.	privet species	F	F	F		Ø	FC*	F	FC	FC*	FC			F		FC	FC*	FC*	FC	FC	FC*	FC*
L. amurense	privet, Amur									FC			FC				FC					
L. japonicum	privet, Japanese				F*		FC	FC		FC		FC	FC				FC	FC			FC*	FC
L. lucidum	privet, waxleaf				F		FC			FC		FC	FC				FC	FC			FC	FC
L. ovalifolium	privet, California						F			F		FC					F	F				F
L. sinense	privet, Chinese											FC	FC								FC*	
Liquidambar styraciflua	sweetgum, American		F		F		F	F	F	FC		FC		F		FC	FC	F				
Liriodendron tulipifera	poplar, tulip; tuliptree		F					F	F			FC		F		FC						
Lonicera spp.	honeysuckle species		F	F			FC*	F	F	F*	FC				F*	FC*	F*	FC*	FC	FC	FC*	FC*
L. japonica	honeysuckle, Jap.				F*					F			FC									F
L. tatarica	honeysuckle, Tatarian												FC									

(continued)

Table 8.3
(Continued)

Key:
F = registered for field (or landscape) use,
C = registered for container use,
FC = registered for field and container use,
Ø = label prohibits use on this species,
F*, C*, FC* = registered for some species (sp.) or cultivars (cvs.); see label,
Ø* = prohibited for some cultivars in a species; see label Genus species.

See footnote for additional symbol keys.

Genus Species	Common Name	Bensulide*	DCPA	Dichlobenil	Dithiopyr	Imazaquin	Isoxaben	Metolachlor	Napropamide	Oryzalin	Oxadiazon G	Pendimethalin*	Prodiamine*	Pronamide*	Simazine	Trifluralin	Benefin + Oryzalin	Isoxaben + Trifluralin	Oxadiazon + Pendimethalin	Oxyfluorfen + Oxadiazon	Oxyfluorfen + Oryzalin	Trifluralin + Isoxaben + Oxyfluorfen
Loropetalum chinense	loropetulum						FC*			FC			FC*					FC*			FC*	
Magnolia spp.	magnolia species		F	F				F		F	FC		F	F		FC			FC	FC		
M. grandiflora	magnolia, southern						FC					FC					F	FC				FC
M. × *soulangiana*	magnolia, saucer											FC										
M. stellata	magnolia, star						FC*					FC						FC*			Ø*	
Mahonia spp.	mahonia species						FC*								F	FC*		FC*		FC	FC*	FC*
M. aquifolium	grapeholly, Oregon						FC*			F	FC	FC					F	FC*	FC		FC	FC*
Malus spp.	apple species				F*		F	F	F	F		FC				F	F	F	FC			
Malus spp.	crabapple species		F	F				F	F	F	FC	FC	FC	F		FC	F		FC			
Malus × *domestica*	apple, common						F										F	F				
Malus floribunda	crabapple, Japanese			F													F					
Morus alba	mulberry, white						F			F						FC	F	F				
Myoporum spp.	myoporum species						FCe			F	FC*					FC*	F*	FCe				
Myrica cerifera	myrtle, wax	F				F	FC	FC				FC				FC		FC				
Myrica pensylvanica	bayberry, northern	F						F														
Myrtus communis	myrtle, compact	F								FC	FC	FC					FC		FC			
Nandina domestica	nandina			F	F*		FC	F	FC	FC	FC	FC	FC			FC	FC	FC	FC		FC*	
Nerium oleander	oleander				F*		FC	F	F	FC	FC	FC	FC			FC	FC	FC	FC	FC	FC	FC
Nyssa sylvatica	tupelo, black											FC				FC						
Ochna serrulata	bird's-eye bush										FC											
Olea spp.	olive						F		F	F		FC*	FC*			F	F*	F				

(continued)

Table 8.3

Preemergence Herbicides Used in Woody Ornamental (*Continued*)

Key:
F = registered for field (or landscape) use,
C = registered for container use,
FC = registered for field and container use,
Ø = label prohibits use on this species,
F*, C*, FC* = registered for some species (sp.) or cultivars (cvs.); see label,
Ø* = prohibited for some cultivars in a species; see label Genus species.

See footnote for additional symbol keys.

Genus Species	Common Name	Bensulide*	DCPA	Dichlobenil	Dithiopyr	Imazaquin	Isoxaben	Metolachlor	Napropamide	Oryzalin	Oxadiazon G	Pendimethalin*	Prodiamine*	Pronamide*	Simazine	Trifluralin	Benefin + Oryzalin	Isoxaben + Trifluralin	Oxadiazon + Pendimethalin	Oxyfluorfen + Oxadiazon	Oxyfluorfen + Oryzalin	Trifluralin + Isoxaben + Oxyfluorfen
Osmanthus spp.	osmanthus species			F			FC*	F	F							FC*		FC*	FC			
O. fragrans	tea olive, fragrant						FC*					FC						FC*			FC	
O. heterophyllus	tea olive, holly									F	FC		FC*				F				FC*	
Oxydendrum arboreum	sourwood				F		FC					FC	FC			FC		FC				
Pachysandra terminalis	pachysandra	F	F		F	F	FC	FC	FC	FC	FC	FC				FC	F	FC	FC	FC		
Paxistima canbyi	paxistima (pachistima)		F	F																		
Philadelphus spp.	mockorange		F	F	F*					FC	FC*	FC		F		FC	FC		FC*			
Photinia spp.	photinia spp.			F				F	FC											FC		
P. glabra	photinia, Japanese																					
P. × *fraseri*	Fraser photinia; red-tip				F	F	FC			FC	FC	FC	FC			FC	FC	FC	FC		FC	FC
Physocarpus opulifolius	ninebark						FC*e	F										FC*e				
Picea spp.	spruce species		F					F	F	F*	FC		Fa	F			F*		FC	FC		
P. abies	spruce, Norway				F		FC			F	FC(i)	FC			F	FC	F	FC			FC*	FC
P. glauca	spruce, white				F		FC			F	FC(i)	FC			F	FC	F	FC			FC*	FC*
P. glauca 'Conica'	spruce, dw. Alberta				F		F			F		FC*				FC	F	F				F
P. omorika	spruce, Serbian																					
P. pungens	spruce, Colorado				F		FC*		F	FC*		FC			F	FC	F*	FC*			FC*	FC*
P. rubens	spruce, red														F							
P. sitchensis	spruce, Sitka																					
Pieris spp.	pieris species					Ø	FC*				FC				F	FC*		FC*		FC		
P. japonica	pieris, Japanese		F		F		FC	F		FC	FC(i)	FC	FC			FC	FC	FC			FC*	

(continued)

Table 8.3
(Continued)

Key:
F = registered for field (or landscape) use,
C = registered for container use,
FC = registered for field and container use,
Ø = label prohibits use on this species,
F*, C*, FC* = registered for some species (sp.) or cultivars (cvs.); see label,
Ø* = prohibited for some cultivars in a species; see label Genus species.

See footnote for additional symbol keys.

Genus Species	Common Name	Bensulide*	DCPA	Dichlobenil	Dithiopyr	Imazaquin	Isoxaben	Metolachlor	Napropamide	Oryzalin	Oxadiazon G	Pendimethalin*	Prodiamine*	Pronamide*	Simazine	Trifluralin	Benefin + Oryzalin	Isoxaben + Trifluralin	Oxadiazon + Pendimethalin	Oxyfluorfen + Oxadiazon	Oxyfluorfen + Oryzalin	Trifluralin + Isoxaben + Oxyfluorfen
Pinus spp.	pine species		F	F*			F*	F	FC	FC	FC	FC*	FC*	F			FC	F*	FC	FC		FC*
P. banksiana	pine, Jack																					
P. bungeana	pine, lacebark																					
P. clausa	pine, sand																					
P. contorta	pine, lodgepole						F			F					F	FC		F				F
P. echinata	pine, shortleaf																					
P. elliottii	pine, slash				F							FCh	FC								FC	
P. mugo	pine, mugo				F		FC					FC			F	FC		FC			FC	FC
P. nigra	pine, Austrian				F		FC					FC	FC		F	FC		FC				
P. palustris	pine, longleaf				F								FC									
P. ponderosa	pine, ponderosa																					F
P. radiata	pine, Monterey	F					F			F		FC	FC		F*	FC		F			FC	F
P. resinosa	pine, red									F		FC			F	FC						
P. strobus	pine, Eastern white				F		FC	FC	F	F		FC	FC		F	FC		FC			FC	FC
P. sylvestris	pine, Scotch				F		FC			F		FC	FC		F	FC		FC			FC	FC
P. taeda	pine, loblolly				F							FC	FC			FC						
P. thunbergii	pine, Jap. black				F		FC	FC		F			FC			FC		FC			FC	FC
P. virginiana	pine, Virginia				F							FC	FC									
Pistacia spp.	pistachio						F		F	F			FC			F	F	F				
Pittosporum tobira	pittosporum, Jap.	F	F	F	F		FC	FC	FC	F	FC	FC	FC			FC	FC	FC	FC	FC	FC	FC
Pittosporum rhombifolium	Queensland pittosporum		F	F				F	FC	FC			FC				FC			FC		

(continued)

Table 8.3
PREEMERGENCE HERBICIDES USED IN WOODY ORNAMENTAL *(CONTINUED)*

Key:
F = registered for field (or landscape) use,
C = registered for container use,
FC = registered for field and container use,
Ø = label prohibits use on this species,
F*, C*, FC* = registered for some species (sp.) or cultivars (cvs.); see label,
Ø* = prohibited for some cultivars in a species; see label Genus species.

See footnote for additional symbol keys.

Genus Species	Common Name	Bensulide*	DCPA	Dichlobenil	Dithiopyr	Imazaquin	Isoxaben	Metolachlor	Napropamide	Oryzalin	Oxadiazon G	Pendimethalin*	Prodiamine*	Pronamide*	Simazine	Trifluralin	Benefin + Oryzalin	Isoxaben + Trifluralin	Oxadiazon + Pendimethalin	Oxyfluorfen + Oxadiazon	Oxyfluorfen + Oryzalin	Trifluralin + Isoxaben + Oxyfluorfen
Platanus × *acerifolia*	planetree, London													F		FC						
Platanus occidentalis	sycamore		F		F		F			F		FC		F		FC	F	F				
P. racemosa	sycamore, California		F				F			F				F		FC	F	F				
Platycladus orientalis	arborvitae, Oriental				F		FC*			FC*				F		FC*	FC	FC*	FC		FC*	FC*
Podocarpus spp.	podocarpus species		F				F	F	FC	F						FC	F	F			FC*	F
P. macrophyllus	podocarpus, Chinese						FC			FC	FC	FC	FC			FC	FC	FC				Ø
Populus spp.	poplar species		F	F				F	F		FC	FC*		F					FC			
P. deltoides	cottonwood, eastern		F	F			F		F	F		FC				FC	F	F				
Potentilla spp.	potentilla; cinquefoil		F		F*		FC*	F		F*	FC*					FC	F*	FC*	FC	FC, Ø*		FC*
P. fruticosa	cinquefoil, bush				F		FC			FC	FC	FC					FC	FC			FC*	FC
Prunus spp.	prunes & stone fruits								F	F*			FC			F*	F*			FC		F
Prunus spp.	cherry, peach and plum species			F*			F*	F	F	F*			FC*	FC	F*		F*	F*	F*		FC	F
P. avium	cherry, sweet								F	F						F						F
P. caroliniana	cherrylaurel, Carolina						Fce			F						FC	F	Fce	FC			
P. cerasifera	plum, cherry											FC										
P. × *cistena*	cherry, purpleleaf sand																				FC	
P. glandulosa	almond, dw. flowering			F	F		FC			FC						FC		FC			Ø*	
P. laurocerasus	cherrylaurel, common						FCe			F	FC	FC	FC				F	FCe	FC			
P. maritima	plum, beach								F													
P. persica	peach, common			F*	F		F		F	F		FC	FC			F	F	F				F

(continued)

Table 8.3
(Continued)

Key:
F = registered for field (or landscape) use,
C = registered for container use,
FC = registered for field and container use,
Ø = label prohibits use on this species,
F*, C*, FC* = registered for some species (sp.) or cultivars (cvs.); see label,
Ø* = prohibited for some cultivars in a species; see label Genus species.

See footnote for additional symbol keys.

Genus Species	Common Name	Bensulide*	DCPA	Dichlobenil	Dithiopyr	Imazaquin	Isoxaben	Metolachlor	Napropamide	Oryzalin	Oxadiazon G	Pendimethalin*	Prodiamine*	Pronamide*	Simazine	Trifluralin	Benefin + Oryzalin	Isoxaben + Trifluralin	Oxadiazon + Pendimethalin	Oxyfluorfen + Oxadiazon	Oxyfluorfen + Oryzalin	Trifluralin + Isoxaben + Oxyfluorfen
P. sargentii	cherry, sargent										FC											
P. serrulata	cherry, Jap. flowering										FC	FC										
P. subhirtella	cherry, Higan																					
P. × yedoensis	cherry, Yoshino						F			F						FC	F	F				F
Pseudotsuga menziesii	Douglasfir				F			F	FC	Ø	FC	FC	Fa	F	F	FC	Ø		FC			
Pyracantha spp.	pyracantha species	F		F				F	FC					F		FC				FC	FC*	
P. coccinea	firethorn, scarlet									FC	FC	FC	FC				FC		FC			
P. fortuneana	firethorn, Chinese						FC*			FC*			FC				FC*	FC*			FC*	FC*
P. koidzumii	firethorn, Formosa				F*					FC*			FC				FC*					
Pyrus spp.	pear species			F*				F	F	F*							F*					
P. calleryana 'Bradford'	pear, Bradford callery											FC	FC	F								
P. communis	pear						F		F	F						F	F	F				
Quercus spp.	oak species		F	F	F*			F		FC	FC	FC*	FC	F			FC		FC	FC		
Q. alba	oak, white											FC										
Q. coccinea	oak, scarlet															FC						
Q. illicifolia	oak, bear						F									FC		F				F
Q. nigra	oak, water											FC										
Q. palustris	oak, pin				F		F		F	F		FC				FC		F				F
Q. phellos	oak, willow				F		FC	FC		F		FC				FC		FC				FC
Q. rubra	oak, red				F		FC			FC		FC	FCc		F	FC		FC				FC
Q. shumardii	oak, Shumard												FCd								FC	

(continued)

Table 8.3

PREEMERGENCE HERBICIDES USED IN WOODY ORNAMENTAL *(CONTINUED)*

Key:
F = registered for field (or landscape) use,
C = registered for container use,
FC = registered for field and container use,
Ø = label prohibits use on this species,
F*, C*, FC* = registered for some species (sp.) or cultivars (cvs.); see label,
Ø* = prohibited for some cultivars in a species; see label Genus species.

See footnote for additional symbol keys.

Genus Species	Common Name	Bensulide*	DCPA	Dichlobenil	Dithiopyr	Imazaquin	Isoxaben	Metolachlor	Napropamide	Oryzalin	Oxadiazon G	Pendimethalin*	Prodiamine*	Pronamide*	Simazine	Trifluralin	Benefin + Oryzalin	Isoxaben + Trifluralin	Oxadiazon + Pendimethalin	Oxyfluorfen + Oxadiazon	Oxyfluorfen + Oryzalin	Trifluralin + Isoxaben + Oxyfluorfen
Q. virginiana	oak, live				F		FC					FC				FC		FC				FC
Raphiolepis indica	hawthorn, Indian					F	FC*	F	FC	FC		FC	FC			FC	FC	FC*		FC	FC*	FC*
R. umbellata	hawthorn, Yeddo							F	FC		FC		FC							FC		
Rhododendron spp.	azalea, rhododendron	F	F	F*	F*	Ø	FC*	F	FC	FC	FC (i)	FC	FC*	F		FC*	FC	FC*	FC	FC*, Ø*	FC*, Ø*	FC*
R. carolinianum	rhododendron, Carolina				F		FC*, Ø*											FC*, Ø*				Ø
R. catawbiense	rhododendron, Catawba				F		FC*, Ø*	FC		FC								FC*, Ø*				FC, Ø*
R. indica	azalea, Indica						FC*	FC		FC*								FC*				FC*
R. maximum	rhodo., rosebay				F		FC			FC								FC				FC
R. molle	azalea, Chinese																					
R. obtusum	azalea, Hiryu						FC	FC										FC				
Rhus typhina	sumac, staghorn						FCe				FC							FCe				
Robinia spp.	locust species		F	F				F								FC*						
Rosa spp.	rose species		F	F			FC*	F	FC	F*	FC	FCg				FC*	F*	FC*	FC			
R. banksiae	rose, Lady Bank's				F								FC									
Rubus spp.	bramble						F*			F*						F*		F*				
Salix spp.	willow species		F	F			F*	F						F		FC		F*				
S. babylonica	willow, Babylon weeping						F			F		FC					F	F				
Samanea saman	rain-tree (ohai)										FC											
Sequoiadendron giganteum	sequoia, giant						F			F		FC				FC	F	F				

(continued)

Table 8.3
(Continued)

Key:
F = registered for field (or landscape) use,
C = registered for container use,
FC = registered for field and container use,
Ø = label prohibits use on this species,
F*, C*, FC* = registered for some species (sp.) or cultivars (cvs.); see label,
Ø* = prohibited for some cultivars in a species; see label Genus species.

See footnote for additional symbol keys.

Genus Species	Common Name	Bensulide*	DCPA	Dichlobenil	Dithiopyr	Imazaquin	Isoxaben	Metolachlor	Napropamide	Oryzalin	Oxadiazon G	Pendimethalin*	Prodiamine*	Pronamide*	Simazine	Trifluralin	Benefin + Oryzalin	Isoxaben + Trifluralin	Oxadiazon + Pendimethalin	Oxyfluorfen + Oxadiazon	Oxyfluorfen + Oryzalin	Trifluralin + Isoxaben + Oxyfluorfen
Skimmia spp.	Skimmia						FC*									FC*		FC*				
Sorbus americana	mountainash, Am.			F										F								
Spiraea spp.	spirea species		F	F	F*		FC*	F				FC*	FC*	F		FC*		FC*		FC	FC*	FC*
S. × vanhouttei	spirea, Vanhoutte						FC			F		FC				FC	F	FC				FC
Symphoricarpos albus	snowberry							F														
Syringa spp.	lilac species		F	F			FC*	F			FC	FC*		F		FC*		FC*	FC	FC	FC*	FC*
S. vulgaris	lilac, common						F			FC		FC				FC	F	F				F
S. × persica	lilac, Persian																					
Taxodium distichum	cypress, bald				F			F			FC	FC				FC			FC			
Taxus spp.	yew species		F	F				F	F		FC			F	F				FC	FC		
T. baccata	yew, English																					
T. canadensis	yew, Canadian																					
T. cuspidata	yew, Japanese				F*		F	FC		F		FC	FC			FC	F	F				F
T. × media	yew, Anglojap				v					F		FC	FC			FC	F				FC*	
Ternstroemia gymnanthera	cleyera, Japanese			F			FC	F		FC	FC		FC			FC	FC	FC			FC	FC
Thuja spp.	arborvitae species		F	F				F	C	F*	FC	FC*		F	F		F*		FC	FC		
T. occidentalis	arborvitae, American				F*		FC*	FC		F, Ø*		FC	FC			FC	F, Ø*	FC*			FC*	FC*
Tilia spp.	basswood; linden			F					F*	FC*	FC*	FC		F		FC	FC*					
Trachelospermum asiaticum	jasmine, Japanese star				F	F*						FC	FC			FC			FC*		FC*, Ø*	
Trachelospermum jasminoides	jasmine, star or confederate					F			FC	F	FC						F		FC*		FC	

(continued)

Table 8.3
Preemergence Herbicides Used in Woody Ornamental (*Continued*)

Key:
F = registered for field (or landscape) use,
C = registered for container use,
FC = registered for field and container use,
Ø = label prohibits use on this species,
F*, C*, FC* = registered for some species (sp.) or cultivars (cvs.); see label,
Ø* = prohibited for some cultivars in a species; see label Genus species.

See footnote for additional symbol keys.

Genus Species	Common Name	Bensulide*	DCPA	Dichlobenil	Dithiopyr	Imazaquin	Isoxaben	Metolachlor	Napropamide	Oryzalin	Oxadiazon G	Pendimethalin*	Prodiamine*	Pronamide*	Simazine	Trifluralin	Benefin + Oryzalin	Isoxaben + Trifluralin	Oxadiazon + Pendimethalin	Oxyfluorfen + Oxadiazon	Oxyfluorfen + Oryzalin	Trifluralin + Isoxaben + Oxyfluorfen
Tsuga spp.	hemlock species							F	F					F	F							
T. canadensis	hemlock, Canadian				F		FC	FC		Ø	FC(i)	FC	FC			FC	Ø	FC	FC		FC	
T. caroliniana	hemlock, Carolina																					
Ulmus spp.	elm species		F	F							FC	FC*		F					FC*			
U. americana	elm, American														F							
U. parvifolia	elm, Chinese				F		F			F						FC	F	F				
U. pumila	elm, Siberian								F						F							
Vaccinium spp.	blueberry				F*		F			F						F		F			FC*	F
Viburnum spp.	viburnum species		F		F*	Ø	FC*	FC	F	FC*	FC	FC*		F		FC	FC*	FC*	FC	FC	FC*	FC*
V. davidii	viburnum, David						FC			F	FC(i)						F	FC			FC	FC
V. japonicum	viburnum, Japanese						F			F			FC				F	F				F
V. odoratissimum	viburnum, sweet												FC									
V. plicatum	viburnum, Jap. snowball						FC			F			FC				F	FC			FC*	FC
V. rigidum	viburnum, Canary Island												FC									
V. suspensum	vib., Sandankwa	F								F		FC	FC				F					
V. tinus	laurustinus						F*			FC*		FC	FC				FC	F*				
V. trilobum	viburnum, Am. cranberrybush				F		FC			F*			FC				F*	FC				FC
V. wrightii	viburnum, Wright				F								FC									
Vitex spp.	chaste tree											FC										
Vitis spp.	grapes						F		F	F			FC			F*	F	F			FC*	

(continued)

Table 8.3
(Continued)

Key:
F = registered for field (or landscape) use,
C = registered for container use,
FC = registered for field and container use,
Ø = label prohibits use on this species,
F*, C*, FC* = registered for some species (sp.) or cultivars (cvs.); see label,
Ø* = prohibited for some cultivars in a species; see label Genus species.

See footnote for additional symbol keys.

Genus Species	Common Name	Bensulide*	DCPA	Dichlobenil	Dithiopyr	Imazaquin	Isoxaben	Metolachlor	Napropamide	Oryzalin	Oxadiazon G	Pendimethalin*	Prodiamine*	Pronamide*	Simazine	Trifluralin	Benefin + Oryzalin	Isoxaben + Trifluralin	Oxadiazon + Pendimethalin	Oxyfluorfen + Oxadiazon	Oxyfluorfen + Oryzalin	Trifluralin + Isoxaben + Oxyfluorfen
Weigela spp.	weigela species		F	F				F								FC				FC	FC*	
W. florida	weigela, old fashioned						FC*			F*		FC	FC				F*	FC*			FC*	
Wisteria spp.	wisteria species							F*	F*		FC*	FCg							FC*			
Xylosma congestum	Xylosma	F			F		F		FC	F	FC	FC				FC	F	F		FC	FC	F
Yucca spp.	yucca species					F*	FC*, Ø*	FC		FC*		FC*	FC*			FC*	FC*	FC*, Ø*				Ø*

a. EC landscape only
b. FC RegalKade; F only Barricade
c. Barricade only
d. RegalKade only
e. Gallery only
f. Snapshot only
g. GRANULAR ONLY
h. SPRAY ONLY
i. Some species, cultivars have shown sensitivity
* see text for other trade names of these products
Authors: J. C. Neal, J. A. Briggs and T. Whitwell. 2005. North Carolina State University, Raleigh, North Carolina and Clemson University, Clemson, South Carolina.

Napropamide is another preemergent that controls annual grasses and a select few broadleaf weeds in woody trees, shrubs, and especially ground covers. Sprayable and granular formulations are available with rates ranging from 4 to 6 lb ai/a (4.48 to 6.72 kg ai/ha). Most annual grasses are controlled along with chickweed, groundsel, knotweed, cudweed, little mallow, and annual sow thistle. Optimal control occurs when napropamide is applied under the mulch layer and watered-in with 1 to 2 in. (2.5 to 5 cm) of rainfall within two to three days after application.

Pendimethalin, prodiamine, oryzalin, and trifluralin are all dinitroanaline herbicides that inhibit root and shoot elongation. These products are used primarily to control annual grasses, but select small seeded broadleaf weeds are also controlled. All four products are labeled for woody trees, shrubs, ground covers, herbaceous perennials, and some flowers and bulbs. Rates range from 2 to 4 lb ai/acre (2.24 to 4.48 kg ai/ha) for pendimethalin, oryzalin, and trifluralin, with prodiamine ranging from 0.75 to 1.5 lb ai/acre (0.84 to 1.68 kg ai/ha). Annual broadleaf weeds like carpetweed, chickweed, knotweed, oxalis, shepherd's purse, spurge, and lambsquarter can be controlled by these products. They are generally applied at a split rate and reapplied eight weeks later to maximize their residual control. The products are volatile and will photodecompose if not watered in soon after application. Applications made under the mulch layer will maximize control and provide longer residual.

Dithiopyr controls annual grasses and many broadleaf weeds in established woody trees, shrubs, ground covers, and herbaceous perennials. Perennial weeds are not controlled, but dithiopyr will control the following broadleaf weeds: chickweed, corn speedwell, Carolina geranium, lespedeza, bittercress, yellow woodsorrel, henbit, and parsley-piert. Sprayable and granular formulations are available with rates ranging from 0.25 to 0.75 lb ai/acre (0.28 to 0.84 kg ai/ha). Dithiopyr should be applied as a directed spray only and should not be allowed to contact ornamentals, or injury will occur. Two applications are needed for season-long control, and applications need to be irrigated to incorporate the herbicide.

Metolachlor is labeled for woody trees, shrubs, ground covers, and some herbaceous ornamentals, for preemergence control of annual grasses, some broadleaf weeds, and yellow nutsedge. Rate ranges from 2 to 4 lb ai/acre (2.24 to 4.48 kg ai/ha), and the herbicide is available only in a liquid formulation. Metolachlor can cause foliar burn on sensitive ornamentals; therefore, irrigation immediately following an application is advised. Metolachlor can be broadcast over the top of tolerant ornamentals. A higher rate should be used for yellow nutsedge control. It can be tank mixed with compatible pre- and postemergence herbicides.

Oxadiazon can be used on a number of woody trees, shrubs, and ground covers. Crabgrass, goosegrass, and other annual grasses, along with oxalis, bittercress, velvetleaf, and other annual broadleaf weeds, can be controlled with oxadiazon. Oxadiazon is formulated as a granular or liquid, but the widest range of labeled ornamentals are noted with the granular formulation, with rates ranging from 2 to 4 lb ai/acre (2.24 to 4.48 kg ai/ha). Several landscape plants are sensitive to oxadiazon, including andromeda, azalea, cotoneaster, canadien hemlock, hibiscus, rhododendron, viburnum, and Norway and white spruce. Unlike most preemergence herbicides discussed, oxadiazon should be placed on top of the mulch layer due its requirement of light for weed control. Oxadiazon also can be found in other granular formulations as combinations with either prodiamine and oxyflurofen, to broaden the weed-control spectrum. (See "Preemergence Combination Herbicides" for further details.)

Preemergence Broadleaf Herbicides

Oxyflurofen, like oxadiazon, is formulated as a granular with other preemergence herbicides like pendimethalin, oxadiazon, and oryzalin, to broaden the weed control spectrum (Tables 8.2 and 8.3). These products are labeled for use in a limited number of woody trees, shrubs, and ground covers. (See "Preemergence Combination Herbicides.") Oxyflurofen alone can be found as a sprayable, but is currently not labeled for landscape applications, only for field nursery purposes. However, the combination products containing oxyflurofen are labeled for landscape ornamentals. Annual grasses like crabgrass, barnyardgrass, and annual bluegrass, plus many broadleaf weeds like clover, oxalis, bittercress, spurge, smartweed, and others, can be controlled with oxyflurofen. Applications to wet foliage must be avoided, as severe injury may occur. As with oxadiazon, applications need to be placed on top of the mulch layer, since light is required for weed control.

Isoxaben is labeled for a wide range of woody trees, shrubs, ground covers, herbaceous perennials, and flowering ornamentals. Alone, isoxaben is available only as a sprayable formulation, but can be found in a granular combination product with trifluralin. Isoxaben controls many broadleaf weeds, including chickweed, oxalis, knotweed, hairy bittercress, henbit, white clover, lawn burweed, groundsel, and prostrate spurge, but does not control perennial weeds. Rates range from 0.5 to 1.0 lb ai/acre (0.56 to 1.12 kg ai/ha). Best results occur if isoxaben is applied under the mulch layer and watered in within 21 days after application. Several ornamental species are highly sensitive to isoxaben; the label should be consulted for details. Isoxaben should not be used on bedding plants, or on areas where bedding plants are to be planted within one year following application, or on ground covers until they are established and well rooted. Isoxaben can be applied over the top of ornamentals, but injury may result if it is applied to plants with newly formed buds. It is often tank-mixed with glyphosate or oryzalin to enhance weed control.

Dichlobenil provides control of annual and perennial broadleaf weeds, as well as some annual grassy weeds in a limited number of woody trees, shrubs, and ground covers. Only a granular formulation is available, at rates ranging from 4 to 6 lb ai/acre (4.48 to 6.72 kg ai/ha). Applications need to be made when air temperatures are cool, 45°F (7.2°C), since dichlobenil is highly volatile; if the herbicide is applied at high temperatures, vapors will cause severe damage to ornamentals. Dichlobenil controls most spring germinating annuals (broadleaves and grasses) preemergently and most winter annuals postemergently. It also has activity on several perennial weeds like dandelion, horsetail, red sorrel, fescue, orchardgrass, and wild carrot. For perennial weed control, apply between mid-fall and mid-winter. When dichlobenil is applied at the highest rate, yellow nutsedge, mugwort, quackgrass, and bindweed can be suppressed. Certain species of azaleas, hollies, and pines are sensitive to injury, and applications to firs, hemlocks, and spruce specifically should be avoided. Applications to wet foliage should also be avoided, as injury is likely to occur, and for best results, dichlobenil must be placed on top of the mulch.

Preemergence Combination Herbicides

An oxyflurofen and oryzalin granular combination utilizes both products' strengths and controls a wide spectrum of grasses and broadleaf weeds (Tables 8.2 and 8.3). This combination is applied at 3.0 lb ai/acre (3.36 kg ai/ha). The major weeds controlled are large crabgrass, foxtails, annual ryegrass, bittercress, cudweed, and prickly lettuce. This product needs to be watered in immediately to activate the herbicide and to wash away any product attached to the foliage. It should not be applied to daylilies, liriope, yucca, or hostas. It should be used strictly on woody ornamentals, but is not recommended either for ground covers, flowers, or herbaceous bedding plants or on deciduous plants less than 12 in (30.5 cm) tall. Applications around several landscape plants is prohibited, including *Euonymus japonica* Silver King, *Ilex crenata* Helleri, *Trachelospermum asiaticum* Oblanceolatum, hydrangea, lantana, heather, and hardy star jasmine.

The combination of oxyflurofen and oxadiazon is useful in situations where dinitroanalines are not desirable. Applications should be made at 3 lb ai/acre (3.36 kg ai/ha) to control a wide variety of annual grasses and broadleaf weeds. Contact to wet foliage may cause injury. This herbicide should not be applied to daylilies, liriope, or hostas, as their leaves will funnel granules to the base, causing injury. Annual bluegrass, goosegrass, bittercress, chickweed, fireweed, groundsel, oxalis, horseweed, and many other weeds are controlled with this combination. Watering in is essential in order to activate the herbicide and wash away any adhered granule from the foliage of desired plantings. Deciduous plantings less than 12 in (30.5 cm) tall do not need to be treated with this product.

Oxadiazon and prodiamine granular combination controls many annual grasses including annual bluegrass, crabgrass, goosegrass, crowfootgrass, and some broadleaf weeds including chickweed and spurge. Application rate is 2.4 lb ai/acre (2.69 kg ai/ha). This combination must be watered in immediately following application in order to activate the herbicide. Granules will burn foliage if they adhere; therefore, thorough irrigation is needed following application. Oxadiazon can also be found in combination with napropamide in a granular formulation. Use rates range from 6 to 9 lb ai/acre (6.7 to 10 kg ai/ha). The herbicide controls annual grasses and many broadleaf weeds, including bittercress, chickweed, fillaree, fireweed, groundsel, liverwort, sowthistle, speedwell, and woodsorrel. Immediate irrigation incorporates the herbicide and washes off any material that may have adhered to the foliage.

The combination of isoxaben and trifluralin is widely used for its broad-spectrum control of annual grasses and broadleaf weeds. Use rates range from 2.5 to 5 lb ai/acre (2.8 to 5.6 kg ai/ha) in established labeled plantings. This product can be applied to new transplants, after the soil has settled, by packing or irrigation, with no cracks present, at a rate of 1.5 to 3.0 lb ai/acre (1.68 to 3.36 kg ai/ha). The product should not be applied to bedding plants or to areas where bedding plants will be planted within one year. No more than 600 lb of product/acre (672 kg/ha) can be applied in one calendar year. A similar three-way granular combination preemergence containing trifluralin, isoxaben, and oxyfluorfen controls many broadleaf and annual grasses.

Oryzalin and benefin are found in a granular combination for control of annual grasses like crabgrass, foxtail, and barnyardgrass and a few broadleaf weeds like chickweed, henbit, and carpetweed. Use rates range from 4 to 6 lb ai/acre (4.48 to 6.72 kg ai/ha). This product should not be applied to emerged tulips, and some injury has been reported on pansies at the high labeled rate. Injury has also been reported on melampodium, begonia, celosia, phlox, monarda, and salvia (perennial). Transplanted plants must be settled with no cracks before application, and irrigation is needed to activate the herbicide.

POSTEMERGENCE HERBICIDES USED IN LANDSCAPES

NONSELECTIVE HERBICIDES

Glyphosate is a systemic nonselective herbicide that can be slow to activate, especially at temperatures below 50°F (10°C). Postemergence control of most herbaceous and woody weeds is provided with glyphosate in most woody trees and shrubs as a directed spray only (Table 8.4). Avoid contact with foliage, green stems, and bark of desirable trees and shrubs, as severe injury will likely result. Use rates range from 1 to 5 lb ai/acre (1.12 to 5.6 kg ai/ha). Glyphosate is also available in formulations that are safe for use around waterways, including ponds.

Table 8.4
OPTIMUM APPLICATION RATES AND TIMING OF GLYPHOSATE 4L (BRIGGS AND WHITWELL, 2001)

Weed Species	Glyphosate 4L Rate Which Provides ⩾90% Control One Season Later	Application Timing for Best Control
Perennial grasses (quackgrass, johnsongrass, fescue, etc.)	1%	At time of first flowering
Bermudagrass	2%	At time of first flowering
Composites (asters, goldenrod, etc.)	1%	At time of first flowering
Poison ivy	2%	Two weeks on either side of full bloom (early summer)
Honeysuckle	1 to 1.5%	Full bloom and up to a month after (early summer)
Kudzu	1.5 to 2%	Full bloom and up to a month after (late summer)
Blackberry	1 to 1.5%	Fall and early winter
Sericea lespedeza	1%	Full bloom (midsummer)
Mugwort	1.5 to 2%	Full flower (late summer to early fall)
Trumpetcreeper	1.5%	Late summer to mid-fall before frost
Passion flower (maypop)	1%	Early bloom to first fruit
Virginia creeper	1%	Late summer or first sign of fall color
Clematis vine	1%	After bloom until frost
Wisteria	1.5 to 2% (not on label)	Six to eight weeks after bloom (mid to late summer)
Greenbriar	3% (not on label)	Five fully expanded leaves (early spring)
English ivy	2 or 3% (not on label)	Three to five fully expanded new leaves (early spring)
Japanese knotweed	2% (not on label)	Late summer to early fall but before frost

Diquat is a contact nonselective that is rapidly absorbed by green tissue only. It is used as a directed spray to control most small annual broadleaf and grassy weeds and to suppress perennial weeds. Due to its quick activity, many perennial weeds will not be completely controlled and will regrow. Use rates range from 0.25 to 1 lb ai/acre (0.28 to 1.12 kg ai/ha) in woody trees and shrub landscape beds. Drifted spray will result in undesirable ornamental plant injury. Unlike glyphosate, cool temperatures have little effect on the efficacy of diquat. Diquat is available for aquatic applications. A new product is available that combines the strengths of glyphosate and diquat in a wettable granular formulation.

Similar to diquat, paraquat is a contact nonselective that is rapidly absorbed by the plant, desiccating all green tissue. Paraquat can be used in and around shade and ornamental trees. It controls most annual weeds, but will only suppress perennial weeds. Paraquat should be used as a directed spray around trees at 0.64 to 1.0 lb ai/acre (0.72 to 1.12 kg ai/ha). Drift and any contact with desirable green foliage or stems should be avoided, as injury will occur. The addition of an nonionic surfactant or crop oil is required. This product should not be used around home gardens, schools, or recreational areas. Paraquat is a restricted-use pesticide with a danger label.

Glufosinate is another contact nonselective that has quick activity, with visual effects seen within two to four days after application. It controls a broad spectrum of annual and perennial grasses and broadleaf weeds. Similar to diquat, glufosinate will only suppress perennial weeds, which will regenerate from their underground portions. It is used as a directed spray only at rates ranging from 0.75 to 1.5 lb ai/acre (0.84 to 1.68 kg ai/ha) in woody trees or shrub landscape beds. Contact to desirable foliage, tree bark, and green tissue can cause injury. Glufosinate has shown better efficacy at higher spray volumes (e.g., 90 GPA, 842 L/ha).

Pelargonic acid is another contact nonselective that is extremely rapid in its activity. If the product is applied on hot days, activity can be seen within hours of application, due to its mode of action of disrupting cell membranes and causing rapid cell desiccation. It controls most young, succulent annual weeds, but will only burn back perennial and older annual weeds. Use rates range from 6.5 to 13.0 fl oz/gal (50 to 100 mL/L). Plargonic acid is to be used as a directed spray only in flower, woody tree, and shrub landscape beds.

Selective Grass Herbicides

Selective herbicides for various weed control in woody and herbaceous ornamentals are listed in Tables 8.5 through 8.7. Fenoxaprop-ethyl controls most annual and a few perennial grasses. Crabgrass, goosegrass, barnyardgrass, foxtails, panicums, and Johnsongrass can all be controlled by fenoxaprop-ethyl. Bermudagrass can be suppressed with multiple applications. Rate ranges from 0.016 to 0.47 lb ai/acre (0.018 to 0.53 kg ai/ha); the low rate should be used to control untillered annual grasses. Fenoxaprop-ethyl can be safely used on a wide variety of woody trees, shrubs, ground covers, herbaceous perennials, flowers, and bulbs.

Clethodim and sethoxydim are both available for a wide variety of woody trees, shrubs, ground covers, herbaceous perennials, flowers, and bulbs. Annual grasses like crabgrass, barnyardgrass, crowfoot grass, foxtails, and field sandbur are all controlled by these two products. Perennial grasses like bermudagrass and Johnsongrass can also be controlled with multiple applications. Clethodim is the more active of the two on annual bluegrass. Use rate for clethodim is 0.06 to 0.25 lb ai/acre (0.067 to 0.28 kg ai/ha), while sethoxydim rate ranges from 0.3 to 0.5 lb ai/acre (0.336 to 0.56 kg ai/ha) . Both of these products will require the use of an 80 percent nonionic surfactant at a rate of 1 pt/50 gal (473 mL/189 L) or 0.25% v/v and are rainfast within one to two hours.

Fluazifop is similar in activity to the previously mentioned products. Annual and perennial grasses can be controlled by this herbicide. Fluazifop is labeled for a wide variety of ornamentals. Use rates range from 0.25 to 0.4 lb ai/acre (0.28 to 0.448 kg ai/ha), and the addition a nonionic surfactant at 0.25% v/v is required. Fluazifop has good to excellent activity on the control of bermudagrass, especially with repeat applications. Annual bluegrass, however, has considerable tolerance to fluazifop.

Selective Broadleaf Herbicides

Clopyralid is a selective broadleaf herbicide available for selected trees and shrubs (Tables 8.5 through 8.7). Some ornamentals are very sensitive to clopyralid, especially those in the legume family (locust, mimosa, and redbud) and composite family (daisies and sunflowers). Clopyralid is effective on clovers, vetch, goldenrod, and thistles. Use rates range from 0.09 to 0.25 lb ai/a (0.1 to 0.28 kg ai/ha). This product is restricted for commercial use only.

Simazine controls most broadleaf weeds and some annual grasses in woody ornamentals, especially junipers. Use rates range from 2 to 4 lb ai/acre (2.24 to 4.48 kg ai/ha), but applications should not be made within one year after transplanting. Several major broadleaf weeds are controlled, including chickweed, horseweed, wild mustard, pigweed, lambsquarters, groundsel, and ragweed. Annual bluegrass is the major annual grassy weed controlled, with crabgrass, panicum, and foxtail all being somewhat tolerant to simazine, especially at low rates.

Table 8.5
Postemergence Herbicides Used in Ornamentals

Key:
Ratings based on label:
G = good control (80–100%),
F = fair control (50–80%)

Genus Species	Common Name	Selective										Non-Selective				
		Bentazon	Clethodim	Clopyralid	Dicholbenil*	Fenoxaprop	Fluazifop	Halosulfuron	Imazaquin	Pronamide	Sethoxydim	Diquat	Glufosinate	Glyphosate	Paraquat	Pelargonic Acid
Grasses (or Grasslike)																
Aegilops cylindrica	Goatgrass, jointed									G				G		
Agrostis spp.	Bentgrass									G	G			F		
Allium canadense	Onion, wild								G				G			
Allium vineale	Garlic, wild								G							
Alopecurus carolinianus	Foxtail, Carolina													G		
Apera spica-venti)	Windgrass												G			
Avena fatua	Oats, wild		G				G			G	G		G	G		
Axonopus affinis	Carpetgrass												G			
Brachiaria platyphylla	Signalgrass, broadleaf		G				G				G			G		
Brachiaria subquadripara	Alexandergrass, smallflowered												G			
Bromus carinatus	Brome, California		G													
Bromus catharticus	Rescuegrass															
Bromus inermus	Brome, smooth				G								G	G		
Bromus japonicus	Brome, Japanese													G		
Bromus rigidus	Brome, rigput		G													
Bromus secalinus	Cheat or Chess		G											G		
Bromus tectorum	Brome, downy		G				G			G	G		G	G		
Cenchrus spp.	Sandbur species						G						G			
Cenchrus incertus	Sandbur, field		G			G	G		G		G		G	G		
Cynodon dactylon	Bermudagrass		G			F	G				G		G	G		
Cyperus compressus	Sedge, annual								F							
Cyperus esculentus	Nutsedge, yellow	F			G			G	F				G	G		
Cyperus globosus	Sedge, globe								F							

(continued)

Table 8.5
POSTEMERGENCE HERBICIDES USED IN ORNAMENTALS *(CONTINUED)*

Key:
Ratings based on label:
G = good control (80–100%),
F = fair control (50–80%)

Genus Species	Common Name	Selective										Non-Selective				
		Bentazon	Clethodim	Clopyralid	Dicholbenil*	Fenoxaprop	Fluazifop	Halosulfuron	Imazaquin	Pronamide	Sethoxydim	Diquat	Glufosinate	Glyphosate	Paraquat	Pelargonic Acid
Grasses (or Grasslike)																
Cyperus iria	Sedge, rice flat								F							
Cyperus rotundifolia	Nutsedge, purple							G	F				G	G		
Dactylis glomerata	Orchardgrass				G					G	G			G		
Dactyloctenium aegyptium	Crowfootgrass		G													
Digitaria abscendens	Crabgrass, hairy		G		G				F				G	G		
Digitaria bicornis	Crabgrass, tropical				G		G		F				G	G		
Digitaria ciliaris	Crabgrass, Southern		G		G		G		F				G	G		
Digitaria ischaemum	Crabgrass, smooth		G		G	G	G		F		G		G	G		
Digitaria sanguinalis	Crabgrass, large		G		G	G	G		G		G		G	G		
Echinochloa colunum	Junglerice		G				G				G			G		
Echinochloa crus-galli	Barnyardgrass		G			G	G				G		G	G		
Eleusine indica	Goosegrass		G			G	G				G		G	G		
Elytrigia repens	Quackgrass (from seed)		G		G		G			G	G		G	G		
Elytrigia repens	Quackgrass (from rhizome)		G		G		G				G		G	G		
Equisetum spp.	Horsetail (Equisetum)				G								G			
Eragrostis spp.	Lovegrass												G			
Eragrostis cilianensis	Stinkgrass		G								G		G	G		
Eremochloa ophiuroides	Centipedegrass										G					
Eriochloa contracta	Cupgrass, prairie						G						G			
Eriochloa gracilis	Cupgrass, Southwestern		G				G						G			
Eriochloa villosa	Cupgrass, wooly		G				G				G		G	G		
Festuca arundinaceae	Fescue, tall (from seed)				G				F	G	G		G	G		
Glyceria obtusa	Mannagrass, Eastern													G		

(continued)

Table 8.5
(Continued)

Key:
Ratings based on label:
G = good control (80–100%),
F = fair control (50–80%)

		Selective										Non-Selective				
Genus Species	**Common Name**	**Bentazon**	**Clethodim**	**Clopyralid**	**Dicholbenil***	**Fenoxaprop**	**Fluazifop**	**Halosulfuron**	**Imazaquin**	**Pronamide**	**Sethoxydim**	**Diquat**	**Glufosinate**	**Glyphosate**	**Paraquat**	**Pelargonic Acid**
Grasses (or Grasslike)																
Holcus lanatus	Velvetgrass, common									G				G		
Holcus mollis	Velvetgrass, German										G			G		
Hordeum jubatum	Barley, foxtail		G							G						
Hordeum leporinum	Barley, wild (hare)				G								G			
Hordeum pusillum	Barley, little													G		
Imperata cylindrica	Cogongrass													**G**		
Kyllinga brevifolius	Kyllinga, green							F	F							
Leptochloa spp.	Sprangletop					G								G		
Leptochloa fascicularis	Sprangletop, bearded		G											G		
Leptochloa filiformis	Sprangletop, red		G								G			G		
Leptochloa panicoides	Sprangletop, Amazon		G											G		
Leptochloa uninervia	Sprangletop, Mexican		G											G		
Lolium multiflorum	Ryegrass, annual		G				G			G	G		G	G		
Lolium perenne	Ryegrass, perennial								F	G			G	G		
Lolium remotum	Ryegrass, hardy		G										G	G		
Melinis repens	Natalgrass				G											
Muhlenbergia frondosa	Muhly, wirestem		G				G				G			G		
Oryza satiza	Red rice		G				G							G		
Panicum capillare	Witchgrass		G		G	G	G				G			G		
Panicum dichotomiflorum	Panicum, fall		G			G	G				G		G	G		
Panicum fasciculatum	Panicum, browntop					G					G			G		
Panicum maximum	Guineagrass, seedling					G	G						G	G		
Panicum miliaceum	Millet, wild Proso		G			G	G				G					

(continued)

Table 8.5

POSTEMERGENCE HERBICIDES USED IN ORNAMENTALS *(CONTINUED)*

Key:
Ratings based on label:
G = good control (80–100%),
F = fair control (50–80%)

Genus Species	Common Name	Selective										Non-Selective				
		Bentazon	**Clethodim**	**Clopyralid**	**Dichlobenil***	**Fenoxaprop**	**Fluazifop**	**Halosulfuron**	**Imazaquin**	**Pronamide**	**Sethoxydim**	**Diquat**	**Glufosinate**	**Glyphosate**	**Paraquat**	**Pelargonic Acid**
Grasses (or Grasslike)																
Panicum repens	Torpedograss					G	F				G		G	F		
Panicum texanum	Panicum, Texas		G		G	G	G				G			G		
Pascopyrum smithii	Wheatgrass, Western													G		
Paspalum dilatatum	Dallisgrass												G	G		
Paspalum notatum	Bahiagrass								F		G		G	G		
Paspalum urvillei	Vaseygrass												G	G		
Pennisetum clandestinum	Kikuyugrass						G							G		
Pennisetum purpureum	Napiergrass													G		
Phalaris arundinacea	Canarygrass, reed													G		
Phalaris canariensis	Canarygrass		G													
Phleum pratense	Timothy				G									G		
Poa annua	Bluegrass, annual		G		G				F	G	G		G	G		
Poa bulbosa	Bluegrass, bulbous									G				G		
Poa pratensis	Bluegrass, Kentucky				G					G			G	G		
Polypogon monspeliensis	Rabbitfootgrass		G				G									
Rottboellia cochinchinensis	Itchgrass		G				G							G		
Setaria faberi	Foxtail, giant		G		G	G	G				G		G	G		
Setaria glauca/viridis	Foxtail (yellow, green)		G		G	G	G				G		G	G		
Sorghum bicolor	Shattercane		G				G				G		G	G		
Sorghum halapense	Johnsongrass (rhizome)		G			F	G				G		G	G		
Sorghum halapense	Johnsongrass (seedling)		G			G	G				G			G		
Sporobolus indicus	Smutgrass															

(continued)

Table 8.5
(CONTINUED)

Key:
Ratings based on label:
G = good control (80–100%),
F = fair control (50–80%)

Genus Species	Common Name	Selective										Non-Selective				
		Bentazon	Clethodim	Clopyralid	Dichlobenil*	Fenoxaprop	Fluazifop	Halosulfuron	Imazaquin	Pronamide	Sethoxydim	Diquat	Glufosinate	Glyphosate	Paraquat	Pelargonic Acid
Grasses (or Grasslike)																
Taeniatherum caput-medusae	Medusahead													G		
Urochloa mutica	Paragrass												G	G		
Broadleaf Weeds																
Abrus precatorius	Rosarypea				G											
Acalypha virginica	Copperleaf, Virginia												G	G		
Acanthospermum hispidum	Starbur, bristly															
Acroptilon repens	Knapweed, Russian			F	G									G		
Alchemilla arvensis	Parsley-piert								G							
Amaranthus spp.	Pigweed spp.				G				G				G	G		
Amaranthus palmeri	Amaranth, palmer															
Amaranthus rudis	Waterhemp, common															
Ambrosia artemisifolia	Ragweed, common	G		G	G								G	G		
Ambrosia grayi	Bursage, woolyleaf													G		
Ambrosia trifida	Ragweed, giant	G		G	G								G	G		
Amsinckia spp.	Fiddleneck				G								G	G		
Anagallis arvensis	Pimpernel, scarlet			G												
Anoda caristata	Anoda, spurred	G												G		
Anthemis cotula	Mayweed, stinking (chamomile)			G										G		
Antriscus caucalis	Chervil													G		
Apocynum cannabinum	Dogbane, hemp												G	G		
Arabis canadensis	Sickelpod			G										G		
Arbutilon theophrasti	Velvetleaf												G	G		

(continued)

Table 8.5
Postemergence Herbicides Used in Ornamentals (*Continued*)

Key:
Ratings based on label:
G = good control (80–100%),
F = fair control (50–80%)

Genus Species	Common Name	Selective										Non-Selective				
		Bentazon	Clethodim	Clopyralid	Dicholbenil*	Fenoxaprop	Fluazifop	Halosulfuron	Imazaquin	Pronamide	Sethoxydim	Diquat	Glufosinate	Glyphosate	Paraquat	Pelargonic Acid
Broadleaf Weeds																
Arctium minus	Burdock, common			G									G			
Artemesia spp.	Artemesia (sagebrush, wormwood, sage wort)				G											
Artemesia vulgaris	Mugwort (from rhizome)				G								G			
Artemesia vulgaris	Mugwort (from seed)				G											
Asclepias spp.	Milkweed species													G		
Aster spp.	Aster species			G	G								G			
Atriplex hortensis	Orach, red															
Barbarea vulgaris	Rocket, yellow				G								G	G		
Bassia hyssopifolia	Bassia, five-hook													G		
Berteroa incana	Alyssum, hoary															
Bidens bipinnata	Spanishneedles				G									G		
Brassica kaber	Mustard, wild	G			G								G	G		
Brassica rapa	Mustard, birdsrape (wild turnip)												G			
Brunnichia ovata	Redvine													F		
Bryum spp.	Moss															
Buglossoides arvensis	Gromwell, field (corn)												G			
Calandrinia ciliata	Redmaids (Dessert rock purslane)															
Calystegia sepium	Bindweed, hedge															
Camelina microcarpa	Falseflax, smallseed													G		
Capsella bursa-pastoris	Shepherdspurse	G			G								G	G		
Cardamine hirsuta	Bittercress, hairy								G					G		
Cardamine oligosperma	Bittercress, lesser													G		

(continued)

Table 8.5
(CONTINUED)

Key:
Ratings based on label:
G = good control (80–100%),
F = fair control (50–80%)

		Selective										Non-Selective				
Genus Species	**Common Name**	**Bentazon**	**Clethodim**	**Clopyralid**	**Dicholbenil***	**Fenoxaprop**	**Fluazifop**	**Halosulfuron**	**Imazaquin**	**Pronamide**	**Sethoxydim**	**Diquat**	**Glufosinate**	**Glyphosate**	**Paraquat**	**Pelargonic Acid**
Broadleaf Weeds																
Cardamine pennsylvanica	Bittercress, Pennsylvania													G		
Cardiospermum halicacabum	Balloonvine	G														
Carduus nutans	Thistle, musk	F		G									G			
Cassia occidentalis	Senna, coffee	G		G	G											
Centaurea cyanus	Cornflower			G												
Centaurea diffusa	Knapweed, diffuse			G										G		
Centaurea maculosa	Knapweed, spotted			G										G		
Centaurea solstitialis	Starthistle, yellow			G										G		
Cerastium vulgatum	Chickweed, mouseear								G	G						
Chenopodium spp.	Goosefoot spp.				G											
Chenopodium album	Lambsquarters	G			G								G	G		
Chorispora tenella	Mustard, blue													G		
Chrysanthemum leucanthemum	Daisy, oxeeye			G												
Cirsium arvense	Thistle, Canada	F		G	G								G	G		
Cirsium vulgare	Thistle, bull			G	G											
Citrullus lanatus	Citron melon				G											
Commelina communis	Dayflower	G														
Convolvulus arvensis	Bindweed, field												G	G		
Conyza bonariensis	Fleabane, hairy													G		
Conyza canadensis	Horseweed (marestail)			G									G	G		
Coreopsis spp.	Coreopsis, Plains, tickseed													G		
Coronopus spp.	Swinecress															

(continued)

Table 8.5
Postemergence Herbicides Used in Ornamentals *(Continued)*

Key:
Ratings based on label:
G = good control (80–100%),
F = fair control (50–80%)

Genus Species	Common Name	Selective										Non-Selective				
		Bentazon	Clethodim	Clopyralid	Dichlobenil*	Fenoxaprop	Fluazifop	Halosulfuron	Imazaquin	Pronamide	Sethoxydim	Diquat	Glufosinate	Glyphosate	Paraquat	Pelargonic Acid
Broadleaf Weeds																
Crepis tectorum	Hawksbeard, narrowleaf			G												
Croton glandulosus	Croton, tropic															
Cuscuta spp.	Dodder															
Cynara cardunculus	Thistle, artichoke													G		
Datura stramonium	Jimsonweed			G									G			
Daucus carota	Carrot, wild			G	G											
Descurainia sophia	Flixweed															
Desmodium tortuosum	Beggarweed, Florida															
Deuscurainia pinnata	Tansymustard												G	G		
Diodia virginiana	Buttonweed, Virginia			F					F							
Dipsacus fullonum	Teasel, common			G												
Eclipta prostrata	Eclipta								G					G		
Epilobium spp.	Willoweed; fireweed															
Erechtites hieracifolia	Fireweed (Am. Burnweed)			G												
Erigeron annuus	Fleabane												G	G		
Erigeron strigosus	Fleabane, rough													G		
Erodium botrys	Filaree, broadleaf												G	G		
Eupatorium capillifolium	Dogfennel			G	G											
Erodium cicutarium	Filaree, redstem												G	G		
Euphorbia	Spurge, annual				G									G		
Euphorbia esula	Spurge, leafy				G								G	F		
Euphorbia heterophylla	Poinsettia, wild	G														
Euphorbia hirta	Spurge, Garden				G											

(continued)

Table 8.5
(Continued)

Key:
Ratings based on label:
G = good control (80–100%),
F = fair control (50–80%)

Genus Species	Common Name	Selective										Non-Selective				
		Bentazon	Clethodim	Clopyralid	Dicholbenil*	Fenoxaprop	Fluazifop	Halosulfuron	Imazaquin	Pronamide	Sethoxydim	Diquat	Glufosinate	Glyphosate	Paraquat	Pelargonic Acid
Broadleaf Weeds																
Euphorbia humistrata	Spurge, prostrate				G									G		
Euphorbia hyssopifolia	Spurge, hyssop				G											
Euphorbia maculata	Spurge, spotted				G									G		
Euphorbia nutans	Spurge, nodding				**G**											
Fatuoa villosa	Mulberry weed															
Foeniculum spp.	Anise (fennel)													G		
Galinsoga ciliata	Galinsoga, hairy	G		G												
Galium aparine	Bedstraw															
Geranium carolinianum	Geranium, Carolina								G					G		
Geranium molle	Dovetail								G							
Gisekia spp.	Gisekia				G											
Glechoma hederacea	Ivy, ground															
Gnaphalium spp.	Cudweed			G	G				F							
Helianthus annuus	Sunflower	G		G										G		
Helianthus ciliaris	Blueweed, Texas													G		
Helianthus tuberosus	Artichoke, Jerusalem			G	G									G		
Heterotheca subaxillaris	Camphorweed				G											
Hieracium aurantiacum	Hawkweed, yellow			G												
Hieracium pratense	Hawkweed, orange			G												
Holosteum umbellatum	Spurry, umbrella													G		
Hydrocotyl spp.	Pennywort; Dollarweed			F					F							
Hypochoeris radicata	Dandelion, false				G									G		
Indigofera hirsuta	Indigo, hairy															

(continued)

Table 8.5
Postemergence Herbicides Used in Ornamentals *(Continued)*

Key:
Ratings based on label:
G = good control (80–100%),
F = fair control (50–80%)

Genus Species	Common Name	Selective										Non-Selective				
		Bentazon	Clethodim	Clopyralid	Dicholbenil*	Fenoxaprop	Fluazifop	Halosulfuron	Imazaquin	Pronamide	Sethoxydim	Diquat	Glufosinate	Glyphosate	Paraquat	Pelargonic Acid
Broadleaf Weeds																
Inula spp.	Inula species			G												
Ipomoea coccinea	Morning glory, red, scarlet													G		
Ipomoea hederacea	Morning glory, ivyleaf													G		
Ipomoea hederacea var. integriuscula	Morning glory, entireleaf													G		
Ipomoea purpurea	Morning glory, annual, tall													G		
Iva xanthifolia	Marshelder			G												
Jacquemontia tamnifolia	Morning glory, smallflower															
Kochia scoparia	Kochia												G	G		
Krigia spp.	Dandelion, dwarf													G		
Lactuca serriola	Lettuce, prickly			G									G	G		
Lamium amplexicaule	Henbit				G				G					G		
Lamium purpureum	Deadnettle, purple				G				G							
Lepidium latifolium	Pepperweed, perennial													G		
Lepidium perfoliatum	Pepperweed, yellowflower															
Lepidium virginicum	Pepperweed, Va.				G									G		
Lespedeza spp.	Lespedeza													G		
Lysimachia angusta	Loosestrife															
Malva parviflora	Cheeseweed			G									G	G		
Malva neglecta	Mallow, common												G			
Marchantia spp.	Liverwort															
Matricaria matricarioides	Pineappleweed			G	G											
Medicago lupulina	Medic, black			G					G							

(continued)

Table 8.5
(Continued)

Key:
Ratings based on label:
G = good control (80–100%),
F = fair control (50–80%)

		Selective										Non-Selective				
Genus Species	**Common Name**	**Bentazon**	**Clethodim**	**Clopyralid**	**Dicholbenil***	**Fenoxaprop**	**Fluazifop**	**Halosulfuron**	**Imazaquin**	**Pronamide**	**Sethoxydim**	**Diquat**	**Glufosinate**	**Glyphosate**	**Paraquat**	**Pelargonic Acid**
Broadleaf Weeds																
Medicago polymorpha	Burclover															
Melilotus spp.	Clover, sweet			G									G			
Melochia corchorifolia	Redweed	G														
Mollugo verticillata	Carpetweed				G				G							
Montia perfoliata	Miners lettuce				G											
Morrenia odorata	Milkweed vine				G											
Murdannia nudiflora	Doveweed															
Oenethera spp.	Eveningprimrose				G				G							
Ostrya virginiana	Hophornbeam													G		
Oxalis corniculata	Woodsorrel, creeping												G			
Oxalis stricta	Woodsorrel, yellow				G								G			
Oxytropis sericea	Locoweed, white			G												
Passiflora incarnata	Maypops				G											
Phyllanthus tenellus	Phyllanthus, long stalked															
Physalis spp.	Groundcherry species															
Phytolacca americana	Pokeweed, common															
Plantago spp.	Plantain (from seed)				G								G			
Plantago lanceolata	Plantain, narrowleaf (buckhorn)			G	G								G			
Plantago major	Plantain, broadleaf			F	G								G			
Polygonum spp.	Knotweed spp.				G									G		
Polygonum aviculare	Knotweed, prostrate			G	G									G		
Polygonum convolvulus	Buckwheat, wild	G		G									G			
Polygonum lapathifolium	Ladysthumb, pale smartweed	G		F										G		

(continued)

Table 8.5
Postemergence Herbicides Used in Ornamentals (*Continued*)

Key: Ratings based on label: G = good control (80–100%), F = fair control (50–80%)		Selective										Non-Selective				
Genus Species	**Common Name**	**Bentazon**	**Clethodim**	**Clopyralid**	**Dicholbenil***	**Fenoxaprop**	**Fluazifop**	**Halosulfuron**	**Imazaquin**	**Pronamide**	**Sethoxydim**	**Diquat**	**Glufosinate**	**Glyphosate**	**Paraquat**	**Pelargonic Acid**
Broadleaf Weeds																
Polygonum pennsylvanicum	Smartweed, Pennsylvania	G		G	G								G	G		
Portulaca oleracea	Purslane, common	G			G								G	G		
Probiscidea louisianica	Devilsclaw	G														
Prosopis spp.	Mesquite			F												
Pteridium aquilinum	Bracken fern													G		
Pueraria lobata	Kudzu			F												
Pyllanthus urinaria	Chamberbitter (niruri)															
Ranunculus spp.	Buttercup			F					G					G		
Raphanus raphanistrum	Radish, wild				G											
Rhus radicans	Ivy, poison												G	G		
Richardia scabra	Pusley, Florida				G									G		
Rosa multiflora	Rose, wild												G	G		
Rumex acetosella	Sorrel, red			G					G							
Rumex crispus, obtusifolius	Dock, curly or broadleaf			G	G								G	G		
Sagina procumbens	Pearlwort, birdseye															
Salsola iberica	Thistle, Russian				G								G	G		
Scleranthus annuus	Knawel								G							
Senecio jacobaea	Ragwort, tansy			G												
Senecio vulgaris	Groundsel, common	F		G	G									G		
Sesbania exaltata	Sesbania, hemp	G												G		
Sibara virginica	Sibara															
Sida spinosa	Sida, prickly	G			G									G		
Sisymbrium altissimum	Mustard, tumble													G		

(continued)

Table 8.5
(Continued)

Key:
Ratings based on label:
G = good control (80–100%),
F = fair control (50–80%)

Genus Species	Common Name	Bentazon	Clethodim	Clopyralid	Dicholbenil*	Fenoxaprop	Fluazifop	Halosulfuron	Imazaquin	Pronamide	Sethoxydim	Diquat	Glufosinate	Glyphosate	Paraquat	Pelargonic Acid
		Selective										Non-Selective				
Broadleaf Weeds																
Sisymbrium irio	Rocket, London												G	G		
Solanum carolinense	Horsenettle													G		
Solanum eleagnifolium	Nightshade, silverleaf												G	G		
Solanum ptycanthum	Nightshade, Eastern black			G									G			
Solanum nigrum	Nightshade, black			G									G	G		
Solanum rostratum	Buffalo bur			F									G			
Solanum sarrachoides	Nightshade, hairy			G									G			
Solanum triflorum	Nightshade, cutleaf			G									G			
Solidago spp.	Goldenrod			G									G			
Soliva pterosperma	Burweed, lawn	G		G					G							
Sonchus arvensis	Sowthistle, perennial			F												
Sonchus oleraceus	Sowthistle, annual			G									G	G		
Spergula arvensis	Spurry, corn			G												
Spergularia rubra	Spurry, red sand															
Stachys floridanum	Betony, Florida															
Stellaria media	Chickweed, common				G				G	G			G	G		
Taraxacum officinale	Dandelion (seedling)			G	G				F				G	G		
Thlaspi arvense	Pennycress, field												G	G		
Toxicodendron toxicarium	Oak, poison												G	G		
Tragopogon pratensis	Salsify, meadow			G												
Tribulus terrestris	Puncturevine													G		
Trifolium aureum	Clover, hop			G									G			
Trifolium pratense	Clover, red			G									G	G		

(continued)

Table 8.5
POSTEMERGENCE HERBICIDES USED IN ORNAMENTALS *(CONTINUED)*

Key:
Ratings based on label:
G = good control (80–100%),
F = fair control (50–80%)

Genus Species	Common Name	Selective										Non-Selective				
		Bentazon	Clethodim	Clopyralid	Dicholbenil*	Fenoxaprop	Fluazifop	Halosulfuron	Imazaquin	Pronamide	Sethoxydim	Diquat	Glufosinate	Glyphosate	Paraquat	Pelargonic Acid
Broadleaf Weeds																
Trifolium repens	Clover, white			G					G				G	G		
Utrica dioica (gracilis, procera)	Nettle, stinging												G			
Urtica urens	Nettle, burning												G			
Verbascum thapsus	Mullein, common													G		
Verbena spp.	Vervain												G			
Veronica spp.	Speedwell spp.			G												
Veronica arvensis	Speedwell, corn													G		
Veronica peregrina	Speedwell, Purslane													G		
Veronica persica	Speedwell, birdseye (Persian)															
Veronica serpyllifolia	Speedwell, thymeleaf															
Vicia sativa	Vetch, common			G												
Viola spp.	Violet; pansy								F							
Xanthium strumarium	Cocklebur, common			G									G	G		
Youngia japonica	Hawksbeard, Asiatic															

*See text for other trade names of this product
Authors: J. C. Neal, J. A. Briggs, and T. Whitwell. 2005. North Carolina State University, Raleigh, North Carolina, and Clemson University, Clemson, South Carolina.

Table 8.6

POSTEMERGENCE HERBICIDES USED IN WOODY ORNAMENTALS

Key:
D = registered for directed applications.
OT = registered for over-the-top use.
D* = registered for directed applications on certain species within the genus; consult label for details.
OT* = registered for over-the-top applications on certain species within genus; consult label for details.

		Selective										Non-Selective				
Genus Species	**Common Name**	Bentazon	Clethodim	Clopyralid	Dichobenil*	Fenoxaprop	Fluazifop	Halosulfuron	Imazaquin	Pronamide	Sethoxydim	Diquat	Glufosinate	Glyphosate	Paraquat	Pelargonic Acid
Abelia spp.	Abelia species	D	OT					D				D	D	D		D
A. × grandiflora	Abelia, glossy	D					OT	D			OT	D	D	D		D
Abeliophyllum distichum	Abelialeaf, Korean	D						D				D	D	D		D
Abies spp.	Fir species	D					OT*	D		OT	OT*	D	D	D	D	D
A. balsamea	Fir, balsam	D		OT				D			OT	D	D	D		D
A. concolor	Fir, white	D					OT	D			OT	D	D	D		D
A. fraseri	Fir, fraser	D		OT			D	D			OT	D	D	D		D
A. grandis	Fir, grand	D		OT				D			OT	D	D	D		D
A. koreana	Fir, Korean	D						D				D	D	D		D
A. procera	Fir, noble	D		OT			OT	D			OT	D	D	D		D
Acacia spp.	Acacia species	D					OT*; D*	D			OT*	D	D	D		D
Acer spp.	Maple species	D	OT		OT			D		OT		D	D	D	D	D
A. ginnala	Maple, amur	D					OT*	D				D	D	D		D
A. griseum	Maple, paperbark	D						D				D	D	D		D
A. negundo	Boxelder	D			OT			D				D	D	D		D
A. palmatum	Maple, Japanese	D				OT	OT	D			OT	D	D	D		D
A. pennsylvanicum	Maple, striped	D						D				D	D	D		D
A. platanoides	Maple, Norway	D					OT	D				D	D	D		D
A. rubrum	Maple, red	D		D		OT	D	D			OT	D	D	D		D
A. saccharinum	Maple, silver	D					OT	D			OT	D	D	D		D
A. saccharum	Maple, sugar	D					OT	D				D	D	D		D
Actinidia deliciosa	Kiwi	D						D				D	D	D	D	D
Aesculus spp.	Buckeye species	D						D		OT*		D	D	D		D
Akebia quintata	Akebia, fiveleaf	D						D				D	D	D		D

(continued)

Table 8.6

POSTEMERGENCE HERBICIDES USED IN WOODY ORNAMENTALS *(CONTINUED)*

Key:
D = registered for directed applications.
OT = registered for over-the-top use.
D* = registered for directed applications on certain species within the genus; consult label for details.
OT* = registered for over-the-top applications on certain species within genus; consult label for details.

		Selective										Non-Selective				
Genus Species	**Common Name**	**Bentazon**	**Clethodim**	**Clopyralid**	**Dichobenil***	**Fenoxaprop**	**Fluazifop**	**Halosulfuron**	**Imazaquin**	**Pronamide**	**Sethoxydim**	**Diquat**	**Glufosinate**	**Glyphosate**	**Paraquat**	**Pelargonic Acid**
Amelanchier laevis	serviceberry	D						D			OT	D	D	D	D	D
Arctostaphylos uva-ursi	bearberry (kinnikinick)	D			OT		OT	D				D	D	D		D
Aucuba spp.	golddust plant	D	OT					D				D	D	D		D
Aucuba japonica	aucuba, Japanese	D					OT	D				D	D	D		D
Berberis spp.	barberry species	D	OT*		OT	OT		D		OT	OT*	D	D	D		D
B. × gladwynensis	barberry	D						D				D	D	D		D
B. julianae	barberry, wintergreen	D						D				D	D	D		D
B. × mentorensis	barberry, mentor	D					OT	D				D	D	D		D
B. thunbergii	barberry, Japanese	D	OT				OT	D			OT	D	D	D		D
B. verruculosa	barberry, warty	D						D				D	D	D		D
Betula spp.	birch species	D			OT			D		OT	OT*	D	D	D	D	D
B. nigra	birch, river	D	OT				D	D			OT	D	D	D		D
B. papyrifera	birch, paper	D	OT					D			OT	D	D	D		D
B. pendula	birch, Eur. white	D	OT				OT	D			OT	D	D	D		D
Bougainvillea spp.	bougainvillea	D					OT	D			OT*					
Buddleia spp.	butterfly bush	D						D				D	D	D		D
Buxus spp.	boxwood species	D			OT			D		OT		D	D	D		D
B. harlandii	boxwood, Harland	D						D				D	D	D		D
B. microphylla	boxwood	OT*		OT		OT*	OT*	D			OT*	D	D	D		D
B. sempervirens	boxwood	D	OT				OT	D			OT	D	D	D		D
Callistemon spp.	bottlebrush	D	OT*				OT*; D*	D			OT*	D	D	D		D
Calluna spp.	heather species	D			OT		OT*	D				D	D	D		D
Camellia spp.	camellia species	D			OT			D				D	D	D		D
C. japonica	camellia, Japanese	D	OT				OT	D			OT	D	D	D		D

(continued)

Table 8.6
(CONTINUED)

Key:
D = registered for directed applications.
OT = registered for over-the-top use.
D* = registered for directed applications on certain species within the genus; consult label for details.
OT* = registered for over-the-top applications on certain species within genus; consult label for details.

		Selective										Non-Selective				
Genus Species	**Common Name**	**Bentazon**	**Clethodim**	**Clopyralid**	**Dichobenil***	**Fenoxaprop**	**Fluazifop**	**Halosulfuron**	**Imazaquin**	**Pronamide**	**Sethoxydim**	**Diquat**	**Glufosinate**	**Glyphosate**	**Paraquat**	**Pelargonic Acid**
C. sasanqua	camellia, Sasanqua	D					OT	D			OT	D	D	D		D
Campsis × tagliabuana	trumpetcreeper	D						D				D	D	D		D
Caragana arborescens	peashrub, Siberian	D			OT			D				D	D	D		D
Carissa grandiflora	plum, Natal	D					OT	D			OT*	D	D	D		D
Carya illinoinensis	pecan	D			OT			D			OT	D	D	D	D	D
Cassia spp.	senna	D					OT*; D*	D			OT*	D	D	D		D
Castanea spp.	chestnut species	D						D				D	D	D	D	D
C. mollissima	chestnut, Chinese	D						D				D	D	D		D
Ceanothus spp.	New Jersey tea	D			OT		D*	D			OT*	D	D	D		D
Cedrus spp.	cedar species	D						D		OT		D	D	D	D	D
C. atlantica	cedar, Atlas	D						D				D	D	D		D
C. deodara	cedar, deodar	D						D				D	D	D		D
Celtis occidentalis	hackberry	D			OT		OT	D			OT	D	D	D		D
Cercis canadensis	redbud, eastern	D	OT				OT	D		OT		D	D	D	D	D
Chaenomeles speciosa	quince, flowering	D			OT		OT	D		OT		D	D	D		D
Chamaecyparis spp.	falsecypress	D			OT*	OT	OT*	D				D	D	D	D	D
C. obtusa	falsecypress, Hinoki	D					OT*	D				D	D	D		D
C. pisifera	falsecypress, Sawara	D					OT*; D*	D			OT	D	D	D		D
Chamaedorea spp.	palm	D						D				D	D	D		D
Cheirodendron spp.	lapalapa	D						D				D	D	D		D
Cistus spp.	rockrose	D			OT		OT*	D			OT*	D	D	D		D
Citrus spp.	citrus, ornamental	D					OT*	D			OT*	D	D	D	D	D
Clethra spp.	clethra species	D						D				D	D	D		D

(continued)

Table 8.6
POSTEMERGENCE HERBICIDES USED IN WOODY ORNAMENTALS (*CONTINUED*)

Key:
D = registered for directed applications.
OT = registered for over-the-top use.
D* = registered for directed applications on certain species within the genus; consult label for details.
OT* = registered for over-the-top applications on certain species within genus; consult label for details.

		Selective										Non-Selective				
Genus Species	Common Name	Bentazon	Clethodim	Clopyralid	Dichobenil*	Fenoxaprop	Fluazifop	Halosulfuron	Imazaquin	Pronamide	Sethoxydim	Diquat	Glufosinate	Glyphosate	Paraquat	Pelargonic Acid
Cornus spp.	dogwood species	D			OT		OT*	D		OT	OT*	D	D	D	D	D
C. florida	dogwood, flowering	D	OT	D		OT	OT	D			OT	D	D	D		D
C. kousa	dogwood, Kousa	D						D				D	D	D		D
C. sericea	dogwood, redosier	D					OT	D			OT	D	D	D		D
Corylus spp.	filbert (hazelnut)	D			OT			D				D	D	D	D	D
Cotinus coggygria	smokebush	D						D				D	D	D		D
Cotoneaster spp.	cotoneaster species	D			OT	OT	OT*	D		OT	OT*	D	D	D		D
C. apiculatus	cotoneaster, cranberry	D					OT	D			OT	D	D	D		D
C. buxifolius	cotoneaster, boxleaf	D										D	D	D		D
C. dammeri	cotoneaster, bearberry	D					OT*	D			OT	D	D	D		D
C. horizontalis	cotoneaster, rock	D						D				D	D	D		D
C. microphyllus	cotoneaster, little-leaf	D					OT	D				D	D	D		D
C. salicifolius	cotoneaster, willowleaf	D					OT*	D				D	D	D		D
Crataegus spp.	hawthorn species	D				OT		D		OT		D	D	D		D
Cryptomeria japonica	cryptomeria, Japanese	D						D				D	D	D	D	D
Cupressocyparis leylandii	cypress, Leyland	D					D	D			OT	D	D	D	D	D
Cupressus spp.	cypress species	D						D				D	D	D	D	D
C. glabra	cypress, smooth	D						D				D	D	D		D
C. sempervirens	cypress, Italian	D					OT	D				D	D	D		D
Cytisus spp.	broom species	D						D			OT	D	D	D		D
Daphne spp	daphne	D						D				D	D	D		D
Deutzia spp.	deutzia	D			OT		OT*	D				D	D	D		D
Elaeagnus spp.	elaeagnus species	D						D			OT*	D	D	D		D
E. angustifolia	Russian-olive	D	OT		OT		OT	D			OT	D	D	D		D
E. pungens	elaeagnus, thorny	D						D				D	D	D		D

(continued)

Table 8.6
(Continued)

Key:
D = registered for directed applications.
OT = registered for over-the-top use.
D* = registered for directed applications on certain species within the genus; consult label for details.
OT* = registered for over-the-top applications on certain species within genus; consult label for details.

		Selective										Non-Selective				
Genus Species	**Common Name**	**Bentazon**	**Clethodim**	**Clopyralid**	**Dichobenil***	**Fenoxaprop**	**Fluazifop**	**Halosulfuron**	**Imazaquin**	**Pronamide**	**Sethoxydim**	**Diquat**	**Glufosinate**	**Glyphosate**	**Paraquat**	**Pelargonic Acid**
Erica spp.	heather species	D						D				D	D	D		D
E. cinerea	heather, bell	D						D				D	D	D		D
Eriobotrya japonica	loquat	D						D			OT	D	D	D		D
Eucalyptus spp.	eucalyptus species	D					OT*; D*	D			OT*	D	D	D	D	D
E. camaldulensis	Eucalyptus	D						D				D	D	D		D
E. cinerea	Argyle apple	D						D				D	D	D		D
Euonymus spp.	euonymus species	D	OT		OT	OT		D		OT		D	D	D		D
E. alatus	euonymous, winged	D					OT*	D			OT	D	D	D		D
E. alatus 'Compactus'	burning bush, dwarf	D					D	D				D	D	D		D
E. fortunei	wintercreeper	D					OT	D				D	D	D		D
E. japonicus	euonymous, Jap.	D					OT*	D			OT*	D	D	D		D
E. kiautschovicus	euonymous, spreading	D					OT	D			OT	D	D	D		D
Fagus spp.	beech species	D						D		OT		D	D	D	D	D
F. sylvatica	beech, European	D						D				D	D	D		D
Fatsia japonica	fatsia	D						D				D	D	D		D
Fatshedera lizei	tree-ivy	D					OT	D				D	D	D		D
Ficus benjamana	fig	D					OT	D			OT	D	D	D		D
Ficus spp.	fig species	D	OT*				OT*	D			OT*	D	D	D	D	D
Forsythia spp.	forsythia species	D			OT	OT	OT	D		OT		D	D	D		D
F. X intermedia	forsythia, border	D					OT	D				D	D	D		D
F. suspensa	forsythia, weeping	D					OT	D				D	D	D		D
F. viridissima	forsythia, greenstem	D						D			OT*	D	D	D		D
Fraxinus spp.	ash species	D	OT		OT		OT*	D		OT		D	D	D	D	D
F. americana	ash, white	D					OT	D			OT	D	D	D		D

(continued)

Table 8.6
Postemergence Herbicides Used in Woody Ornamentals *(Continued)*

Key:
D = registered for directed applications.
OT = registered for over-the-top use.
D* = registered for directed applications on certain species within the genus; consult label for details.
OT* = registered for over-the-top applications on certain species within genus; consult label for details.

		Selective										Non-Selective				
Genus Species	**Common Name**	**Bentazon**	**Clethodim**	**Clopyralid**	**Dichobenil***	**Fenoxaprop**	**Fluazifop**	**Halosulfuron**	**Imazaquin**	**Pronamide**	**Sethoxydim**	**Diquat**	**Glufosinate**	**Glyphosate**	**Paraquat**	**Pelargonic Acid**
F. pennsylvanica	ash, green	D					OT	D			OT	D	D	D		D
Gardenia spp.	gardenia species	D	OT		OT		OT*	D			OT*	D	D	D		D
Gardenia jasminoides	cape jasmine	D					OT*	D	OT		OT	D	D	D		D
Gelsemium sempervirens	jessamine, Carolina	D					OT	D			OT	D	D	D		D
Ginkgo biloba	ginkgo	D						D		OT		D	D	D	D	D
Gleditsia triacanthos	honeylocust, thornless	D					OT	D		OT	OT	D	D	D	D	D
Hamamelis virginiana	witchhazel	D						D				D	D	D	D	D
Hedera helix	ivy, English	OT	OT		OT	OT	OT	D			OT*	D	D	D		D
Hemigraphis alternata	red-ivy	D						D				D	D	D		D
Hibiscus syriacus	rose-of-sharon	D					OT	D				D	D	D		D
Hydrangea spp.	hydrangea	D				OT	OT*	D			OT*	D	D	D		D
Hypericum spp.	St.-John's-wort species	D						D			OT*	D	D	D		D
Ilex spp.	holly species	D	OT		OT*			D		OT	OT*	D	D	D		D
I. X attenuata 'Fosteri'	holly, Foster's hybrid	D					OT	D				D	D	D		D
I. aquifolium	holly, English	D						D				D	D	D		D
I. aquifolium x cornuta	holly, English x Chinese	D						D				D	D	D		D
I. cassine	dahoon	D						D				D	D	D		D
I. cornuta	holly, Chinese	OT*					OT*	D	OT*		OT*	D	D	D		D
I. crenata	holly, Japanese	OT*				OT	OT	D	OT*		OT*	D	D	D		D
I. crenata 'Green Lustre'	holly, 'Green Luster'	D						D				D	D	D		D
I. glabra	inkberry	D					OT*	D				D	D	D		D
I. latifolia	holly, lusterleaf	D						D				D	D	D		D
I. meserveae	holly, Meserve hybrids	D				OT	OT	D				D	D	D		D
I. opaca	holly, American	D				OT	OT	D				D	D	D	D	D
I. pernyi	holly, Perny	D						D				D	D	D		D

(continued)

Table 8.6
(Continued)

Key:
D = registered for directed applications.
OT = registered for over-the-top use.
D* = registered for directed applications on certain species within the genus; consult label for details.
OT* = registered for over-the-top applications on certain species within genus; consult label for details.

		Selective										Non-Selective				
Genus Species	**Common Name**	**Bentazon**	**Clethodim**	**Clopyralid**	**Dichobenil***	**Fenoxaprop**	**Fluazifop**	**Halosulfuron**	**Imazaquin**	**Pronamide**	**Sethoxydim**	**Diquat**	**Glufosinate**	**Glyphosate**	**Paraquat**	**Pelargonic Acid**
I. verticillata	holly, winterberry	D						D				D	D	D		D
I. vomitoria	holly, yaupon	D					OT*	D	OT*		OT	D	D	D		D
Illicium spp.	anise-tree	D						D				D	D	D		D
I. floridanum	anise-tree, Florida	D	OT					D				D	D	D		D
Itea virginica	sweetspire, Virginia	D						D				D	D	D		D
Jasminum spp.	jasmine species	D	OT					D			OT*	D	D	D		D
Juglans spp.	walnut species	D			OT*			D		OT	OT	D	D	D	D	D
J. nigra	walnut, black	D						D			OT	D	D	D		D
Juniperus spp.	juniper species	D	OT		OT		OT*; D*	D		OT	OT*	D	D	D		D
J. chinensis	juniper, Chinese	D				OT*	D*	D	OT*		OT*	D	D	D		D
J. conferta	juniper, shore	D		OT		OT	D*	D	OT*		OT*	D	D	D		D
J. davurica	juniper, Dahurian	D				OT*		D				D	D	D		D
J. excelsa	juniper, Greek	D						D				D	D	D		D
J. horizontalis	juniper, creeping	D		OT*		OT*	OT*; D*	D	OT*		OT*	D	D	D		D
J. procumbens	juniper, Japanese garden	D						D				D	D	D		D
J. sabina	juniper, Savin	D					D*	D			OT*	D	D	D		D
J. squamata	juniper, singleseed	D		OT*				D	OT*			D	D	D		D
J. virginiana	redcedar, eastern	D			OT		OT; D*	D			OT*	D	D	D	D	D
Justicia brandegeana	shrimp plant	D					D	D			OT	D	D	D		D
Kalmia latifolia	mountainlaurel	D			OT			D		OT		D	D	D		D
Koelreuteria paniculata	golden raintree	D			OT		OT	D			OT	D	D	D	D	D
Kolkwitzia amabilis	beautybush	D			OT			D				D	D	D		D

(continued)

Table 8.6
POSTEMERGENCE HERBICIDES USED IN WOODY ORNAMENTALS *(CONTINUED)*

Key:
D = registered for directed applications.
OT = registered for over-the-top use.
D* = registered for directed applications on certain species within the genus; consult label for details.
OT* = registered for over-the-top applications on certain species within genus; consult label for details.

		Selective										Non-Selective				
Genus Species	**Common Name**	**Bentazon**	**Clethodim**	**Clopyralid**	**Dichobenil***	**Fenoxaprop**	**Fluazifop**	**Halosulfuron**	**Imazaquin**	**Pronamide**	**Sethoxydim**	**Diquat**	**Glufosinate**	**Glyphosate**	**Paraquat**	**Pelargonic Acid**
Laburnum spp.	golden-chain tree	D	OT*					D				D	D	D	D	D
Lagerstroemia indica	crapemyrtle	D	OT				OT	D	OT*		OT	D	D	D	D	D
Lantana spp.	lantana	D	OT*				OT*; D*	D			OT*	D	D	D		D
Larix spp.	larch species	D						D			OT*	D	D	D	D	D
Leucothoe spp.	leucothoe species	D			OT			D				D	D	D		D
L. axillaris	leucothoe, coast	D					D	D				D	D	D		D
Ligustrum spp.	privet species	D	OT		OT	OT	OT*	D		OT	OT*	D	D	D		D
L. amurense	privet, Amur	D					OT*	D				D	D	D		D
L. japonicum	privet, Japanese	D					D	D			OT	D	D	D		D
L. lucidum	privet, waxleaf	D					OT	D			OT*	D	D	D		D
L. ovalifolium	privet, California	D					OT	D				D	D	D		D
L. sinense	privet, Chinese	D						D				D	D	D		D
Liquidambar styraciflua	sweetgum, American	D	OT				OT	D		OT	OT	D	D	D	D	D
Liriodendron tulipifera	poplar, tulip; tuliptree	D						D		OT	OT	D	D	D	D	D
Lonicera spp.	honeysuckle species	D	OT*		OT		OT*	D			OT*	D	D	D		D
L. japonica	honeysuckle, Jap.	D					D	D			OT	D	D	D		D
L. tatarica	honeysuckle, Tatarian	D						D			OT*	D	D	D		D
Loropetalum chinense	loropetulum	D						D				D	D	D		D
Maclura																
Magnolia spp.	magnolia species	D			OT	OT		D		OT		D	D	D	D	D
M. grandiflora	magnolia, southern	D					D	D			OT	D	D	D		D
M. X soulangiana	magnolia, saucer	D						D				D	D	D		D
M. stellata	magnolia, star	D					OT	D				D	D	D		D
Mahonia spp.	mahonia species	D					OT*	D				D	D	D		D

(continued)

Table 8.6
(CONTINUED)

Key:
D = registered for directed applications.
OT = registered for over-the-top use.
D* = registered for directed applications on certain species within the genus; consult label for details.
OT* = registered for over-the-top applications on certain species within genus; consult label for details.

Genus Species	Common Name	Selective										Non-Selective				
		Bentazon	Clethodim	Clopyralid	Dichobenil*	Fenoxaprop	Fluazifop	Halosulfuron	Imazaquin	Pronamide	Sethoxydim	Diquat	Glufosinate	Glyphosate	Paraquat	Pelargonic Acid
M. aquifolium	grapeholly, Oregon	D	OT				OT	D			OT	D	D	D		D
Malus spp.	apple species	D			OT			D			OT	D	D	D	D	D
Malus spp.	crabapple species	D	OT*		OT		OT*	D		OT	OT*	D	D	D	D	D
Malus × domestica	apple, common	D			OT			D				D	D	D	D	D
Malus floribunda	crabapple, Japanese	D			OT		OT	D		OT		D	D	D	D	D
Morus alba	mulberry, white	D	OT					D				D	D	D		D
Myoporum spp.	myoporum species	D					OT*	D			OT*	D	D	D		D
Myrica cerifera	myrtle, wax	D					OT	D	OT			D	D	D		D
Myrica pensylvanica	bayberry, northern	D	OT					D				D	D	D		D
Myrtus communis	myrtle, compact	D						D			OT*	D	D	D		D
Nandina domestica	nandina	D	OT		OT	OT	D	D			OT	D	D	D		D
Nerium oleander	oleander	D	OT			OT	OT*; D*	D			OT	D	D	D		D
Nyssa sylvatica	tupelo, black	D						D				D	D	D	D	D
Ochna serrulata	bird's-eye bush	D						D			OT	D	D	D		D
Olea spp.	olive	D					OT*	D			OT	D	D	D	D	D
Osmanthus spp.	osmanthus species	D						D				D	D	D		D
O. fragrans	tea olive, fragrant	D	OT				OT	D			OT	D	D	D		D
O. heterophyllus	tea olive, holly	D			OT			D			OT	D	D	D		D
Oxydendrum arboreum	sourwood	D						D				D	D	D	D	D
Pachysandra terminalis	pachysandra	OT	OT				OT	D	OT		OT*	D	D	D		D
Paxistima canbyi	paxistima (pachistima)	D			OT			D				D	D	D		D
Philadelphus spp.	mockorange	D			OT		OT*	D		OT		D	D	D		D
Photinia spp.	photinia spp.	D	OT		OT	OT		D			OT	D	D	D		D
P. glabra	photinia, Japanese	D						D				D	D	D		D

(continued)

Table 8.6
Postemergence Herbicides Used in Woody Ornamentals *(Continued)*

Key:
D = registered for directed applications.
OT = registered for over-the-top use.
D* = registered for directed applications on certain species within the genus; consult label for details.
OT* = registered for over-the-top applications on certain species within genus; consult label for details.

		Selective										Non-Selective				
Genus Species	**Common Name**	**Bentazon**	**Clethodim**	**Clopyralid**	**Dichobenil***	**Fenoxaprop**	**Fluazifop**	**Halosulfuron**	**Imazaquin**	**Pronamide**	**Sethoxydim**	**Diquat**	**Glufosinate**	**Glyphosate**	**Paraquat**	**Pelargonic Acid**
P. × fraseri	Fraser photinia; red-tip	D					OT	D	OT		OT	D	D	D		D
Physocarpus opulifolius	ninebark	D					OT*	D			OT*	D	D	D	D	D
Picea spp.	spruce species	D						D		OT		D	D	D	D	D
P. abies	spruce, Norway	D		OT			OT	D			OT	D	D	D		D
P. glauca	spruce, white	D		OT			OT*	D			OT	D	D	D		D
P. glauca 'Conica'	spruce, dw. Alberta	D					OT	D				D	D	D		D
P. omorika	spruce, Serbian	D					OT	D				D	D	D		D
P. pungens	spruce, Colorado	D		OT			OT	D			OT	D	D	D		D
P. rubens	spruce, red	D						D				D	D	D		D
P. sitchensis	spruce, Sitka	D						D				D	D	D		D
Pieris spp.	pieris species	D						D				D	D	D		D
P. japonica	pieris, Japanese	D					OT	D			OT	D	D	D		D
Pinus spp.	pine species	D			OT		OT*	D		OT	OT*	D	D	D	D	D
P. banksiana	pine, Jack	D						D			OT	D	D	D		D
P. bungeana	pine, lacebark	D						D				D	D	D		D
P. clausa	pine, sand	D					OT	D				D	D	D		D
P. contorta	pine, lodgepole	D		OT				D			OT*	D	D	D		D
P. echinata	pine, shortleaf	D					OT	D				D	D	D		D
P. elliottii	pine, slash	D					OT	D			OT	D	D	D		D
P. mugo	pine, mugo	OT*		OT*			OT	D			OT	D	D	D		D
P. nigra	pine, Austrian	D					OT	D			OT	D	D	D		D
P. palustris	pine, longleaf	D					OT	D			OT	D	D	D		D
P. ponderosa	pine, ponderosa	D		OT			OT	D			OT	D	D	D		D
P. radiata	pine, Monterey	D						D			OT	D	D	D		D
P. resinosa	pine, red	D					OT	D			OT	D	D	D		D

(continued)

Table 8.6
(Continued)

Key:
D = registered for directed applications.
OT = registered for over-the-top use.
D* = registered for directed applications on certain species within the genus; consult label for details.
OT* = registered for over-the-top applications on certain species within genus; consult label for details.

		Selective										Non-Selective				
Genus Species	**Common Name**	**Bentazon**	**Clethodim**	**Clopyralid**	**Dichobenil***	**Fenoxaprop**	**Fluazifop**	**Halosulfuron**	**Imazaquin**	**Pronamide**	**Sethoxydim**	**Diquat**	**Glufosinate**	**Glyphosate**	**Paraquat**	**Pelargonic Acid**
P. strobus	pine, Eastern white	D		OT		OT	OT	D			OT	D	D	D		D
P. sylvestris	pine, Scotch	D		OT			OT	D			OT	D	D	D		D
P. taeda	pine, loblolly	D					OT	D			OT	D	D	D		D
P. thunbergii	pine, Japanese black	D						D			OT	D	D	D		D
P. virginiana	pine, Virginia	D					OT	D			OT	D	D	D		D
Pistacia spp.	pistachio	D						D			OT	D	D	D	D	D
Pittosporum tobira	pittosporum, Japanese	D	OT		OT		OT*	D			OT	D	D	D		D
Pittosporum rhombifolium	pittosporum, Queensland	D	OT		OT			D				D	D	D		D
Platanus × acerifolia	planetree, London	D					OT	D		OT		D	D	D	D	D
Platanus occidentalis	sycamore			D			OT	D		OT	OT	D	D	D	D	D
Platanus racemosa	sycamore, California						OT	D		OT		D	D	D	D	D
Platycladus orientalis	arborvitae, Oriental	D					OT*	D		OT	OT*	D	D	D	D	D
Podocarpus spp.	podocarpus species	D	OT				OT*	D				D	D	D		D
P. macrophyllus	podocarpus, Chinese	D					OT; D*	D			OT	D	D	D		D
Populus spp.	poplar species	D			OT			D		OT	OT*	D	D	D	D	D
P. deltoides	cottonwood, eastern	D			OT			D				D	D	D		D
Potentilla spp.	potentilla; cinquefoil	D	OT				OT*; D*	D			OT*	D	D	D		D
P. fruticosa	cinquefoil, bush	D		D			OT*	D				D	D	D		D
Prunus spp.	prunes & stone fruits	D						D			OT*	D	D	D	D	D
Prunus spp.	cherry, peach and plum species	D			OT		OT*	D		OT*	OT*	D	D	D	D	D
P. avium	cherry, sweet	D			OT			D			OT	D	D	D		D
P. caroliniana	cherrylaurel, Carolina	D					OT*	D			OT*	D	D	D		D
P. cerasifera	plum, cherry	D						D				D	D	D		D

(continued)

Table 8.6
Postemergence Herbicides Used in Woody Ornamentals *(Continued)*

Key:
D = registered for directed applications.
OT = registered for over-the-top use.
D* = registered for directed applications on certain species within the genus; consult label for details.
OT* = registered for over-the-top applications on certain species within genus; consult label for details.

		Selective										Non-Selective				
Genus Species	Common Name	Bentazon	Clethodim	Clopyralid	Dichobenil*	Fenoxaprop	Fluazifop	Halosulfuron	Imazaquin	Pronamide	Sethoxydim	Diquat	Glufosinate	Glyphosate	Paraquat	Pelargonic Acid
P. X cistena	cherry, purpleleaf sand	D						D			OT	D	D	D		D
P. glandulosa	almond, dw. flowering	D			OT			D				D	D	D		D
P. laurocerasus	cherrylaurel, common	D						D				D	D	D		D
P. maritima	plum, beach	D						D				D	D	D		D
P. persica	peach, common	D						D			OT	D	D	D		D
P. sargentii	cherry, sargent	D						D				D	D	D		D
P. serotina																
P. serrulata	cherry, Jap. flowering	D						D				D	D	D		D
P. subhirtella	cherry, Higan	D						D				D	D	D		D
P. X yedoensis	cherry, Yoshino	D						D				D	D	D		D
Pseudotsuga menziesii	Douglasfir	D		OT			OT	D		OT	OT	D	D	D	D	D
Pyracantha spp.	pyracantha species	D	OT		OT	OT	OT*	D		OT	OT*	D	D	D		D
P. coccinea	firethorn, scarlet	D					OT*	D				D	D	D		D
P. fortuneana	firethorn, Chinese	D						D				D	D	D		D
P. koidzumii	firethorn, Formosa	D					OT*	D				D	D	D		D
Pyrus spp.	pear species	D						D			OT*	D	D	D	D	D
P. calleryana 'Bradford'	pear, Bradford callery	D					OT	D		OT		D	D	D		D
P. communis	pear	D			OT			D			OT	D	D	D		D
Quercus spp.	oak species	D	OT		OT			D		OT		D	D	D	D	D
Q. alba	oak, white	D						D				D	D	D		D
Q. coccinea	oak, scarlet	D						D				D	D	D		D
Q. illicifolia	oak, bear	D						D				D	D	D		D
Q. nigra	oak, water	D						D			OT	D	D	D		D
Q. palustris	oak, pin	D					OT	D				D	D	D		D
Q. phellos	oak, willow	D		D				D			OT	D	D	D		D

(continued)

Table 8.6
(Continued)

Key:
D = registered for directed applications.
OT = registered for over-the-top use.
D* = registered for directed applications on certain species within the genus; consult label for details.
OT* = registered for over-the-top applications on certain species within genus; consult label for details.

		Selective										Non-Selective				
Genus Species	**Common Name**	**Bentazon**	**Clethodim**	**Clopyralid**	**Dichobenil***	**Fenoxaprop**	**Fluazifop**	**Halosulfuron**	**Imazaquin**	**Pronamide**	**Sethoxydim**	**Diquat**	**Glufosinate**	**Glyphosate**	**Paraquat**	**Pelargonic Acid**
Q. rubra	oak, red	OT		D				D				D	D	D		D
Q. shumardii	oak, Shumard	D						D				D	D	D		D
Q. virginiana	oak, live	D					OT	D				D	D	D		D
Raphiolepis indica	hawthorn, Indian	D	OT				OT	D	OT		OT	D	D	D		D
Raphiolepsis umbellata	hawthorn, Yeddo	D					OT	D				D	D	D		D
Rhododendron spp.	azalea, rhododendron		OT		OT*	OT	OT*; D*	D		OT	OT*	D	D	D		D
R. carolinianum	rhododendron, Carolina							D				D	D	D		D
R. catawbiense	rhododendron, Catawba			OT*			OT*	D				D	D	D		D
R. indica	azalea, Indica						D	D				D	D	D		D
R. maximum	rhodo., rosebay							D				D	D	D		D
R. molle	azalea, Chinese							D				D	D	D		D
R. obtusum	azalea, Hiryu			OT*			OT*; D*	D				D	D	D		D
Rhus typhina	sumac, staghorn	D						D				D	D	D		D
Ribes																
Robinia spp.	locust species	D			OT			D			OT*	D	D	D		D
Rosa spp.	rose species	D	OT		OT	OT	OT	D				D	D	D		D
R. banksiae	rose, Lady Bank's	D					OT	D				D	D	D		D
Rubus spp.	bramble	D			OT			D			OT*	D	D	D		D
Salix spp.	willow species	D			OT	OT	OT*; D*	D		OT	OT*	D	D	D	D	D
S. babylonica	willow, Babylon weeping	D					OT	D				D	D	D		D
Samanea saman	rain-tree (ohai)	D						D				D	D	D		D
Sequoiadendron giganteum	sequoia, giant	D						D				D	D	D	D	D

(continued)

Table 8.6
Postemergence Herbicides Used in Woody Ornamentals (Continued)

Key:
D = registered for directed applications.
OT = registered for over-the-top use.
D* = registered for directed applications on certain species within the genus; consult label for details.
OT* = registered for over-the-top applications on certain species within genus; consult label for details.

Genus Species	Common Name	Selective										Non-Selective				
		Bentazon	Clethodim	Clopyralid	Dichobenil*	Fenoxaprop	Fluazifop	Halosulfuron	Imazaquin	Pronamide	Sethoxydim	Diquat	Glufosinate	Glyphosate	Paraquat	Pelargonic Acid
Skimmia spp.	Skimmia	D						D				D	D	D		D
Sorbus americana	mountainash, Am.	D			OT		OT	D		OT	OT	D	D	D		D
Spiraea spp.	spirea species	D	OT*	D*	OT		OT*	D		OT	OT*	D	D	D		D
S. × vanhouttei	spirea, Vanhoutte	D					D	D			OT	D	D	D		D
Symphoricarpos albus	snowberry	D						D				D	D	D		D
Syringa spp.	lilac species	D			OT		OT* D*	D		OT	OT*	D	D	D		D
S. vulgaris	lilac, common	D						D			OT*	D	D	D		D
S. × persica	lilac, Persian	D						D				D	D	D		D
Taxodium distichum	cypress, bald	D						D				D	D	D	D	D
Taxus spp.	yew species	D			OT	OT		D		OT		D	D	D	D	D
T. baccata	yew, English	D						D			OT	D	D	D		D
T. canadensis	yew, Canadian	D						D				D	D	D		D
T. cuspidata	yew, Japanese	OT*					OT	D			OT*	D	D	D		D
T. × media	yew, Anglojap	OT*		OT			OT*	D				D	D	D		D
Ternstroemia gymnanthera	cleyera, Japanese	D	OT		OT		OT	D				D	D	D		D
Thuja spp.	arborvitae species	D			OT			D		OT		D	D	D	D	D
T. occidentalis	arborvitae, American	OT		OT			OT*	D			OT*	D	D	D		D
Tilia spp.	basswood; linden	D	OT		OT		OT*	D		OT	OT*	D	D	D	D	D
Trachelospermum asiaticum	jasmine, Japanese star	D	OT				OT	D	OT*		OT	D	D	D		D
Trachelospermum jasminoides	jasmine, star or confederate	D	OT				OT	D	OT		OT	D	D	D		D
Tsuga spp.	hemlock species	D						D		OT		D	D	D	D	D
T. canadensis	hemlock, Canadian	D					OT	D			OT	D	D	D		D

(continued)

Table 8.6
(CONTINUED)

Key:
D = registered for directed applications.
OT = registered for over-the-top use.
D* = registered for directed applications on certain species within the genus; consult label for details.
OT* = registered for over-the-top applications on certain species within genus; consult label for details.

		Selective										Non-Selective				
Genus Species	Common Name	Bentazon	Clethodim	Clopyralid	Dichobenil*	Fenoxaprop	Fluazifop	Halosulfuron	Imazaquin	Pronamide	Sethoxydim	Diquat	Glufosinate	Glyphosate	Paraquat	Pelargonic Acid
T. caroliniana	hemlock, Carolina	D						D				D	D	D		D
Ulmus spp.	elm species	D			OT			D		OT		D	D	D	D	D
U. americana	elm, American	D						D				D	D	D		D
U. parvifolia	elm, Chinese	D						D			OT	D	D	D		D
U. pumila	elm, Siberian	D						D				D	D	D		D
Vaccinium spp.	blueberry	D			OT		D*	D			OT	D	D	D		D
Viburnum spp.	viburnum species	D				OT	OT*; D*	D		OT	OT*	D	D	D		D
V. davidii	viburnum, David	D						D				D	D	D		D
V. japonicum	viburnum, Japanese	D					OT	D			OT	D	D	D		D
V. odoratissimum	viburnum, sweet	D						D				D	D	D		D
V. plicatum	viburnum, Jap. snowball	D					OT*	D				D	D	D		D
V. rigidum	viburnum, Canary Island	D						D				D	D	D		D
V. suspensum	viburnum, Sandankwa	D					OT	D			OT	D	D	D		D
V. tinus	laurustinus	D	OT					D			OT	D	D	D		D
V. trilobum	viburnum, American cranberrybush	D					OT*	D			OT	D	D	D		D
V. wrightii	viburnum, Wright	D						D				D	D	D		D
Vitex spp.	chaste tree	D						D				D	D	D		D
Vitis spp.	grapes	D			OT			D			OT	D	D	D	D	D
Weigela spp.	weigela species	D			OT	OT		D				D	D	D		D
W. florida	weigela, old fashioned	D					OT*	D				D	D	D		D
Wisteria spp.	wisteria species	D	OT					D			OT*	D	D	D		D
Xylosma congestum	xylosma	D						D				D	D	D		D
Yucca spp.	yucca species	D					OT*	D	OT*			D	D	D		D

*see text for other trade names of these products
Authors: J. C. Neal, J. A. Briggs, and T. Whitwell. 2005. North Carolina State University, Raleigh, North Carolina, and Clemson University, Clemson, South Carolina.

Table 8.7
Herbicides Used in Herbaceous Ornamentals

Key:
y = registered for some species of this genus.
e = registered for some species of this genus, apply only after established.
d = registered for some species of this genus, directed application only
x = NOT registered/prohibited from use on some species.

		Preemergence															Postemergence					
Genus	Common Name	Dithiopyr	Isoxaben	Metolachlor	Napropamide	Oryzalin	Oxadiazon	Pendimethalin 2 G	Pendimethalin 2.86 G	Pendimethalin 3.3 EC	Trifluralin	Benefin + Oryzalin	Isoxaben + Trifluralin	Oxadiazon + Pendimethalin	Oxadiazon + Prodiamine	Oxyfluorfen + Oxadiazon	Fenoxaprop	Bentazon	Clethodim	Fluazifop	Pelargonic Acid	Sethoxydim
Achillea	Yarrow	d	e	y		e		e	y		e	e	e	e			e	d	y	y	d	y
Acorus	Sweet flag	d						e	y									d	y		d	y
Agapanthus	African Lily (Lily-of-Nile)	d	e	y	y	e	e	e	y	e	e	e	e					d		y	d	y
Ageratum	Ageratum			y	y			e			e			e				d	y	y	d	
Ajuga	Bugleweed	d	x	y	y	e	e	e	y	e	e	e	x	e			e	y	y	d	d	y
Alcea	Hollyhock																	d		y	d	
Allium	Allium			y														d		y	d	
Alternanthera	Joseph's Coat																	d			d	
Alyssum; Lobularia	Alyssum			y				e			e		e				e	d	y	y	d	y
Amaranthus	Love-Lies-bleeding																	d			d	
Ammophila	Beach Grass	d	e					e	y	e	e		e					d			d	
Anagallis	Pimpernel																e	d			d	
Anemone	Anemone							e										d			d	
Antirrhinum	Snapdragon	d		y		e		e			e	e	e				e	y	y	y	d	y
Aquilegia	Columbine	d	e	y				e						e			e	d		d	d	y
Arctotheca	Capeweed (Cape Marigold)		e			e		e	y	e	e	e	e					d		y	d	y
Arctotis	African Daisy				y						e							d			d	
Arenaria	Sandwort												x					d		y	d	y
Arisaema	Jack-in-the-Pulpit																	d			d	y
Armeria	Sea Pink										e							d			d	y
Artemesia	Artemesia		e					e			e			e				d			d	

(continued)

Table 8.7
(Continued)

Key:
y = registered for some species of this genus.
e = registered for some species of this genus, apply only after established.
d = registered for some species of this genus, directed application only
x = NOT registered/prohibited from use on some species.

		Preemergence															Postemergence					
Genus	**Common Name**	Dithiopyr	Isoxaben	Metolachlor	Napropamide	Oryzalin	Oxadiazon	Pendimethalin 2 G	Pendimethalin 2.86 G	Pendimethalin 3.3 EC	Trifluralin	Benefin + Oryzalin	Isoxaben + Trifluralin	Oxadiazon + Pendimethalin	Oxadiazon + Prodiamine	Oxyfluorfen + Oxadiazon	Fenoxaprop	Bentazon	Clethodim	Fluazifop	Pelargonic Acid	Sethoxydim
Artemesia	Dusty Miller (Wormwood)							e			e		e	e				d			d	
Asclepias	Butterflyweed (Milkweed)			y				e	y									d			d	y
Asparagus	Asparagus Fern	d	e		y	e		e	y	e	e		e					d	y	y	d	y
Aster	Aster		e	y	y	e		e	y		e		e	e				d			d	y
Astilbe	Astilbe (False Spriea)	d				e		e			e	e	e	e			e	d			d	
Athyrium	Fern, Lady or Japanese					e					e							d			d	
Aurinia	Alyssum (Basket of Gold, per.)																	d		y	d	
Begonia	Begonia, Fibrous	d	e					e			e	x	e				e	d		d	d	y
Brassica	Cabbage or Kale, Ornamenta		x			e		e					x					y			d	
Caladium	Caladium					e		e				e						d			d	
Calamagrostis	Calamagrostis (Reed Feather Grass)													e				d			d	
Calendula	Pot Marigold																	d		y	d	y
Callistephus	Aster, China	d	e			e		e					e					d			d	
Calluna	Heather		e		y			e		e	e		e					d		y	d	
Campanula	Bellflower			y		e		e			e	e					e	d		y	d	y
Canna	Canna			y				e		e				e				d			d	y
Capsicum	Pepper, Ornamental																	d			d	y
Carex	Carex	d	e	y				e			e		e					d			d	

(continued)

Table 8.7
Herbicides Used in Herbaceous Ornamentals *(Continued)*

Key:
y = registered for some species of this genus.
e = registered for some species of this genus, apply only after established.
d = registered for some species of this genus, directed application only
x = NOT registered/prohibited from use on some species.

		Preemergence															Postemergence					
Genus	Common Name	Dithiopyr	Isoxaben	Metolachlor	Napropamide	Oryzalin	Oxadiazon	Pendimethalin 2 G	Pendimethalin 2.86 G	Pendimethalin 3.3 EC	Trifluralin	Benefin + Oryzalin	Isoxaben + Trifluralin	Oxadiazon + Pendimethalin	Oxadiazon + Prodiamine	Oxyfluorfen + Oxadiazon	Fenoxaprop	Bentazon	Clethodim	Fluazifop	Pelargonic Acid	Sethoxydim
Catharanthus	Periwinkle (bedding plant)		e					e					e				e	d		d	d	y
Celosia	Cockscomb	d						e										d			d	y
Centaurea	Cornflower	d									e						e	d			d	y
Cerastium	Snow-in-Summer		e								e		e				e	d		y	d	x
Chrysanthemum	Chrysanthemum		e	y	y	e	e	e			e	e	e				e	d	y	y	d	y
Cirsium	Cirsium, Pink Beauty																	d			d	
Clarkia	Godetia																e	d			d	y
Coleus	Coleus	d															e	d	y	y	d	y
Consolida	Larkspur																	d			d	
Convallaria	Lily-of-the-Valley																	d			d	y
Coreopsis	Coreopsis (Calliopsis)	d	e	y		e	e	e			e	e	e	e			e	d	y	y	d	y
Coronilla	Crown Vetch							e			e							d		y	d	y
Cortaderia	Pampas Grass	d	e	y		e		e	y	e	e	e	e		y			d		d	d	
Cosmos	Cosmos										e						e	d			d	y
Crocus	Crocus			y				e										d			d	
Cuphea	Heather, False		e			e							e					d	y		d	y
Dahlia	Dahlia		e		y		e	e			e		e					d	y		d	y
Daucus	Queen Anne's Lace			y														d			d	y
Delosperma; Drosantheumum	Iceplant	d	e	y	y	e	y	e	y	e	e	e	e	e				d		d	d	y
Delphinium	Delphinium	d		y													e	d			d	y

(continued)

Table 8.7
(Continued)

Key:
y = registered for some species of this genus.
e = registered for some species of this genus, apply only after established.
d = registered for some species of this genus, directed application only
x = NOT registered/prohibited from use on some species.

		Preemergence															Postemergence					
Genus	**Common Name**	**Dithiopyr**	**Isoxaben**	**Metolachlor**	**Napropamide**	**Oryzalin**	**Oxadiazon**	**Pendimethalin 2 G**	**Pendimethalin 2.86 G**	**Pendimethalin 3.3 EC**	**Trifluralin**	**Benefin + Oryzalin**	**Isoxaben + Trifluralin**	**Oxadiazon + Pendimethalin**	**Oxadiazon + Prodiamine**	**Oxyfluorfen + Oxadiazon**	**Fenoxaprop**	**Bentazon**	**Clethodim**	**Fluazifop**	**Pelargonic Acid**	**Sethoxydim**
Deschampsia	Hair Grass, Tufted	d	e					e	y	e	e		e					d			d	
Dianthus	Carnation	d					y				e		e					d	y	y	d	y
Dianthus	Pink	d	e			e					e		e				e	d	y	y	d	y
Dianthus	Sweet William	d		y		e	e	e			e	e					e	d	y	y	d	y
Dicentra	Bleeding Heart					e		e		e	e	e	x				e	d	y	y	d	y
Digitalis	Foxglove					e	e	e			e		x	e				d			d	y
Dimorphotheca	Cape Marigold					e					e	e						d			d	y
Doronicum	Leopard's Bane					e	y	e						e			e	d			d	
Echinacea	Coneflower, Purple		x			e		e			e	e	x				e	d			d	y
Endymion	Hyacinth, Wood			y														d			d	
Erica	Heath	d	e								e		e					d			d	
Erysimum	Wallflower, Siberian												e				e	d			d	
Eschscholzia	Poppy, California							e			e						e	d			d	y
Euphorbia	Snow-on-Mountain		x								e		x					d		y	d	
Festuca	Fescue, Blue	d				e		e	y	e	e	e	e	e				d			d	y
Fragaria	Strawberry, Ornamental				y			e	y		e							d		y	d	
Freesia	Freesia							e		e								d			d	
Gaillardia	Gaillardia (Blanketflower)	d	e	y		e	y	e			e		e	e			e	d			d	y
Gazania	Gazania		e	y	y	e	y	e	y	e	e	e	e	e			e	d	y	d	d	y
Geranium	Cranesbill (True Geranium)		e	y	y						e		x					d	y	y	d	y
Gerbera	Gerbera Daisy																	d			d	y

(continued)

Table 8.7
Herbicides Used in Herbaceous Ornamentals (*Continued*)

Key:
y = registered for some species of this genus.
e = registered for some species of this genus, apply only after established.
d = registered for some species of this genus, directed application only
x = NOT registered/prohibited from use on some species.

		Preemergence															Postemergence					
Genus	**Common Name**	**Dithiopyr**	**Isoxaben**	**Metolachlor**	**Napropamide**	**Oryzalin**	**Oxadiazon**	**Pendimethalin 2 G**	**Pendimethalin 2.86 G**	**Pendimethalin 3.3 EC**	**Trifluralin**	**Benefin + Oryzalin**	**Isoxaben + Trifluralin**	**Oxadiazon + Pendimethalin**	**Oxadiazon + Prodiamine**	**Oxyfluorfen + Oxadiazon**	**Fenoxaprop**	**Bentazon**	**Clethodim**	**Fluzaifop**	**Pelargonic Acid**	**Sethoxydim**
Geum	Geum (Avens)		e	y		e		e			e	e						d			d	y
Gilia	Gilia																e	d			d	y
Gladiolus	Gladiolus		y	y	y	e	e	e			e	e		e				d	y	d	d	y
Goniolimon	Statice, German							e										d			d	
Gypsophila	Baby's Breath		e			e	e	e			e	e	x	e			e	d			d	y
Hedera	Ivy	d	e	y	y	e	y	e	y	e	e	e	e	e			e	y	y	y	d	y
Helianthus	Sunflower										e							d			d	
Helichrysum	Strawflower		e										e					d			d	
Heliotropium	Heliotrope		e			e	e											d			d	
Hemerocallis	Daylily	d	e	y		e		e	y		e	e	e				e	d	y	y	d	y
Herniaria	Rupturewort							e			e							d		d	d	y
Hesperis	Dame's rocket																e	d			d	y
Heuchera	Coral Bell		e			e					e		e					d	y		d	
Hosta	Hosta (Plantain-Lily)	d	e	y	y	e		e	y	e	e	e	e	e			e	d	y	y	d	y
Hyacinthus	Hyacinth		y	y		y					y	y	y					d			d	
Hypericum	St.-John's-wort		e	y	y	e	e	e	y	e	e	e	e	e		y		d			d	y
Iberis	Candytuft	d	x			e	e						x				e	d	y	y	d	y
Impatiens	Impatiens (Balsam)	d	e	y		e		e			e	e	e					y	y		d	y
Ipomea	Morningglory (Sweet Potato Vine)		e					e										d			d	
Iris	Iris, Bulbous	d	y	y		y		e			y	e	y				e	d	y	y	d	y

(*continued*)

Table 8.7
(Continued)

Key:
y = registered for some species of this genus.
e = registered for some species of this genus, apply only after established.
d = registered for some species of this genus, directed application only
x = NOT registered/prohibited from use on some species.

		Preemergence															Postemergence					
Genus	**Common Name**	**Dithiopyr**	**Isoxaben**	**Metolachlor**	**Napropamide**	**Oryzalin**	**Oxadiazon**	**Pendimethalin 2 G**	**Pendimethalin 2.86 G**	**Pendimethalin 3.3 EC**	**Trifluralin**	**Benefin + Oryzalin**	**Isoxaben + Trifluralin**	**Oxadiazon + Pendimethalin**	**Oxadiazon + Prodiamine**	**Oxyfluorfen + Oxadiazon**	**Fenoxaprop**	**Bentazon**	**Clethodim**	**Fluazifop**	**Pelargonic Acid**	**Sethoxydim**
Iris	Iris, Rhizomatous	d	e	y		e		e			e	e	e				e	d	y	y	d	y
Kniphofia	Poker Plant																	d			d	
Lantana	Lantana	d	e		y		e	e			e		e					d	y	d	d	y
Lathyrus	Sweet Pea										e							d			d	
Lavendula	Lavender					e		e		e	e		e	e				d			d	y
Layia	Tidy Tips																e	d			d	y
Leucanthemum (Chrysanthemum)	Daisy, Shasta		e	y	y	e		e			e	e	e				e	d	y	y	d	y
Liatris	Gayfeather					e	e	e			e	e	e	e			e	d		y	d	y
Lilium	Lily		y	y				e		e				e				d			d	
Limonium; Statice	Statice		e	y		e	e	e			e		e				e	d			d	y
Linum	Flax, Scarlet or Blue																e	d			d	y
Liriope	Liriope	d	e	y	y	e		e	y	e	e	e	e				e	y	y	y	d	y
Lobelia	Cardinal Flower					e			y		e		e				e	d			d	y
Lonicera	Honeysuckle	d	e	y	y	e	e				e	e	e	e	y	y		d	y	d	d	y
Lupinus	Lupine			y				e			e							d			d	y
Lysimachia	Moneywort (Loosestrife)		e										e					d	y	y	d	y
Lythrum	Loosestrife, Purple			y				e	y									d	y		d	y
Matricaria	Chamomile																e	d			d	
Matthiola	Stock		e								e		e					d			d	y
Mertensia	Bluebells																e	d			d	

(continued)

Table 8.7
HERBICIDES USED IN HERBACEOUS ORNAMENTALS *(CONTINUED)*

Key:
y = registered for some species of this genus.
e = registered for some species of this genus, apply only after established.
d = registered for some species of this genus, directed application only
x = NOT registered/prohibited from use on some species.

		Preemergence															Postemergence					
Genus	Common Name	Dithiopyr	Isoxaben	Metolachlor	Napropamide	Oryzalin	Oxadiazon	Pendimethalin 2 G	Pendimethalin 2.86 G	Pendimethalin 3.3 EC	Trifluralin	Benefin + Oryzalin	Isoxaben + Trifluralin	Oxadiazon + Pendimethalin	Oxadiazon + Prodiamine	Oxyfluorfen + Oxadiazon	Fenoxaprop	Bentazon	Clethodim	Fluazifop	Pelargonic Acid	Sethoxydim
Mirabilis	Four O'Clock										e							d			d	
Miscanthus	Miscanthus (Maiden Grass)		e					e	y		e		e	e				d			d	
Molucella	Bells of Ireland																	d			d	
Monarda	Bee Balm	d									e		e					d			d	y
Morea	Fortnight Lily		e	y				e	y				e					d			d	
Muscari	Hyacinth, Grape			y														d			d	
Myosotis	Forget-Me-Not										e						e	d			d	
Narcissus	Daffodil	d	y	y	y	y		e			y	y	y					d			d	y
Nemophila	Baby-Blue-Eyes																e	d			d	y
Nephrolepsis	Fern, Boston							e										d		y	d	
Nicotiana	Nicotiana (Flowering Tobacco)									e							d	y		d	y	
Nipponanthemum (Chrys. nip.)	Daisy, Montauk			y	y	e		e				e	e				e	d	y	y	d	y
Oenothera	Eveningprimrose (Sundrops)		e	y							e		e				e	d		d	d	y
Ophiopogon	Mondo Grass	d	e	y		e		e	y	e	e	e	e					d	y	d	d	y
Opuntia	Prickly Pear																	d		d	d	
Ornithogalum	Star-of-Bethlehem			y				e		e								d			d	
Osteospermum	African Daisy, Trailing	d	e		y	e	y	e	y	e	e	e	e				e	d	y	y	d	y
Pachysandra	Pachysandra	d	e	y	y	e	e	e	y	e	e		e	e	y	y		y	y	y	d	y
Paeonia	Peony						e	e		e				e			e	d			d	

(continued)

Table 8.7
(Continued)

Key:
y = registered for some species of this genus.
e = registered for some species of this genus, apply only after established.
d = registered for some species of this genus, directed application only
x = NOT registered/prohibited from use on some species.

		Preemergence															Postemergence					
Genus	**Common Name**	**Dithiopyr**	**Isoxaben**	**Metolachlor**	**Napropamide**	**Oryzalin**	**Oxadiazon**	**Pendimethalin 2 G**	**Pendimethalin 2.86 G**	**Pendimethalin 3.3 EC**	**Trifluralin**	**Benefin + Oryzalin**	**Isoxaben + Trifluralin**	**Oxadiazon + Pendimethalin**	**Oxadiazon + Prodiamine**	**Oxyfluorfen + Oxadiazon**	**Fenoxaprop**	**Bentazon**	**Clethodim**	**Fluzaifop**	**Pelargonic Acid**	**Sethoxydim**
Papaver	Poppy										e						e	d			d	y
Pelargonium	Geranium, Zonal	d	e	y	y	e		e	y	e		e	e	e			e	d	y	y	d	y
Pennisetum	Fountain Grass	d	e			e	e	e	y	e	e		e	e				d		d	d	y
Penstemon	Beardtongue		e		y			e	y		e		e	e			e	d			d	
Petunia	Petunia	d	e	y	y	e		e			e	e	e				e	y	y	y	d	y
Phalaris	Ribbon Grass	d	e					e	y	e	e		e					d			d	
Phlox	Phlox		e	y			e				e		x				e	d	y		d	y
Physostegia	False Dragonhead			y														d			d	y
Platycodon	Balloonflower							e										d			d	
Polygonum	Pink Clover																	d		d	d	
Portulaca	Moss-Rose	d				e		e			e	e						d	y	y	d	y
Potentilla	Cinquefoil	y						e	y	e				e				d	y	d	d	y
Primula	Primrose																	d			d	
Ranunculus	Ranunculus					e						e						d			d	
Rosa	Rose	d	e	y	y	e	e	e			e	e	e	e			e	d	y	y	d	
Rosmarinus	Rosemary	d	e			e		e	y		e	e	e					d		y	d	y
Rudbeckia	Black-Eyed Susan	d	e			e		e			e	e	e	e			e	d			d	y
Rumohra	Fern, Leatherleaf			y				e		e								d		y	d	
Salvia	Salvia (Sage, ann.)		e			e		e			e	e	e				x	d	y	y	d	y
Salvia	Salvia (Sage, per.)	d	e			e					e	e	e	e			x	d	y	y	d	y
Sanguisorba	Burnet																e	d			d	

(continued)

Table 8.7

Herbicides Used in Herbaceous Ornamentals *(Continued)*

Key:
y = registered for some species of this genus.
e = registered for some species of this genus, apply only after established.
d = registered for some species of this genus, directed application only
x = NOT registered/prohibited from use on some species.

		Preemergence															Postemergence					
Genus	**Common Name**	Dithiopyr	Isoxaben	Metolachlor	Napropamide	Oryzalin	Oxadiazon	Pendimethalin 2 G	Pendimethalin 2.86 G	Pendimethalin 3.3 EC	Trifluralin	Benefin + Oryzalin	Isoxaben + Trifluralin	Oxadiazon + Pendimethalin	Oxadiazon + Prodiamine	Oxyfluorfen + Oxadiazon	Fenoxaprop	Bentazon	Clethodim	Fluzaifop	Pelargonic Acid	Sethoxydim
Santolina	Lavendercotton														y	y		d		y	d	y
Sanvitalia	Zinnia, Creeping																e	d			d	
Saponaria	Soapwort																e	d			d	
Saxifrage	Saxifrage																	d	y		d	
Scabiosa	Mourning-Bride										e			e				d			d	
Scilla	Squill			y														d			d	
Sedum	Sedum (Stonecrop)	d	x	y	y	e	e	e			e	e	x					d	y	y	d	y
Sempervivum	Hens and Chickens																	d		y	d	
Senecio	Dusty Miller	d	e			e		e					e					d	y		d	
Silene	Catchfly																e	d			d	y
Solanum	Nightshade		e					e	y				e					d			d	
Stachys	Lamb's Ear			y			e				e							d			d	y
Stokesia	Stoke's Aster					e		e	y		e	e						d			d	y
Strelitzia	Bird of Paradise				y	e						e						d		y	d	
Tagetes	Marigold	d	e	y		e		e			e	e	e					y	y	y	d	y
Teucrium	Germander										e							d			d	
Thymus	Thyme, Ornamental																e	d			d	
Tradescantia	Spiderwort																	d			d	y
Trientalis	Starflower																e	d			d	
Tropaeolum	Nasturtium										e							d			d	
Tulipa	Tulip	d	y	y		y		e			y		y					d			d	y

(continued)

Table 8.7
(Continued)

Key:
y = registered for some species of this genus.
e = registered for some species of this genus, apply only after established.
d = registered for some species of this genus, directed application only
x = NOT registered/prohibited from use on some species.

		Preemergence															Postemergence					
Genus	**Common Name**	**Dithiopyr**	**Isoxaben**	**Metolachlor**	**Napropamide**	**Oryzalin**	**Oxadiazon**	**Pendimethalin 2 G**	**Pendimethalin 2.86 G**	**Pendimethalin 3.3 EC**	**Trifluralin**	**Benefin + Oryzalin**	**Isoxaben + Trifluralin**	**Oxadiazon + Pendimethalin**	**Oxadiazon + Prodiamine**	**Oxyfluorfen + Oxadiazon**	**Fenoxaprop**	**Bentazon**	**Clethodim**	**Fluzaifop**	**Pelargonic Acid**	**Sethoxydim**
Verbena	Verbena (Vervain)	d	e			e		e	y		e		e					d	y		d	y
Veronica	Speedwell			y		e	e				e			e				d			d	y
Vinca	Periwinkle (perennial gr cvr)	d	e	y	y	e	y	e	y		e	e	x		y		e	d	y	y	d	y
Viola	Pansy	d		y		e		e				e						d	y		d	y
Yucca	Yucca		x	y		e		e		e	e	e	x					d		y	d	
Zantedeschia	Calla lily							e		e								d			d	
Zinnia	Zinnia			y	y	e		e			e	e	e				e	d	y	y	d	Y

Authors: A. F. Senesac and I. Tsontakis-Bradley, Cornell Cooperative Extension of Suffolk County, Long Island Horticultural Research and Extension Center, Riverhead, New York.

Selective Nutsedge Herbicides

Halosulfuron can be safely used on established woody ornamentals to control yellow and purple nutsedge and green kyllinga at rates ranging from 0.031 to 0.062 lb ai/acre (0.035 to 0.7 kg ai/ha). Halosulfuron cannot be applied over the top, but must be applied as a directed spray only (Tables 8.5 through 8.7). Applications need to be made in spring on young nutsedge with three to eight leaves, with reapplication six weeks later for season-long control. Contact to foliage can injure several woody ornamentals, especially arborvitae, hemlock, and taxus. Do not apply to herbaceous perennials or annual color beds.

Bentazon is labeled to control yellow nutsedge, annual sedges, and certain seedling broadleaf weeds like mallow, purslane, smartweed, wild mustard, and thistle. Use rates range from 1 to 2 lb ai/acre (1.12 to 2.24 kg ai/ha), and bentazon can used on selected woody trees, shrubs, and ground covers. Applications should begin in early summer when sedges are vigorously growing. Repeat applications are sometimes necessary. Bentazon can be applied over the top of only a limited number of ornamentals, but can be used as directed spray on many ornamentals. Applications around sycamore, taxus, and rhododendron should be avoided, as injury from root uptake has been reported. While most herbaceous perennials are very susceptible to injury from bentazon, impatiens, marigold, petunia, and snapdragon are exceptions.

Imazaquin controls broadleaf weeds, sedges, and some grasses in select ornamentals. The major weeds controlled with imazaquin are hairy bittercress, chickweed, Carolina geranium, and yellow and purple nutsedge. Application rates range from 0.4 to 0.5 lb ai/acre (0.45 to 0.56 kg ai/ha). Azaleas, hollies, viburnums, ligustrums, abelia, pieris, and birch trees are sensitive to imazaquin, with a number of reported cases of injury. Application should be made only to labeled ornamentals; however, stunting has been reported, especially to labeled woody plants. Do not apply where bedding plants are to be planted.

appendix a

Common and Trade Names of Specialty-Use Herbicides and Plant-Growth Regulators*

Common Name	Trade Name(s)
Aminopyralid	Milestone 2L, Milestone VM 2L
Ammoniated soap of fatty acids	Quick-fire, Herbicidal Soap
Asulam	Asulox 3.34L, Asulam 3.3L
Atrazine	AAtrex, Atrazine Plus, Purge II, Aatrex 90, Atrazine 4L, Bonus S, St. Augustine Weedgrass Control, others
Benefin	Balan 2.5G. 1.5EC, Crabgrass Preventer, others
Benefin + oryzalin	XL 2G
Benefin + trifluralin	Team 2G, Crabgrass Preventer 0.92%, Team Pro
Bensulide	Betasan, Pre-San 12.5 & 7 G, Bensumec 4L, Lescosan, Weedgrass Preventer, Betamec, Squelch, others
Bentazon	Basagran T/O 4L, Lescogran 4L, Nutgrass 'Nihilator
Bentazon + atrazine	Prompt 5L, Laddock S-12
Bispyribac-sodium	Velocity 80WP, 17.6 WDG
Bromoxynil	Buctril 2L, Brominal 4L, Bromox 2E, Moxy 2E
Cacodylic acid	Montar, Weed Ender
Carfentrazone	Quicksilver 1.9 L
Carfentrazone + 2,4-D + MCPP + dicamba	Speed Zone Southern, Speed Zone Northern, Bermuda 2.2L
Carfentrazone + MCPA + MCPP + dicamba	Power Zone
Chlorsulfuron	Corsair 75DF, Telar 75DG
Clethodim	Envoy 0.94 EC, Clethodim 2EC
Clopyralid	Lontrel T&O 3L, Transline 3L
CMA (CAMA)	Calar, Ortho Crabgrass Killer-Formula II, Selectrol
Corn gluten	Dynaweed, WeedzSTOP 100G
Cytokinin	Agriplex PGR for T&O
Dazomet	Basamid
Dichlobenil	Casoron 4G, Dyclomec 4G, Norosac 4G
2,4-D	2,4-D Amine & Ester, Weedone LV4, Dacamine, Weedar 64, AM-40, 2, 4-D LV4, Dymec, Lesco A-4D, Hardball, Esteron 638, Savana, others
2,4-D + clopyralid + dicamba	Millennium Ultra 3.75L
2,4-D + clopyralid + triclopyr	Momentum, Confront 3
2,4-D + dicamba	81 Selective Weedkiller, Four Power Plus, Triple D Lawn Weed Killer, Banvel 2,4-D
2,4-D + dichlorprop (2,4-DP)	2D + 2DP Amine, Turf D + DP, Fluid Broadleaf Weed Control, Weedone DPC Ester & Amine, others

*Refer to the herbicide label for specific site and use registration.

(continued)

2,4-D + dichlorprop (2,4-DP) + dicamba	Super Trimec, Brushmaster
2,4-D + dichlorprop (2,4-DP) + MCPP + spoiler 4.1L	Broadleaf Granular Herbicide, Dissolve, Triamine, Triamine Jet-Spray Triplet SF, Turf Weeder, Weed Whacker
2,4-D + mecoprop (MCPP)	2D Amine + 2MCPP, 2 Plus 2, MCPP-2,4-D, Phenomec, Ortho Weed-B-Gon Lawn Weed Killer, Patron 170, others
2,4-D + MCPP + 2,4-DP	Broadleaf Granular Herbicide, Dissolve, Triamine, Tri-Ester, Jet-Spray 3-Way Weed Control, Turf Weeder, others
2,4-D + MCPP + dicamba + MCPA and/or 2,4-DP	Trimec Southern, Three-Way Selective, Eliminate DG, 33-Plus, Dissolve, Triamine 3.9 lb/gal, TriEster, Triplet, Trex-San, Weed-B-Gon, 2 Plus 2, Bentgrass Selective Weed Killer, Trimec Bentgrass Formula, Strike 3, Broadleaf Trimec, MECAmine-D, Trimec 992, Weed-B-Gon for Southern Lawns, Formula II, Endrun 3.22L, others
2,4-D TIPA + MCPP + dicamba	Triplet Low Odor
DCPA	Dacthal W-75 WP, Dacthal 6F
Dicamba	Vanquish 4 L, K-O-G Weed Control, Bentgrass Selective, Banvel 4S, Oracle, Vision, others
Dicamba + MCPA + MCPP	Encore DSC, Tri-Power Dry, Tri-Power Selective, Trimec Encore
Dichlobenil	Casoron 4G, Barrier 4G
Diclofop	Illoxan 3EC
Diflufenzopyr + Dicamba	Overdrive 70WG
Dikegulac-sodium	Atrimmec
Dimethenamid	Outlook 6L
Diquat	Reward 2LS, Watrol, Vegetrol, Aquatate, Aquatrim II
Dithiopyr	Dimension 1L, Dimension Ultra 40WSP, Lifeguard, Crab and Spurge Preventer, Dimension 270-G
Diuron	Karmex, Diuron
DSMA	Ansar, DSMA Liquid, Methar 30, Namate, DSMA 4
DSMA + 2,4-D	Weed Beater Plus
Ethofumesate	Prograss 1.5EC, 4.0SC
Ethephon	Proxy 2L, Ethephon 2, ProTrim
Fenarimol	Rubigan 1AS, Patchwork 0.78G
Fenoxaprop	Acclaim Extra 0.94L
Fluazifop	Fusilade II T&O, Ornamec 170, Ornamec Over-The-Top
Fluroxypyr	Spotlight 1.5L
Fluroxypyr + 2,4-D + dicamba	Escalade 4.4L, Escalade2 4L
Fluroxypyr + 2,4-D TIPA + dicamba	Escalade Low Odor 4.4L
Fluroxypyr + MCPP	Bastion T
Fluroxypyr + MCPP + 2,4-DP	Chaser Ultra 2 Selective Herbicide
Flurprimidol	Cutless 50WP
Flurprimidol + trinexapac-ethyl	Legacy 1.52 MEC
Foramsulfuron	Revolver 0.19L
Fosamine	Krenite 4S
Gibberellic acid	RyzUp, ProGibb T&O
Gibberellic acid + indolebutyric acid	PGR IV
Glufosinate	Finale 1L
Glyphosate	Roundup Pro 4L, Roundup ProDry, Gly-Flo, GlyphoMate 41, Clear-Out 41 Plus, Glypro, AquaNeat, Razor Pro, Rodeo 5.4L, Kleenup Pro, Weed Wrangler, Prosecutor, Touchdown Pro, Trailblazer, Glyphomate 41 (3.8L), Fireball 1.55L (acid), others

Glyphosate + 2,4-D	Campaign 3.1 L
Glyphosate + Diquat	QuikPRO, Prosecutor Swift Acting
Halosulfuron	Sedgehammer, 75WP, Sandea 75WP, Manage 75WP, Sempra 75WP
Hexazinone	Velpar 2L
Imazapic	Plateau 70DG, Panoramic 2SL
Imazapic + Glyphosate	Journey 2.25L
Imazapyr	Arsenal 2S, Arsenal Powerline 2L, Arsenal Applicators Concentrate 4L
Imazapyr + diuron	Sahara DG
Imazaquin	Image 1.5L, 70DF
Isoxaben	Gallery 75DF
Isoxaben + trifluralin	Preen 1.9G, Snapshot 2.5 TG, Gallery + Team Woodace Preen Plus
Maleic hydrazide	Royal Slo-Gro
MCPA	Weedar MCPA 4 lb/gal, MCPA-4 Amine, others
MCPA + clopyralid + dichlorprop	Chaser Ultra
MCPA + clopyralid + triclopyr	Battleship
MCPA + fluroxypyr + triclopyr	Battleship III
MCPA + MCPP + dicamba	Trimec Encore, Tri-Power, Trimec Encore DSC
MCPA + MCPP + 2,4-DP	Triamine II, Tri-Ester II
MCPA + dicamba + triclopyr	Eliminate, Three-Way Ester II, Horsepower 4.56 lb/gal, CoolPower 3.6 lb/gal, Clover Power, Spurge Power
MCPP	Mecomec 4, Chickweed & Clover Control, Lescopex, MCPP-4 Amine, MCPP-4K, others
MSMA	Daconate 6, Dal-E-Rad, Crab-E-Rad, MSMA 6.6L, Drexar 530, Bueno 6L, 120 Herbicide, Daconate Super, 912 Herbicide, MSMA Turf, Summer Crabicide, Target MSMA, Weed Hoe, others
MSMA + 2,4-D +MCPP + dicamba	Trimec Plus (Quadmec)
Mefluidide	Embark T&O, Embark 2S, Sta-Lo
Mesotrione	Tenacity 4L
Methyl chlorflurenol	Maintain CF
Metribuzin	Sencor 75DF
Metolachlor	Pennant 7.8 lb/gal, Pennant Magnum 7.62L
Metsulfuron	Manor 60 DF, Blade 60DF, Escort 60 DF, Patriot 60 WDG, Metsulfuron Pro
Methyl bromide	Brom-O-Gas, Terr-O-Gas, MB 98, MBC
Napropamide	Devrinol 50 DF, 2G, 10G, Ornamental Herbicide 5G
Napropamide + oxadiazon	PrePair 6G
Norflurazon	Predict
Oryzalin	Surflan AS 4 lb/gal, Oryzalin Pro, Weed Impede, Surflan Coated Granules
Oxadiazon	Ronstar 2G, 50WP, Ronstar Flo 3.17 L
Oxadiazon + benefin	Regalstar 1.5G
Oxadiazon + bensulide	Goosegrass/Crabgrass Control 6.56G
Oxadiazon + dithiopyr	SuperStar
Oxadiazon + pendimethalin	Kansel + (20-2-13) 3G
Oxadiazon + prodiamine	Regalstar II 1.2G
Oxyfluorfen	Goal 2XL
Oxyfluorfen + oxadiazon	OO-Herbicide 3G, Regal OO, LaSar
Oxyfluorfen + pendimethalin	OH2

(continued)

Oxyfluorfen + oryzalin	Rout
Paclobutrazol	Turf Enhancer 50WP, 2SC, Trimmit 2SC, TGR
Paraquat	Gramoxone Max 3L
Pelargonic acid	Scythe, Quik
Pendimethalin	Pendulum (3.3EC, 2G), Pendulum AquaCap (3.8 CS), Hurdle, Turf Weedgrass Control, Halts, Corral 2.68G, ProPendi, Pendiflex 32
Penoxsulam	LockUp, Sapphire
Picloram	Grazon, Tordon
Prodiamine	Barricade 65WDG, Endurance 65 WDG, Factor 65 WDG, RegalKade 0.5G & 0.37G, Stonewall, ProClipse 65WDG, Cavalcade + others
Pronamide	Kerb 50WP
Pyraflufen-ethyl	Octane 2%SC (0.177 lbs/gal)
Quinclorac + Drive XLR8	Drive 75 DF
Quinclorac + 2,4-D + dicamba + sulfentrazone	Q4
Rimsulfuron	TranXit GTA 25DG
Sethoxydim	Sethoxydim G-Pro 1L, Vantage 1.0 lb/gal, Grass Getter, Poast, Poast Plus
Siduron	Tupersan 50WP, 3.5%G, 4.6%G
Simazine	Princep 4 lb/gal, T&O, 80WP, Simazine, Wynstar, Sim-Trol, others
Sulfentrazone	Dismiss Turf Herbicide 4L
Sulfentrazone + 2,4-D + MCPP + dicamba	Surge 2.18L
Sulfentrazone + prodiamine	Echelon 0.3G, 4SC
Sulfometuron-methyl	Oust 75DG, Spyder 75DG, SFM G-Pro 75EG
Sulfosulfuron	Certainty 75WDG, Outrider 75WDG
Triclopyr	Turflon Ester 4L, Garlon 3A (amine), Garlon 4A (ester), Pathfinder 1L (RTU), Tahoe 3A, Tahoe 4E
Triclopyr + 2,4-D	Turflon II Amine, Chaser 3L Ester, Chaser 2 Amine
Triclopyr + clopyralid	Confront 3L, Confront NR
Triclopyr + MCPP + dicamba	3-Way Ester II
Trifloxysulfuron	Monument 75DF
Trifluralin	Treflan 5G, Trifluralin 4EC, Trilin 4EC, 5EC, Preen, Vegetable and Ornamental Weeder
Trifluralin + isoxaben + oxyfluorfen	Showcase 2.5G
Trinexapac-ethyl	Primo 1EC, Triple Play, Primo WSP, Primo MAXX, Governor 0.055% 5-0-10; 0.17%
Xanthomonas campentris	X-Po

All chemicals mentioned are for reference only. Not all are available for turf use. Some may be restricted by some states, provinces, or federal agencies. It is advisable to check the current use status of the pesticide being considered. Always read and follow the manufacturer's label as registered under the Federal Insecticide, Fungicide, and Rodenticide Act. Mention of a proprietary product does not constitute a guarantee or warranty of the product by the authors or the publishers of this book and does not imply approval to the exclusion of other products that also may be suitable.

references and additional readings

Anderson, W. P. 1983. *Weed Science: Principles.* 2d ed. West Publishing Company. St. Paul, MN.

Ashton, F. M., and T. J. Monaco. 1991. *Weed science principles and practices.* John Wiley and Sons, Inc. NY.

Askew, S. D., and J. B. Beam. 2002a. Weed management in cool-season turf with mesotrione. *Proc. Northeast Weed Sci. Soc.* 56:129.

Askew, S. D., J. B. Beam, and S. R. King. 2002b. Nimblewill control in cool-season turf. South. *Weed Sci. Soc. Proc.* 55:64.

Askew, S. D., J. B. Beam, and W. L. Barker. 2003. Isoxaflutole and mesotrione for weed management in cool-season turfgrasses. *Proc. Northeast Weed Sci. Soc.* 57:111.

Baldwin, C. M., H. Liu, L. B. McCarty, and W. L. Bauerle. 2006. Effects of trimexapac-ethyl on the salinity tolerance of two ultra-dwarf bermudagrass cultivars. *HortSci.* 41:808–814.

Banks, P. A., B. Branham, K. Harrison, T. Whitson, and I. Heap. 2004. Determination of the potential impact from the release of glyphosate- and glufosinate-resistant *Agrostis stolonifera* L. in various crop and non-crop ecosystems. Weed Science Society of America. http://www.wssa.net/society/ bentgrass.pdf

Barrett, L. H., and J. A. Jagschitz. 1975. Control of crabgrass and goosegrass with preemergence chemicals in turfgrass. *Proc. Northeast Weed Sci. Soc.* 29:359–364.

Beam, J. B., W. L. Barker, and S. D. Askew. 2006. Selective creeping bentgrass (*Agrostis stolonifera*) control in cool-season turfgrass. *Weed Tech.* 20:340–344.

Berg, R. 1997. *Introductory botany: plants, people, & the environment.* Harcourt, Inc., 6277 SE Harbor Drive, Orlando, FL.

Bettini, P., S. McNally, M. Sevignac, H. Darmency, J. Gasquez, and M. Drone. 1987. Atrazine resistance in *Chenopodium album*: low and high levels of resistance to the herbicide are related to the same chloroplast *psbA* gene mutation. *Plant Physiol.* 84:1442–1446.

Bingaman, B. R., and N. E. Christians. 1995. Greenhouse screening of corn gluten meal as a natural control product for broadleaf and grass weeds. *HortSci.* 30:1256–1259.

Bingham, S. W. 1974. Influence of selected herbicides on rooting of turfgrass sod. pp. 372–377. In E. C. Roberts, (ed.), *Proc. 2nd Intern. Turf. Res. Conf. Amer. Soc. Agron. and Crop Sci. Soc. Amer.* Madison, WI.

Bingham, S. W. 1985. Effectiveness of herbicides for *Eleusine indica* control during *Cynodon dactylon* improvement in golf course fairways. pp. 705–715. In F. Lemaire, (ed.), *Proc. 5th Intern. Turf. Res. Conf.*, Avignon, France.

Bingham, S. W., R. E. Schmidt, and C. K. Curry. 1969. Annual bluegrass control in overseeded bermudagrass putting green turf. *Agron. J.* 61:908–911.

Bingham, S. W., and R. L. Shaver. 1979. Effectiveness of herbicide programs for annual bluegrass (*Poa annua*) control in bermudagrass (*Cynodon dactylon*). *Weed Sci.* 27:367–370.

Bingham, S. W., and R. L. Shaver. 1980. Goosegrass control in bermudagrasses. 1980. pp. 237–245. In J. B. Beard (ed.), *Proc. 3rd Intern. Turf Res. Conf., Am. Soc. Agron.*, Madison, WI.

Blum, R. R., J. Isgrigg III, and F. H. Yelverton. 2000. Purple (*Cyperus rotundus*) and yellow nutsedge (*Cyperus esculentus*) control in bermudagrass (*Cyondon dactylon*) turf. *Weed Tech.* 14:357–365.

Bode, L. E. 1987. Spray application technology. In G. G. McWhorter and M. R. Gebhardt (eds.), *Methods of applying herbicides.* Monograph series of the Weed Science Society of America. Champaign, IL. pp. 85–121.

Bohwmik, P. C. 1986. Fenoxaprop-ethyl for postemergence crabgrass control in Kentucky bluegrass turf. *HortSci.* 21:457–458.

Bohwmik, P. C. and D. C. Riego. 2003. Management alternatives for glyphosate resistant creeping bentgrass. *Proc. Northeast Weed Sci. Soc.* 57:114.

Boyd, J. 2000. Kill off bermudagrass with one less spraying. *Golf Course Manag.* 68(5):68–71.

Brady, N. C., and R. R. Well. 2002. *The nature and properties of soils*, 13th edition. Prentice Hall. Upper Saddle River, NJ.

Branham, B. E., and P. E. Rieke. 1986. Effects of turf cultivation practices on the efficacy of preemergence grass herbicides. *Agron. J.* 78:1089–1091.

Branham, B. E., W. Sharp, E. A. Kohler, T. W. Fermanian, and T. B. Voight. 2005. Selective control of creeping bentgrass (*Agrostis stolonifera* [L.]) in Kentucky bluegrass (*Poa pratensis* [L.]) *Turf. Inter. Turf. Soc. Res. J.* 10:1164–1169.

Brecke, B. J. 1981. Smutgrass (*Sporobolus poiretii*) control in bahiagrass (*Paspalum notatum*) pastures. *Weed Sci.* 291:553–555.

Bregitzer, P., S. E. Halbert, and P. G. Lemaux. 1998. Somaclonal variation in the progeny of transgenic barley. *Theor. Appl. Genet.* 96:421–425.

Briggs, J. A., and T. Whitwell. 2001. Weed control. Grower's Program—South Carolina Horticultural Program.

Buchanan, B. B., W. Gruissem, and R. J. Jones. 2000. Biochemistry & molecular biology of plants. *American Society of Plant Physiologists*, Rockville, MD.

Buchanan, G. A., C. S. Hoveland, and M. C. Harris. 1975. Response of weeds to soil pH. *Weed Sci.* 23:473–477.

Bunnell, B. T., L. B. McCarty, and W. C. Bridges. 2005. 'TifEagle' bermudagrass response to growth factors and mowing heights when grown at various hours of sunlight. *Crop Sci.* 45:575–581.

Busey, P. 2004. Managing goosegrass II: Removal. *Golf Course Manag.* 72(2):132–136.

Busey, P. 2003. Cultural management of weeds in turfgrass: A review. *Crop Sci.* 43:1899–1912.

Callahan, L. M., and E. R. McDonald. 1992. Effectiveness of bensulide in controlling two annual bluegrass (*Poa annua*) subspecies. *Weed Tech.* 6:97–103.

Caress, S. M., and A. C. Steinemann. 2004. A national population study of the prevalence of multiple chemical sensitivity. *Arch. Environ. Health* 59(6): 300–305.

Carroll, M. J., M. J. Mahoney, and P. N. Dernoeden. 1992. Creeping bentgrass (*Agrostis palustris*) quality as influenced by multiple low-rate applications of fenoxaprop. *Weed Tech.* 6:356–360.

Cauchy, P. 2000. Carfentrazone-ethyl. Cereal herbicide. *Phytoma.* 531:55–58

Chernicky, J. P., B. J. Gossett, and T. R. Murphy. 1984. Factors influencing control of annual grasses with sethoxydim and RO-13-8895. *Weed Sci.* 32:174–177.

Chism, W. J., and S. W. Bingham. 1991. Postemergence control of large crabgrass (*Digitaria sanguinalis*) with herbicides. *Weed Sci.* 39:62–66.

Chou, C., and Z. A. Patrick. 1976. Identification and phytotoxic activity of compounds produced during decomposition of corn and rye residues in soil. *J. Chem Ecol.* 2:369–387.

Christians, N. E. 1985. Response of Kentucky bluegrass to four growth stimulates. *J. Amer. Soc. Hort. Sci.* 110:765–769.

Christians, N. 1993a. Making its way to the marketplace: A natural product for the control of annual weeds. *Golf Course Manag.* 61(10): 72, 74, 76.

Christians, N. 1993b. The use of corn gluten meal as a natural preemergence weed control in turf. *Int. Turfgrass Soc. Res. J.* 7:284–290.

Coats, G. E. 1986. Turfgrass weed control. *Miss. Agric. and Forestry Exp. Station Bull. 95.*

Coats, G. E., D. C. Heering, and J. W. Scruggs. 1987. Wild garlic and purple nutsedge control in turf. *Proc. South. Weed Sci.* 40:98.

Cummins, J. 2005. Deregulation of Glyphosate Tolerant Creeping Bentgrass Out of Question, http://www.indsp.org/DGTCBOQ.php.

della-Cioppa, G., S. C. Bauer, B. K. Klein, D. Shah, R. T. Fraley, and G. Kishore. 1986. Translocation of the precursor of 5-enolpyruvylshikimate-3-phosphate synthase into chloroplasts of higher plants *in vitro*. *Proc. Natl Acad. Sci. USA.* 83:6873–6877.

Dernoeden, P. H. 1984. Four-year response of a Kentucky bluegrass–red fescue turf to plant growth retardants. *Agron. J.* 76:807–813.

Dernoeden, P. H. 1985. Controlling crabgrass with reduced rates of herbicides. *Grounds Maint.* 20:12–22, 94.

Dernoeden, P. H. 1987. Tolerance of perennial ryegrass and tall fescue seedlings to fenoxaprop. *Agron. J.* 79:1035–1037.

Dernoeden, P. H. 1989a. Bermudagrass suppression and zoysiagrass tolerance to fenoxaprop. pp. 285–290. In H. Takatoh (ed.). *Proc. 6th Intern. Turf. Res. Conf., Intern. Turf. Soc.* Tokyo, Japan.

Dernoeden, P. H. 1989b. Mature creeping bentgrass and seedling Kentucky bluegrass tolerance to fenoxaprop. pp. 279–283. In H. Takatoh (ed.). *Proc. 6th Intern. Turf. Res. Conf., Intern. Turf. Soc.* Tokyo, Japan.

Dernoedon, P. H. 1990. Comparison of three herbicides for selective tall fescue control in Kentucky bluegrass. *Agron. J.* 82:278–282.

Dernoeden, P. H., C. A. Bigelow, and J. E. Kaminski. 2003. Preemergence herbicides for smooth crabgrass control generally performed poorly in Maryland in 2002. *Proc. Northeast Weed Sci. Soc.* 57:101–102.

Dest, W. M., and K. Guillard. 1987. Nitrogen and phosphorus nutritional influence on bentgrass-annual bluegrass community composition. *J. Amer. Soc. Hort. Sci.* 112:769–773.

Dickens, R. 1979. Control of annual bluegrass (*Poa annua*) in overseeded bermudagrass (*Cynodon* spp.) golf greens. *Weed Sci.* 27: 642–644.

Diesburg, K. L., and N. E. Christians. 1989. Seasonal applications of ethephon, flurprimidol, mefluidide, paclobutrazol, and amidochlor as they affect Kentucky bluegrass shoot morphogenesis. *Crop Sci.* 29:841–847.

DiPaola, J. M., and W. M. Lewis. 1989. Growth regulators and *Poa annua* suppression. *Proc. 27th Ann. N. C. Turf. Conf.* 5:22–23.

Eggens, J. L., and C. H. Wright. 1983. Effects of ethephon on Kentucky bluegrass, annual bluegrass, and creeping bentgrass. pp. 5–6. In *Annual Turfgrass Research Report.* Ontario Agric. Coll., Univ. of Guelph.

Eggens, J. L., and C. P. M. Wright. 1985. Kentucky bluegrass and annual bluegrass response to ethephon. *J. Amer. Soc. Hort. Sci.* 110:609–611.

Eggens, J. L., C. P. M. Wright, D. P. Murr, and K. Carey. 1989. The effect of ethephon on annual bluegrass and creeping bentgrass growth. *Can. J. Plant Sci.* 69:1353–1357.

Emmons, R. D. 1984. *Turfgrass science and management.* Delmar Publishers, Inc. Albany, NY.

Engel, R. E., and R. D. Ilnicki. 1969. Turf weeds and their control. pp. 240–287. In A. A. Hanson and F. V. Juska (eds.). Turfgrass Science. *Am. Soc. Agron.* Madison, WI.

Estes, A. G., and L. B. McCarty. 2005. Selective annual bluegrass control in cool season turf with Velocity (bispyribac-sodium). *South. Weed Sci. Soc. Proc.* 58 (CD format).

Fagerness, M. J., D. C. Bowman, F. H. Yelverton, and T. W. Rufty. 2004. Nitrogen use in Tifway bermudagrass, as influenced by trinexapac-ethyl. *Crop Sci.* 44:595–599.

Fagerness, M. J., F. H. Yelverton, J. Isgrigg, and R. J. Cooper. 2000. Plant growth regulators and mowing height affect ball roll and quality of creeping bentgrass putting greens. *HortSci.* 35:755–759.

Fagerness, M. J., and F. H. Yelverton. 2001. Plant growth regulators and mowing height effects on seasonal root growth of 'Penncross' creeping bentgrass. *Crop Sci.* 41:1901–1905.

Fech, J. C., F. P. Baxendale, J. Powell, and J. Derr. 2000. New pest controls for the new millennium. *Grounds Maint.* 35(4):49–50.

Fei, S., and E. Nelson. 2004. Greenhouse evaluation of fitness-related reproductive traits in Roundup-tolerant transgenic creeping bentgrass (*Agrostis stolonifera* [L.]). 2004. *InVitro Cell. Dev. Biol.—Plant.* 40:266–273.

Ferguson, G. P., G. E. Coats, G. B. Wilson, and D. R. Shaw. 1992. Postemergence control of wild garlic (*Allium vineale*) in turfgrass. *Weed Tech.* 6:144–148.

Figliola, S. S., N. D. Camper, and W. H. Ridings. 1988. Potential biological control agents for goosegrass (*Eleusine indica*). *Weed Sci.* 36:830–835.

Fulwider, J. R., and R. E. Engel. 1959. The effect of temperature and light on germination of seed of goosegrass, *Eleusine indica. Weeds* 7:359–361.

Gaul, M. C., and N. E. Christians. 1988. Selective control of annual bluegrass in cool-season turfs with fenarimol and chlorsulfuron. *Agron. J.* 80:120–125.

Gaussoin, R. E. 2006. Personal communication.

Gaussoin, R. E., and B. E. Branham. 1987. Annual bluegrass and creeping bentgrass germination response to flurprimidol. *HortSci.* 22:441–442.

Gibeault, V. A., and N. R. Goetze. 1972. Annual meadow-grass. *J. Sports Turf Res. Inst.* 48:1–11.

Goloubinoff, P., M. Edelman, and R. B. Hallick. 1984. Chloroplast-encoded atrazine resistance in *Solanum nigrum: psb*A loci from susceptible and resistant biotypes are isogenic except for a single codon change. *Nucleic Acid Res.* 12:9489–9496.

Goss, R. L., S. E. Brauen, and S. P. Orton. 1975. The effects of N, P, K and S on *Poa annua* (L.) in bentgrass putting green turf. *J. Sports Turf Res. Inst.* 51:74–82.

Goss, R. M., J. H. Baird, S. L. Kelm, and R. M. Calhoun. 2002. Trinexapac-ethyl and nitrogen effects on creeping bentgrass grown under reduced light conditions. *Crop Sci.* 42:472–479.

Gressel, J., and L. A. Segel. 1982. Interrelating factors controlling the rate of appearance of resistance: The outlook for the future. In *Herbicide resistance in plants.* H. M. LeBaron and J. Gressel (eds.). John Wiley and Sons, NY.

Grossmann, K. 1998. Quinclorac belongs to a new class of highly selective auxin herbicides. *Weed Sci.* 46:707–716.

Grossmann, K. 2003. Mediation of herbicide effects by hormone interactions. *J. Plant Growth Regul.* 22:109–222.

Guenzi, W. D., and T. M. McCalla. 1962. Inhibition of germination and seedling development by crop residues. *Proc. Soil Sci. Amer.* 26:456–458.

Halisky, P. M., C. R. Funk, and R. E. Engel. 1966. Melting-out of Kentucky bluegrass varieties by *Helminthosporium vagans* as influenced by turf management practices. *Plant Disease Reporter* 50:703–706.

Handly, J. V., J. M. Breuninger, and M. Drinkall. 2001. Characterization of fluroxypyr for broadleaf weed control in turf. *Proceedings: IXth International Turfgrass Research Conference.* 9:53.

Hansen, A. A. 1921. The use of chemical weed killers on golf courses. *USGA Bul.* 1:128–131.

Hart, S., D. W. Lycan, and J. A. Meade. 2003. Fact Sheet—Weed control around the home grounds. Rutgers Cooperative Extension. FS020.

Hart, S. 2001. Weed management in ornamental plantings. Rutgers Coop. Ext. Bulletin E272.

Heap, I. 2004. Criteria for confirmation of herbicide-resistant weeds. *International Survey of Herbicide Resistant Weeds.* www.weedscience.org/in.asp

Hedden, P., and A. L. Phillips. 2000. Gibberellin metabolism: New insights revealed by the genes. *Trends Plant Sci.* 5:523–530.

Higgins, J. M., L. B. McCarty, T. Whitwell, and L. C. Miller. 1987. Bentgrass and bermudagrass putting green tolerance to postemergence herbicides. *HortSci.* 22:248–250.

Higingbottom, J. K., A. G. Estes, and L. B. McCarty. 2005. Goosegrass (*Eleusine indica*) control in common bermudagrass (*Cynodon dactylon*). *South. Weed Sci. Soc. Proc.* 58 (CD format).

Hiltbold, A. E., and G. A Buchanan. 1977. Influence of soil pH on persistence of atrazine in the field. *Weed Sci.* 25:515–520.

Holt, J. S., S. B. Powles, and J. A. M. Holtum. 1993. Mechanisms and agronomic aspects of herbicide resistance. *Ann. Rev. Plant Physiol. Plant Molec. Biol.* 44:203–229.

Howieson, M. J., and N. E. Christians. 2005. Leaf blade and internode lengths of eight grass species after ethephon treatment. *PGRSA Quarterly.* 33(2):66–75.

Isgrigg III, J., F. H. Yelverton, C. Brownie, and L. S. Warren Jr. 2002. Dinitroanaline resistant annual bluegrass in North Carolina. *Weed Sci.* 50:86–90.

Jagschitz, J. A., and J. S. Ebdon. 1985. Influence of mowing, fertilizer and herbicide on crabgrass infestation in red fescue turf. pp. 699–704. In F. Lemaire (ed.). *Proc. 5th Inter. Turf. Res. Conf.* Avignon, France.

Johnson, B. J. 1975. Postemergence control of large crabgrass and goosegrass in turf. *Weed Sci.* 23:404–409.

Johnson, B. J. 1976. Dates of herbicide application for summer weed control in turf. *Weed Sci.* 24:422–424.

Johnson, B. J. 1977. Sequential herbicide treatments for large crabgrass and goosegrass control in bermudagrass. *Agron. J.* 69:1012–1014.

Johnson, B. J. 1980. Goosegrass (*Eleusine indica*) control in bermudagrass (*Cynodon dactylon*) turf. *Weed Sci.* 28:378–381.

Johnson, B. J. 1982a. Frequency of herbicide treatments for summer and winter weed control in turfgrasses. *Weed Sci.* 30:116–124.

Johnson, B. J. 1982b. Oxadiazon treatments on overseeded putting-green turf. *Weed Sci.* 30:335–338.

Johnson, B. J. 1983. Response to ethofumesate of annual bluegrass (*Poa annua*) and overseeded bermudagrass (*Cynodon dactylon*). *Weed Sci.* 31:385–390.

Johnson, B. J. 1984. Influence of nitrogen on recovery of bermudagrass (*Cynodon dactylon*) treated with herbicides. *Weed Sci.* 32:819–823.

Johnson, B. J. 1987a. Effects of core cultivation on preemergence herbicide activity in bermudagrass. *HortSci.* 22:440–441.

Johnson, B. J. 1987b. Turfgrass species response to herbicides applied postemergence. *Weed Tech.* 1:305–311.

Johnson, B. J. 1988a. Fenarimol for control of annual bluegrass in dormant bermudagrass turf. *Georgia Agric. Exp. Stations Research Report 552.*

Johnson, B. J. 1988b. Glyphosate and SC-0224 for bermudagrass (*Cynodon* spp.) cultivar control. *Weed Tech.* 2:20–23.

Johnson, B. J. 1995. Frequency of quinclorac treatments on bermudagrass tolerance and large crabgrass control. *J. of Turf. Manag.* 1(1):49–71.

Johnson, B. J., and R. E. Burns. 1985. Effect of soil pH, fertility, and herbicides on weed control and quality of bermudagrass (*Cynodon dactylon*) turf. *Weed Sci.* 33:366–370.

Johnson, B. J., and R. N. Carrow. 1989. Bermudagrass encroachment into creeping bentgrass as affected by herbicides and plant growth regulators. *Crop Sci.* 29:1220–1227.

Johnson, B. J., R. N. Carrow, and T. R. Murphy. 1990. Foliar-applied iron enhances bermudagrass tolerance to herbicides. *J. Amer. Soc. Hort. Sci.* 115:422–426.

Johnson, B. J. 1997. Growth of 'Tifway' bermudagrass following application of nitrogen and iron with trinexapac-ethyl. *HortSci.* 32:241–242.

Kadir, J., and R. Charudattan. 2000. *Dactylaria higginsii*, a fungal bioherbicide agent for purple nutsedge (*Cyperus rotundus*). *Biological Control. Theory and Application in Pest Management.* 17(2):113–124.

Kageyama, M. E., L. R. Widell, D. G. Cotton, and G. R. McVey. 1989. Annual bluegrass to bentgrass conversion with a turf growth retardant (TGR). pp. 387–390. In H. Takatoh (ed.). *Proc. 6th Intern. Turf. Res. Conf., Intern. Turf. Soc.* Tokyo, Japan.

Kaufman, J. E. 1989. How turfgrass biology affects responses to growth regulators. pp. 83–88. In H. Takatoh (ed.). *Proc. 6th Intern. Turf. Res. Conf., Intern. Turf. Soc.* Tokyo, Japan.

Kelly, S. T., G. E. Coats, and D. S. Luthe. 1999. Mode of resistance of triazine-resistant annual bluegrass (*Poa annua* [L.]) *Weed Tech.* 13:747–752.

Kelley, A. D., and V. F. Burns. 1975. Dissemination of weed seeds by irrigation water. *Weed Sci.* 23:486–493.

Klingman, G. C., and F. M. Ashton. 1982. *Weed Science principles and practices.* John Wiley & Sons, Inc. NY.

Kuhns, L. J., T. Harpster, J. Sellmer, and S. Guiser. 2003. Controlling weeds in nursery and landscape plantings. Penn State Agricultural Research and Cooperative Extension. UJ236.

Kuo, S., S. E. Brauen, and E. J. Jellum. 1992. Phosphorus availability in some acid soils influences bentgrass and annual bluegrass growth. *HortSci.* 27:370.

Larson, A. 1997. Corn gluten meal: Good news for gardeners. *Horticulture and Home Pest News.* 477(16):98–99.

Leach, B. R., and J. W. Lipp. 1927. Additional experiments in grub-proofing turf. *USGA Bul.* 7:28.
LeBaron, H. M., and J. Gressel (eds.). 1982. *Herbicide resistance in plants.* John Wiley and Sons, NY.
Lee, W. D. 1977. Winter annual grass control in Italian ryegrass with ethofumesate. *Weed Sci.* 25:252–255.
Lewis, W. M. 1985. Weeds in turf. pp. 18–34. In A. H. Bruneau, (ed.). *Turfgrass pest management manual.* NC Agric. Ext. Ser. Raleigh, NC.
Lewis, W. M., and J. M. DiPaola. 1989. Ethofumesate for *Poa annua* control in bentgrass. pp. 303–305. In H. Takatoh (ed.). *Proc. 6th Intern. Turf. Res. Conf., Intern. Turf. Soc.* Tokyo, Japan.
Lewis, W. M., J. M. DiPaola, and A. H. Bruneau. 1988. Effects of preemergence herbicides on turfgrass rooting. *Grounds Maint.* 23(2):48–49.
Lickfeldt, D. W., D. S. Gardner, B. E. Branham, and T. B. Voight. 2001. Implications of repeated trinexapac-ethyl applications on Kentucky bluegrass. *Agron. J.* 93:1164–1168.
Liu, X., B. Juang, and G. Banowetz. 2002. Cytoknin effects on creeping bentgrass responses to heat stress: I. Shoot and root growth. *Crop Sci.* 42:457–465.
Loughner, D. L., D. W. Lickfeldt, R. L. Smith, and J. M. Breuninger. 2004. Broadleaf weed control in turf with fluroxypyr based herbicide formunlations. *Proc. of the Western Soc. of Weed Sci.* 57:17.
Lowe, D. B., G. A. Swire-Clark, L. B. McCarty, T. Whitwell, and W. V. Baird. 2001. Biology and molecular analysis of dinitroaniline-resistant *Poa annua* (L.) *Intern. Turf. Soc. Res. J.* 9:1019–1025.
Lycan, D. W., and S. E. Hart. 2006. Cool-season Turfgrass Reseeding Intervals for Bispyribac-sodium. *Weed Tech.* 20:526–529.
Maddy, K. T., F. Schneider, and S. Edmiston. 1989. *Regulation and registration. Principles of weed Control in California,* 2d ed. Fresno, CA.
Mahady, M. M., K. Gard, and D. Mosdell. 2001. Trimmit 2 SC (paclobutrazol) for control of annual bluegrass (*Poa annua*) in perennial ryegrass overseeded common bermudagrass fairways. *Rub of the Green.* 22(12):32, 34, 37, 44, 47, 49.
Maxwell, B. D., and A. M. Mortimer. 1994. Selection for herbicide resistance. In S. B. Powles and J. A. M. Holtum (eds.). *Herbicide resistance in plants, biology and biochemistry.* CRC Press. Boca Raton, FL.
McCarty, L. B. 2005. *Best golf course management practices.* 2d ed. Prentice Hall, Inc. Upper Saddle River, NJ.
McCarty, L. B. 1992. Quinclorac evaluations in warm-season turfgrasses. *South. Weed Sci. Soc. Proc.* 45:136.
McCarty, L. B. 1991. Goosegrass (*Eleusine indica*) control in bermudagrass (*Cynodon* spp.) turf with diclofop. *Weed Sci.* 39:255–261.
McCarty, L. B., J. W. Everest, D. W. Hall, T. R. Murphy, and F. Yelverton. 2008. *Color atlas of turfgrass weeds.* 2d ed. Wiley & Sons, Inc. Hoboken, NJ.
McCarty, L. B., and A. G. Estes. 2005. A new weapon in the fight against *Poa annua. Golf Course Manag.* 73(2)106–109.
McCarty, L. B., and D. L. Colvin. 1991. Carpetgrass response to postemergence herbicides. *Weed Tech.* 5:563–565.
McCarty, L. B., J. M. Higgins, L. C. Miller, and T. Whitwell. 1986. Centipedegrass tolerance to postemergence grass herbicides. *HortSci.* 21:1405–1407.
McCarty, L. B., J. M. Higgins, T. Whitwell, and L. C. Miller. 1989. Tolerance of tall fescue to postemergence grass herbicide. *HortSci.* 24:309–311.
McCarty, L. B., L. C. Miller, and D. L. Colvin. 1991. Bermudagrass (*Cynodon* spp.) cultivar response to diclofop, MSMA, and metribuzin. *Weed Tech.* 5:27–32.
McCarty, L. B., and T. R. Murphy. 1993. Perennial ryegrass (*Lolium perenne*) establishment following diclofop application timings and rates. In R. N. Carrow et al. (ed.), *Int. Turfgrass Soc. Res. Intertec Publ. Corp.*, Overland Park, KS.
McCarty, L. B., and J. S. Weinbrecht. 1997. pp. 327–328. *Cynodon dactylon* X *C. Transvaalensis* cv. Tifway sprigging establishment and weed control following pre-emergence herbicide use. *Inter. Turf. Soc. Res. J.* 8:507–515.
McCarty, L. B. 2002. *Southern Lawns.* Clemson Univ. Ext. Ser. Pub. EC707. 566 pp.
McCarty, L. B., and B. T. Tucker. 2005. Prospects for managing turf weeds without protective chemicals. *Intern. Turf. Soc. Res. J.* 10:34–41.
McCarty, L. B., and F. Yelverton. 2005. Dealing with dallisgrass. *Golf Course Manag.* 73(7):82–88.
McCullough, P., and S. Hart. 2005. Temperature determines efficacy of Velocity herbicide. *Golf Course Manag.* 73(6):85–89.
McCullough, P, H. Liu, and L. B. McCarty. 2005a. Mowing operations influence creeping bentgrass putting green ball roll following plant growth regulator applications. *HortSci.* 40:471–474.
McCullough, P., H. Liu, and L.B. McCarty. 2005b. Trinexapac-ethyl application regimes influence creeping bentgrass putting green performance. *HortSci.* 40:2167–2169.
McCullough, P., L. B. McCarty, H. Liu, and T. Whitwell. 2005c. Response of 'Tifeagle' bermudagrass to fenarimol and gibberellic acid. *Inter. Turf. Soc. Res. J.* 10:1245–1250.
McCullough, P. E., H. Liu, L. B. McCarty, T. Whitwell, and J. E. Toler. 2006a. Nutrient allocation of 'TifEagle' bermudagrass as influenced by trinexapac-ethyl. *J. Plant Nutrition* 29:273–282.
McCullough, P., H. Liu, L. B. McCarty, T. Whitwell, and J. E. Toler. 2006b. Growth and nutrient partitioning of TifEagle bermudagrass as influenced by nitrogen and trinexapac-ethyl. *HortSci.* 41:453–458.
McElroy, J. S., P. D. Hahn, G. K. Breeden, and J. C. Sorochan. 2004. Weed control options for seeded hybrid bluegrass (*Poa pratensis X Poa arachnifera*) establishment. *ASA Annual Meeting Abstracts.* (CD format.)
Meyers, H. G., W. L. Currey, and D. E. Barnes. 1973. Deactivation of Kerb with sewage sludge, topdressing and activated charcoal. *Proc. Florida State Hort. Soc.* 86:442–444.
Miller, L. C., J. P. Krausz, C. L. Parks, and C. S. Gorsuch. 1985. Centipedegrass and its problems. S.C. Coop. Ext. Service. Clemson, SC. Cir. 583.
Miltner, E., A. Bary, and C. Cogger. 2003. Clopyralid and compost: formulation and mowing effects on herbicide content of grass clippings. *Compost Science and Utilization.* 11:289–299.
Monaco, T. J., S. C. Weller, and F. M. Ashton. 2002. *Weed science: Principles and practices,* 4th ed. John Wiley & Sons, Inc. New York, NY. 671 pages.
Monroe, J. H., W. M. Lewis, and J. M. DiPaola. 1990. Aerification effects on preemergence herbicide activity. *Weed Sci. Soc. Am. Abstr.* 30:27.
Morrissette, N. S., A. Mitra, D. Sept, and L. D. Sibley. 2004. Dinitroanilines bind to α-tubulin to disrupt microtubules. *Molecular Biology of the Cell* 15:1960–1968.
Mortimer, A. M. 2005. A review of graminicide resistance. http://www.plantprotection.org/hrac/Bindex.cfm?doc = Monograph1.htm

Murdoch, C. L., and R. K. Nishimoto. 1982. Diclofop for goosegrass control in bermudagrass putting greens. *HortSci.* 17:914–915.

Murphy, T. R. 2004. Herbicide-resistant weeds in Georgia turfgrasses. www.griffin.uga.edu/grf/dept/ cropsci/turf/weedcontrol/HERBRESTa-1.pdf

Murphy, T. R., and B. Nutt. 2003. Tall fescue control with MON 44951. *Summary of Turfgrass, Forage, & Noncropland Management Research.* 1–12.

Murphy, T. R. 1988. Turfgrass weed control for professional managers. Univ. Georgia Coop. Ext. Service. Bulletin 991.

Murphy, T. R., and B. J. Johnson. 1995. Winter annual weed control in bermudagrass. *South. Weed Sci. Soc. Proc.* 48:99–100.

National Research Council, Committee on Plant and Animal Pests. 1968. *Principles of plant and animal pest control, vol. 2, Weed Control.* Natl. Acad. Sci., Washington, DC.

Neal, J. C. 1998. Weed management in annual color beds. North Carolina Cooperative Extension Service. HIL-644.

Neil, J. C. 1995. Yellow nutsedge: Biology and control in cool-season turf. *Turfgrass Trends.* 4(7):15–19.

Neilsen, K. F., T. F. Cuddy, and W. B. Woods. 1960. The influence of the extract of some crops and soil residues on germination and growth. *Can. J. Plant Sci.* 40:188–197.

Nishimoto, R. K., and C. L. Murdoch. 1999. Mature goosegrass (*Eleusine indica*) control in bermudagrass (*Cynodon dactylon*) turf with a metribuzin-diclofop combination. *Weed Tech.* 13:169–171.

Nishimoto, R. K., and L. B. McCarty. 1997. Fluctuating temperature and light influence on seed germination of goosegrass (*Eleusine indica*). *Weed Sci.* 45:426–429.

Nittler, L. W., and T. J. Kenny. 1976. Response of *Festuca rubra* (L.) seedlings to ethephon. *Agron. J.* 68:711–713.

Norcini, J. G., and R. H. Stamps. 1994. Container nursery weed control. Univ. of Florida—Florida Coop. Ext. Ser. Cir. 678.

Padgette, S. R., D. B. Re, G. F. Barry, D. E. Eichholtz, X. Delannay, R. L. Fuchs, G. M. Kishore, and R. T. Fraley. 1996. New weed control opportunities: Development of soybeans with a Roundup ready gene. In Duke, K. (ed.). *Herbicide-resistant crops: Agriculture, economic, environmental, regulatory and technological aspects.* CRC Press. Boca Raton, FL. 1996:53–84.

Polomski, B. 1998. Questions and answers. *Horticulture: The Magazine of American Gardening.* 95(9):12.

Prather, T. S., J. M. DiTomaso, and J. S. Holt. 2000. History, mechanisms, and strategies for prevention and management of herbicide resistant weeds. *Proceedings of the California Weed Science Society.* 52:155–163.

Quan, Y. L., and M. C. Engelke. 1999. Influence of trinexapac-ethyl on 'Diamond' zoysiagrass in a shade environment. *Crop Sci.* 39:202–208.

Rademacher, W. 2000. Growth retardants: Effects on gibberellin biosynthesis and other metabolic pathways. *Annual Review of Plant Physiology and Plant Molecular Biology.* 51:501–531.

Radosevich, S., J. Holt, and C. Ghersa. 1997. *Weed ecology: Implications for management,* 2d ed. John Wiley & Sons, Inc. New York, NY. 589 pages.

Richardson, D. H. S. 1981. 1st ed. *The biology of mosses.* Blackwell Scientific Pub. Boston, MA.

Richardson, M. D. 2002. Turf quality and freezing tolerance of Tifway bermudagrass as affected by late-season nitrogen and trinexapac-ethyl. *Crop Sci.* 42:162–166.

Riddle, G. E., L. L. Burpee, and G. J. Boland. 1991. Virulence of *Sclerotinia sclerotiorum* and *S. minor* on dandelion (*Taraxacum officinale*). *Weed Sci.* 39:109–118.

Robinson, R. J. 1949. Annual weeds and their viable seed population in the soil. *Agron. J.* 41:513–518.

Ross, M. A., and C. A. Lembi. 1999. *Applied weed science.* Prentice Hall. Upper Saddle River, NJ.

Schmidt. R. E. 1992. Biostimulants. *Grounds Maint.* 27:38–56.

Schnick, P. J., and G. J. Boland. 2004. 2.4-D and *Phoma herbarum* to control dandelion (*Taraxacum officinale*). *Weed Sci.* 52:808–814.

Scott, R. 1929. Preventing crabgrass seed. *USGA Bul.* 4:118–119.

Shearman, R. C. 1986. Kentucky bluegrass and annual bluegrass response to ethofumesate. *HortSci.* 21:1157–1159.

Slade, R. E., W. G. Templeman, and W. A. Sexton. 1945. Plant growth substances as selective weed killers. *Nature.* 155:497–498.

Sprague, H. B., and G. W. Burton. 1937. Annual bluegrass (*Poa annua* [L.]) and its requirements for growth. *NJ Agr. Exp. Sta. Bul.* 630.

Sprague, H. B., and E. E. Evaul. 1930. Experiments with turfgrasses. *NJ Agric. Exp. Sta. Bul.* 803.

Steinke, K., and J. C. Stier. 2003. Nitrogen selection and growth regulator applications for improving shaded turf performance. *Crop Sci.* 43:1399–1406.

Stoller, E. W., and R. D. Sweet. 1987. Biology and life cycle of purple and yellow nutsedges (*Cyperus rotundus* and *Cyperus esculentus*). *Weed Tech.* 1:66–73.

Sturkie, G. 1933. Control of weeds in lawns with calcium cyanamid. *J. Am. Soc. Agron.* 25:82–84.

Taiz, L., and E. Zeiger. 2002. *Plant Physiology.* 3d ed. Sinauer Associates, Inc. Sunderland, MA.

Taylor, J. M., G. E. Coats, J. C. Arnold, and K. C. Hutto. 2001. Broadleaf weed control in bermudagrass turf. *South. Weed Sci. Soc. Proc.* 54:67–68

Teuton, T. C., B. J. Brecke, J. B. Unruh, G. E. MacDonald, and J. A. Treadway. 2001. CGA362622 for perennial weed management in warm season turfgrasses. *South. Weed Sci. Soc. Proc.* 54:69.

Tucker, B. J., L. B. McCarty, H. Liu, C. E. Wells, and J. R. Rieck. 2006. Mowing height, nitrogen rate, and biostimulant influence root development of field-grown 'TifEagle' bermudagrass. *HortSci.* 41:805–807.

Turfgrass Producers International. *Guideline specifications to sodding.* New Brunswick, NJ.

Turgeon, A. J. 2008. *Turfgrass management.* 8th ed. Prentice Hall. Upper Saddle River, NJ.

Turgeon, A. J. (ed.) 1994. *Turf weeds and their control.* American Society of America and Crop Science Society of America. Madison, WI.

Unruh, J. B. 1998. Methyl bromide ban is coming soon. *Golf Course Manag.* 66(11):67–70.

Unruh, J. B., and B. J. Brecke. 2001. Seeking alternatives for methyl bromide. *Golf Course Manag.* 69(3):65–72

van Andel, O. M., and D. R. Verkerke. 1978. Stimulation and inhibition by ethephon of stem and leaf growth of some Gramineae at different stages of development. *J. Exp. Bot.* 24:245–257.

Vencill, W. K. 2002. *Herbicide handbook.* 8th ed. Weed Science Society of America. Lawrence, KS.

Walker, R. H., G. R. Wehtje, and J. L. Belcher. 2003. *Poa annua* control with rimsulfuron. *Golf Course Manag.* 71(3):120–123.

Waltz, F. C., J. K. Higingbottom, T. R. Murphy, F. Yelverton, and L. B. McCarty. 2001. Bermudagrass control in centipedegrass with clethodim and adjuvant combinations. *Intern. Turf. Soc. Res. J.* 9:1045–1049.

Watrud, L. S., E. H. Lee, A. Fairbrother, C. Burdick, J. R. Reichman, M. Bollman, M. Storm, G. King, and P. K. Van de Water. 2004. Evidence for landscape-level, pollen-mediated gene flow from genetically modified creeping bentgrass with *CP4 EPSPS* as a marker. *Plant Biology.* 101:14533–14538.

Watson, E. V. 1964. *The structure and life of bryophytes.* 1st ed. Hutchinson & Co. New York.

Weston, L., and J. Barney. 2004. Broadleaf weed control with carfentrazone combinations in turfgrass settings. *Proc. Northeastern Weed Sci. Soc.* 58:116.

Wiecko, G. 2000. Sequential herbicide treatments for goosegrass (*Eleusine indica*) control in bermudagrass (*Cynodon dactylon*) turf. *Weed Tech.* 14:686–691.

Wilen, C. A., and C. L. Elmore. 2001. *Weed management in landscapes.* University of California Statewide IPM Program. UC ANR Pub. 7441.

Willis, G., and B. McCarty. 2006. Standing tall. *Grounds Maint.* 41(4):C1-4.

Yang, Y. S., and S. W. Bingham. 1984. Effects of metribuzin on net photosynthesis of goosegrass (*Eleusine indica*) and bermudagrass (*Cynodon* spp.). *Weed Sci.* 32:247–250.

Yelverton, F. H., J. Isgrigg III, and J. Hinton. 1999. New approaches to management of annual bluegrass in bentgrass putting greens. *South. Weed Sci. Soc. Proc.* 52:72.

Yelverton, F. 2002. Ousting crabgrass—after it germinates. *Golfdom.* 58(6):76–77.

Yelverton, F. 2003. A new herbicide for weeds in bermudagrass and zoysiagrass. *Golf Course Manag.* 71(5):119–122.

Yelverton, F. 2004. New weed control in warm-season grasses. *Golf Course Manag.* 72(1):203–206.

Yelverton, F. 2005. Managing silvery thread moss in creeping bentgrass greens. *Golf Course Manag.* 72(3):103–107.

Zimdahl, R. L. 1999. *Fundamentals of weed science.* 2d ed. Academic Press. New York. 556 pages.

Zhang, X., and R. E. Schmidt. 1999. Antioxidant response to hormone-containing product in Kentucky bluegrass subjected to drought. *Crop Sci.* 39:545–546.

Zhang, X., R. E. Schmidt, E. H. Ervin, and S. Doak. 2002. Creeping bentgrass physiological responses to natural plant growth regulators and iron under two regimes. *HortSci.* 37:898–902.

Ziem, G., and J. McTamney. 1997. Profile of patients with chemical injury and sensitivity. *ENVIRON. Health Perspect.* 105(Suppl 2):417–436.

Index

D

H

N

O